GRAPHIC REPRODUCTION PHOTOGRAPHY

Graphic Reproduction Photography

James Walter Burden
F.R.P.S., M.I.O.P., A.I.I.P., C.G.I.A.

Focal Press · London
Focal/Hastings House · New York

© FOCAL PRESS LIMITED 1973

ISBN (excl. USA) 0 240 50757 6
ISBN (USA only) 0 8038 5778 0

All Rights Reserved. No part of this publication may be reproduced, stored in a retrieval system, or transmitted, in any form or by any means, electronic, mechanical, photocopying, recording or otherwise, without the prior permission of the copyright owner.

First edition 1973
Second imprint 1980

Spanish edition
LA FOTORREPRODUCCIÓN EN LAS ARTES GRÁFICAS
Ediciones Don Bosco, Barcelona

Printed in Great Britain by Thomson Litho Ltd., East Kilbride, and bound by Hunter & Foulis, Edinburgh

CONTENTS

	page
Acknowledgments	10
Foreword by F. PATEMAN BA, FIOP	9

THE BIRTH OF PHOTOGRAPHY — 11
Introduction	11
A moment in time	12
Evolution of the camera obscura	13
Early photochemistry	15
The beginnings of a photographic process	17
The first negative/positive photographic process	19
The first practical photographic process	20
The wet collodion era	21
You press the button, we do the rest	22
Graphic reproduction	23

ORIGINAL PICTURE — 27
Historical introduction	27
Visual interpretation	31
Mechanism of vision	32
Viewing tone and colour	34
Colour	38
Classification of originals	44
Scaling the original	45
Masking to proportion	48
Instructions for graphic reproduction	50
Line originals	51
Tone originals	56
Colour originals	61
Colour transparencies and prints	63

LIGHT BEHAVIOUR — 65
Historical introduction	65
Visible spectrum	66
Light terminology	68
Photometry	72
Intensity of illumination	73
Illumination control	73

Contents

	page
LIGHT ILLUMINANTS	76
Historical introduction	76
Classification of illuminants	77
Illumination requirements	77
Illuminants	80
Lens flare	89
Calculating the flare factor	91
Viewing conditions	92
THE CAMERA	94
Historical introduction	94
The pinhole camera	96
The simple lens	97
Focal length	100
Image formation	103
Magnification of the image	103
Lens aberrations	108
Compound lenses	113
Lens blooming	114
Lens aperture	116
Reversal systems	118
Straight-line reversing systems	120
Graphic reproduction cameras	122
LIGHT SENSITIVE EMULSIONS	134
Historical introduction	134
Chemical introduction	138
Action of light	141
Photochemical requirements	141
Emulsion manufacture	142
Manufacturing process	144
Emulsion structure	150
Emulsion data	152
Latent image theory	154
Latent image formation	155
PROCESSING	161
Historical introduction	161
Chemical introduction	163
The action of the developer	163
The developer's constituents	167
Development theory	173
Variants in development	174
Methods of development	179
Final stages	182

Contents

	page
Activation stabilization system	184
Diffusion or chemical transfer	184
After-treatment	184
Reduction	184
Intensitification	185
Photographic control	185
SYSTEMATIC WORKING	188
Historical introduction	188
Sensitometry	189
Sensitometric terms	191
Logarithms	195
Log of exposure	200
The characteristic curve	201
The manufacturer's characteristic curve	203
Original's density range	204
Exposures on the characteristic curve	206
Correct exposure time	208
Image characteristics	210
Gamma	215
Time/Gamma curve	216
Densitometry	217
Specular and diffuse densities	219
Photo-voltaic or barrier-layer photo-cell	221
Vacuum emission photo-cell	222
The photomultiplier cell	222
Densitometers	223
Reproduction quality	224
The need for a system	225
Standardization	225
Photographic procedure	226
Fundamental relationships	228
Controlled exposure	229
Controlled development	248
An organized glance	249
A matter of economics	252
Quality control strip	252
A line standard	252
A tone standard	254
A colour standard	258
Record of work	266
The moment of truth	269
LINE REPRODUCTION	270
Historical introduction	270

Contents

	page
Assessment of line originals	272
Camera procedure	274
Line results for each reproductive process	284
Line and tint combinations	288
TONE REPRODUCTION	**291**
Historical introduction	291
Types of halftone screens	295
Halftone theories	304
Assessment of the original	320
Camera procedure	324
Tone results for each reproductive process	337
Line and tone combination	340
COLOUR REPRODUCTION	**344**
Historical introduction	344
Colour theory	348
Colour and tonal correction	360
Assessment of coloured originals	388
Camera procedure	390
Colour separation for each reproductive process	395
Colour and line combination	399
ELECTRONIC REPRODUCTION	**403**
Introduction	403
Scanning methods	403
Drum scanners	405
Scanning magnification	406
Computer functions	408
Numerical set-up	409
Colour correction set-up	410
Tonal correction set-up	410
Contact screening and electronic laser screening	411
Page assembly and electronic retouching	414
REPRODUCTION PROCESSES	**416**
Introduction	416
Letterpress printing	416
Gravure printing	423
Lithographic printing	428
Screen process printing	434
The future scene	439
Appendix	440
INDEX	**443**

FOREWORD

For some time now the processes of Graphic Reproduction have been moving from craft processes dependent on 'know-how' and hard-won experience to scientific exactness and quality control. It has become a technology in the modern sense of the term. This book is very welcome at the present time because it sets out the implications of such changes both for the designer, who has to use the processes, and the technician, who has to control them. As the initiator of materials to be printed, the designer has an increasing need to ensure that the material he submits for reproduction should be suitable for the chosen process without the need for elaborate retouching.

Mr. Burden has made clear the main characteristics of both drawn and photographic originals suitable for the main processes of graphic reproduction. For the technician he has described control procedures and illustrated the effect of changes in various parameters on the finished print. A typical programme used to enable a computer to generate a table of control parameters for Graphic Reproduction is given in the text—a refreshing change from the rather mystical approach to computers in some books on printing.

I hope this book becomes the basis of a further advance in the teaching of what has always been an important and fascinating field.

F. PATEMAN

Twickenham College of Technology

ACKNOWLEDGEMENTS

In writing this book I am indebted to the following: Mr. F. Pateman, B.A., F.I.O.P.; Mr. F. V. Paine, M.S.I.A., M.S.T.D.; Mr. J. Parsonson, B.Sc. Hons.; Mr. A. Young; Mr. A. Miles; Mr. D. Page; Mr. D. Simpson, B.Sc., A.F.I.M.A.; Mr. E. E. S. Taylor; Mr. J. Elliot and Mrs. S. Elliot. The table photographs were done by David Searle, the graphic photographs by Paul Green, Eric Glover and Jack Smith, and the pencil sketches by James Joyce. I should like to thank these friends and the many others who encouraged and helped me in this venture.

In addition I wish to thank the following manufacturers for their help and permission in presenting a selection of their excellent technical procedures. Where further investigation is required the reader is directed to the technical literature published by these manufacturers.

Agfa-Gevaert Ltd. (Lens flare testing, use of contact screens, prediction and calculation of exposure times and methods of colour separation.)

Crosfield Electronics Ltd., Dr-ing. Rudolf Hell Scanners Ltd. and Pershke Price Service Organisation.

Du Pont Organisation. (Calibration of contact screens and colour control systems.)

Kodak Ltd. (The use of colour-correction filters, double-overlay making technique and masking alignment chart.)

Ilford Ltd. (Sensitometry and densitometry, contrast control techniques with contact screens, calculation of filter factors and ratios.)

1. THE BIRTH OF PHOTOGRAPHY

Introduction

It can be seen from the beautiful pre-historic cave paintings of animals and contemporary events that even the illiterate cave dwellers felt an innate desire to record and recapture visual sensations that they had experienced.

Prehistoric visual sensations recorded at Lascaux in France

Correlated with the development of education this desire has inspired men to perfect skills, construct equipment and formulate chemical processes capable of capturing almost every visual facet of human experience.

This desire manifested itself for many centuries in the art of drawing and painting. Although this reached high levels of execution during the early civilizations and near perfection during the Renaissance, it was only available to a few people, those fortunate enough to be born with an aptitude for art and the wealthy who bought or commissioned their work. It was during the pursuit of perfection that the camera obscura was employed, its image being an exact replica in perspective and colour of the scene the artist wished to capture. The early camera obscura was a small private room (camera), completely dark (obscura) inside. To the person inside the

The birth of photography

room the experience was similar to watching a present-day motion picture for the first time. Light rays were admitted through a small aperture in the centre of one wall. These light rays formed an inverted, coloured image of the scene outside on the surface of the opposite wall.

An early camera obscura

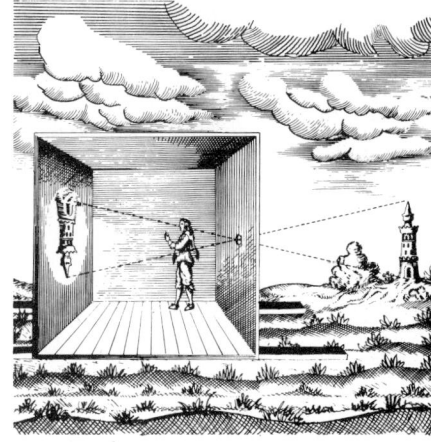

Many people attempted to capture permanently the camera obscura's image, but it was only little more than 150 years ago that these early experiments succeeded, and with them the birth of photography.

The seemingly impossible ambitions of these early inventors have been virtually realized today. Modern colour photography linked with graphic reproduction results in the printing of innumerable reproductions of any image capable of being captured with a lens, and in doing so condenses into a matter of hours, months of laborious hand work required in the past. Modern colour printing is not far short of perfection and satisfies the human desire to capture for posterity the contemporary scene.

A moment in time

A human eye scans, a brain reacts to the stimulation of a visual sensation, hand moves, glass eye blinks, light explodes, time moves on, sensation gone; but captured in a sea of minute chemical crystals an image lies hidden, awaiting a developing solution to bring it back to life. Photography has frozen a moment of time. Photography is absolute because it allows reality to print its own images.

We will join the world of photography by trying briefly to relive the birth of photography, conceived some 150 years

A contemporary scene recorded

Evolution of the camera obscura

ago. The following descriptions only serve to illustrate the events responsible for photography as we know it today. More detailed accounts are given as historical introductions to each subject covered in this book.

Early photographic processes were the result of two separate avenues of thought being united, the first being *optical—The Evolution of the Camera Obscura,* and the second, *chemical—The Darkening Effects of Light on Certain Substances;* so it is logical to re-examine these two basic concepts.

Evolution of the camera obscura

As early as the tenth and eleventh centuries nomadic tribes roaming the vast deserts of Arabia used a darkened tent with a small hole cut in one side to view the sun on the opposite flap. This was an aid to navigation and a comfortable way of viewing the wondrous silhouettes cast during occasional eclipses of the sun. Even during the reign of Richard the Lionheart, when men were more interested in donning armour and fighting battles than developing scientific thought, cameras were being used in the form of camera obscuras. Their dim, rather fuzzy images were employed as guides for drawings and maps.

Many names have been associated with the camera obscura. Roger Bacon (1214–1294), an English friar and alchemist, and Leonardo da Vinci (1452–1519), the famous Italian painter, sculptor and inventor, amongst others, described the use of the camera obscura in their works, and the knowledge of the principles involved would appear to be reasonably widely known during these periods. The early camera obscura,

The birth of photography

being virtually a room providing essentially the same views all the time, was extremely limited in its application. The projected image lacked definition and brilliance. More light could be allowed in by enlarging the hole, but this accentuated the loss of definition.

The optical principle of a lens was well understood in Egyptian times when jewellers utilized the lens as a magnifier to engrave gems which accompanied the noble Pharaohs into their pyramidal tombs. This knowledge was passed on and most learned Europeans of the early Renaissance period understood the magnifying and ignition powers of the lens. From 1300 onwards Italian physicians began to prescribe eyeglasses, but it was not until the year 1550 that Girolamo Cardano, a Milanese physician, positioned a simple spectacle lens in the hole of a camera obscura. By performing this simple action Cardano had begun to cure the lack of definition and brilliance; in retrospect he unveiled the first camera. Of course, at this stage the lens was still housed in a camera obscura in which people stood, viewing or tracing the upsidedown image.

A few years later another Italian, Giovanni Battista della Porta (1523–1590), discomforted and annoyed by having,

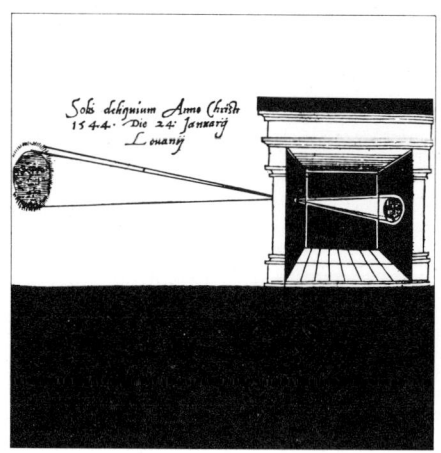

Gemma Frisius. First published illustration of a camera obscura; observing a solar eclipse in January 1544

in his estimation, to stand on his head to see the camera obscura's image properly, bounced the light projected by the lens off a concave mirror which turned the image up the right way. Porta was quick to realize the progress he had made and proclaimed,

'If you cannot paint, you can by this arrangement draw the outlines of the images with a pencil. You have then only to lay on the colours.'

Early photochemistry

Similarly a Venetian nobleman, Daniello Barbaro (1530–1601), in 1568, after introducing a convex lens complete with a diaphragm used to sharpen the projected image, exclaimed to the artistic world,

'If you wish to study the outlines, colours and shadows of things as nature spaces them in distance, make a hole in a window shutter and set in it a thick lens from an old man's eyeglass (not a thin lens made for a young man). Now close all the shutters and doors until no light enters the chamber except through the lens, and opposite it hold a sheet of paper, which you move forward and backward until the scene appears in the sharpest detail. There on the paper you will see the whole view as it really is, with its distances, its colours and shadows and motion, the clouds, the water twinkling, the birds flying. If you partly cover the lens to leave only a small aperture, the image grows sharper. By holding the paper steady you can trace the whole perspective outlines with a pen, shade it, and delicately colour it from nature,'

By now the exponents of the camera obscura were becoming discontented with static views and demanded portable camera. New models, reduced in size and containing many refinements were produced to meet the demands, culminating in a small portable reflex camera for the travelling artist.

A portable reflex camera

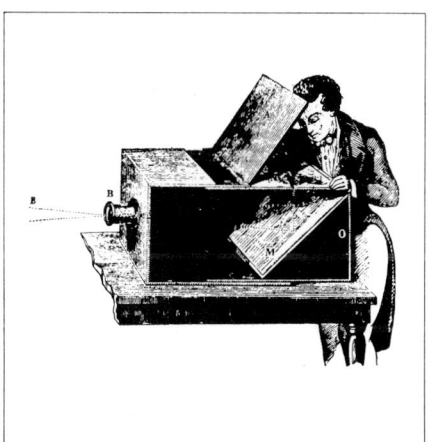

Early photochemistry

The darkening effect of light on certain substances has been noted and observed since ancient times. Aristotle (322–384 B.C.), the eminent Greek scholar and teacher, was fully aware that it was sunlight that tanned the human skin and fulfilled an important role in the natural colouring of plants and flowers. During the Middle Ages, when mysterious

The birth of photography

alchemists and self-proclaimed sorcerers searched in vain for the ingredients of gold, it became known that certain substances when mixed with silver gradually blackened on exposure to light. This information was used by the more diagnostic scientists of the early Renaissance period who recorded the fact that when silver nitrate was mixed with chlorine a light sensitive mixture of silver chloride was formed. The Medieval and early Renaissance alchemists did not, however, mention the fact that the action of light was responsible for the blackening. This was left to an Italian physician, Angelo Sala, who in 1614 wrote, 'When you expose powdered silver nitrate to the sun it turns black as ink.'

This information seems to have lain dormant for nearly 100 years until Dr. Johann Heinrich Schulze (1687–1744), a German chemist and philosopher, picked up the threads and proved in 1725 that light alone was responsible for this phenomenon by producing darkened images exposed on a white chalk background held in bottles. He described his work in the following manner.

'I covered the larger part of the glass (containing the silver nitrate and chalk) with opaque bodies and patterns which permitted the light access to only a small portion of the vessel's contents. In this manner I frequently wrote names and whole sentences on paper and carefully cut out with a very sharp knife the parts covered with ink. The paper perforated in this manner I fastened with wax on the bottle. Before long one could see the sun's rays write through the perforation in the paper, through which they could reach the glass, these very words and sentiments on the chalky sediment.'

With Schulze's work the science of photochemistry began and it was this, reinforced by his bottle images, that inspired

A re-created bottle image

The beginnings of a photographic process

Thomas Wedgwood (1771–1805), the youngest son of Josiah Wedgwood, the famous potter, to try and apply this knowledge. Thomas was well acquainted with the use of the camera obscura, having seen it employed frequently as an aid in producing drawings for pottery designs; so it is not unnatural that he should prove to be the first man to unite the two avenues of work and endeavour to capture the image of the camera obscura on a piece of paper coated with light-sensitive silver nitrate. In this Thomas was unsuccessful; presumably the paper was not sensitive enough and the exposure times too short. His theoretical deductions provided a basis for photography, but his practical evidence had to be confined to silhouettes of insects and leaves produced by

A re-created silhouette

exposing them in contact with pieces of white leather saturated with silver nitrate. Although these silhouettes were beautiful in their delicacy, Thomas Wedgwood's aspirations were once again retarded by his inability to 'fix' the silhouettes and so render them permanent.

Thomas Wedgwood carried this early method of photography to the brink of success, but unfortunately, because of his death at the early age of 34, left the production of a complete photographic process to others.

The beginnings of a photographic process

During the summer of the year 1826 as the dawn light heralded the start of a long sunny day, Joseph Nicéphore de Niépce (1765–1833), a French Army Officer living in retirement at Gras, his family estate in the village of Saint-Loupe-de-Varennes, quickly prepared a pewter plate by pouring on

The birth of photography

Niépce: the first photograph—a view from his attic window

The first negative/positive photographic process

Joseph Nicéphore de Niépce (1765–1833)

a coating of light sensitive bitumen and clamped it into the back of his wooden camera. All day long the lens of this camera pointed out of his attic window swallowing up the surrounding scene. As the light faded into dusk the exposure ended. Niépce rather nonchalantly washed the exposed surface with a solution of lavender oil and petroleum, and slowly an image revealed itself—walls, windows, rooftops, trees, light and shade, all drawn by light alone! Niépce had captured the camera obscura's image.

The first negative/positive photographic process

William Henry Fox Talbot (1800–1877), an English gentleman, mathematician, landowner and one-time MP, wrote the following treatise on the thoughts which first prompted him to discover the delights of photography.

'I reflected on the inimitable beauty of the pictures of nature's painting which the glass lens of the camera throws upon the paper in its focus—fairy pictures, creatures of a moment and destined as rapidly to fade away. It was during these thoughts that the idea occurred to me . . . how charming it would be if it were possible to cause these natural images to imprint themselves durably and remain fixed upon the paper.'

These words and others accompanied photographs in *The Pencil of Nature*, the first book to be illustrated by the camera, published by him in 1844. Fox Talbot, stimulated by what he had learnt from the previous photographic researchers, had miniature wooden cameras made, into which he placed good quality writing paper coated with silver chloride. After selecting pictorial views around Lacock Abbey, his Wiltshire home, he left his cameras allowing the sunlight to

The birth of photography

Fox Talbot: earliest paper negative—a lattice window

do its work. Fox Talbot's first notable success was the paper negative recording the lattice window of his library.

Fox Talbot modified and refined his process, his work culminating in the 'Calotype' process which produced paper negatives and contact positive paper prints of the type that illustrated *The Pencil of Nature*.

The first practical photographic process

Joseph Nicéphore de Niépce, flushed with the success of his early experiments, endeavoured to improve his process which he called 'heliography' (sundrawing) by substituting the pewter plate for copper coated with silver iodide. Niépce's work attracted the attention of Louis Jacques Mandé Daguerre (1787–1851), an astute showman and inventor of the Diorama—a stage set which produced realistic scenes by using skilful lighting effects. In the year 1829 Niépce and Daguerre struck up a partnership with the view of improving the practicability of Niépce's heliography, but before this partnership could flourish Niépce died, leaving Daguerre to carry on alone. Daguerre persevered with the silver iodide coated copper plates and discovered that the exposed image could be developed with mercury vapour and fixed by applying a solution of common salt. The year 1838 saw this manipulation launched as the 'Daguerreotype' process, producing delicate, direct positive images in silver on a copper base.

Despite the handicap of being an image on metal and therefore making copies impossible, Daguerreotypes seized the imagination of Europe and photography became the vogue of the wealthy.

Daguerre: the first daguerreotype—a still life

The wet collodion era

The wet collodion era

During the years 1850 and 1851 British craftsmen and inventors were busy producing and preparing their best work, in the hope of having it selected for the exhibition stands housed in the magnificient Crystal Palace, the centrepiece of the Great Exhibition of British Industry commissioned by Queen Victoria. All this frenzied activity may have bypassed an English sculptor named Frederick Scott Archer (1813–1857), who, owing to early experiences gained while employing the 'Calotype' process as an aid to his work, became so deeply engrossed in photography that he spent most of this period investigating the realms of photographic possibilities instead of creating models.

While on one of these excursions into the world of exposure and development Archer came across 'collodion', a syrupy fluid prepared by dissolving pyroxyline (gun-cotton) in a mixture of ether and alcohol, which he used as a vehicle to hold the light sensitive silver grains in a finely divided state. Archer introduced this discovery into a negative–positive process using glass plates, which he coated, sensitized, exposed and processed while still in a moist or wet state. The 'wet plate' process was born.

This process, more sensitive than the existing Daguerreotype and Calotype processes and free from patent restrictions, was the redemption of photography, freeing it to all would-be photographers, and in doing so started in 1851 the 'Golden Age' of photography which lasted for thirty years; during this time photography, surrounded by the paraphernalia of camera, tripod, darkroom tent, dishes, messy chemicals, water and all, travelled the world, went to war, conveyed current events, preserved family groups, changed from an

The birth of photography

amateur pursuit to a professional business and became a servant to the printing processes, supplying photographic negatives from which master printing blocks could be etched.

You press the button, we do the rest

Dr. Richard Leach Maddox (1816–1902), a physician by profession but photographer by vocation, became so aggravated by the fumes of the wet plate process that he strove to find a cleaner, more healthy process, and the fate of the wet collodion age was sealed.

Dr. Maddox turned to another colloidal substance—gelatine, and in September 1871 the doctor wrote an article entitled, 'An Experiment with Gelatine Bromide' in which he describes 'the result of some careless experiments tried at first on an exceedingly dull afternoon.' In these experiments he produced a photographic dry plate by coating glass plates with a solution of gelatine containing silver nitrate, cadmium bromide and nitric acid. The developer was a solution of pyrogallol to which a few drops of silver nitrate had been added. Dr. Maddox concluded his article with these words. 'As there will be no chance of my being able to continue these experiments, they are placed in their crude state before the readers, and may eventually receive correction and improvement under abler hands. So far as can be judged the process seems quite worth more carefully conducted experiments.'

Others did develop and improve on his ideas; by 1902 Kodak were producing a dry silver bromide gelatine emulsion coated on to rolls of thin flexible celluloid to be used in their complete 'everyone can be a photographer' camera kits,

Kodak: photographic kits—circular snapshots

Graphic reproduction

advertised under the now famous slogan, 'You press the button, we do the rest'. Photography was big business now, everybody wanted to press the photographic button and did so, pointing their cameras at all the epoch-making inventions of the moment—the Zeppelin airship, Wright brothers' petrol engined aeroplane and the first motor cars—capturing all in a world of light, lenses and chemicals.

Graphic reproduction

Today the photographic process is well understood and satisfies many needs. We are surrounded, if not bombarded by printed literature of every description; most of it would seem unreal without illustrations. An eye-catching picture is the most dynamic way of arousing interest. Graphic reproduction is concerned with the photographic production of master printing plates from which numerous facsimile reproductions of an original picture can be printed.

Modern printing may be divided into four major processes:
1. Letterpress Printing from master plates with Relief (raised up) images produced via the photo-engraving process

During letterpress or relief printing the raised surfaces are coated with a film of viscous ink and pressed on to paper. Letterpress printing is widely used for the reproduction of type matter and small to medium sized illustrations; it produces results of excellent quality complemented by the advantage of easy correction, alteration and rearrangement, making it an ideal process for any type of printing where words, figures or illustrations need to be constantly changed, e.g. daily newspapers, business forms and illustrated catalogues. The

The elements of letterpress printing:
1. Forme (lines of type)
2. Inking
3. Laying on the paper
4. Printing
5. Removing the print

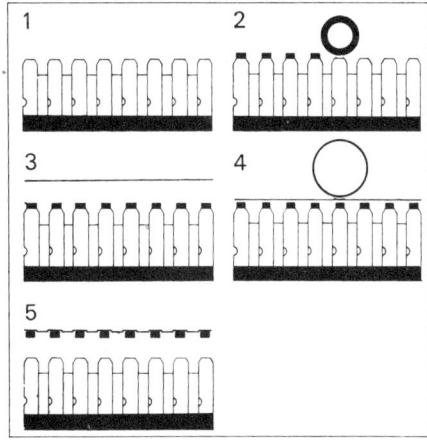

The birth of photography

text may be printed from metal letters held in a forme, whereas in gravure, lithography and screen process, type matter is photographically produced together with the illustrations on the image cylinder, printing plate or stencil.

2. Gravure Printing from master plates containing intaglio (recessed) images achieved by the photogravure process

Gravure printing is produced from images incised or etched below the surface of the plate or image cylinder. The image recesses are filled with a liquid, volatile ink, and the paper, or any other suitable substrate (thin plastic, foil, etc.) is positioned between the image cylinder and a rubber-covered impression roller. As the paper moves through, the ink held in the recesses is transferred to the paper. Varying ink densities are printed because deep recesses impart more ink than shallow ones.

Operation of the intaglio printing process:
1. Forme
2. Inking with the "doctor" removing excessive ink
3. Laying on the paper
4. Printing
5. Removing the print

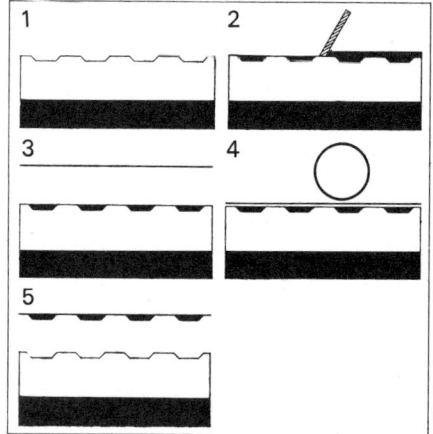

The initial high cost of producing the image cylinder restricts rotary gravure to the printing of long runs, especially in the field of illustrated magazines and packaging materials. The printed result has a vividness all of its own and reproduction on uncoated papers is equal to offset-litho printing and surpasses letterpress; but type matter is poor, because the letters are cut up by the screen lines, giving them a serrated edge. Typematter is usually provided via a photo-typesetting machine, or reproduced photographically from a letterpress proof printed in white ink on a black background. There are various other 'conversion' methods which convert the image from metal letters on to a film base.

3. Lithographic Printing from master plates have planographic (on the surface) images prepared during the photo-lithography process

The birth of photography

Planographic printing processes differ from the letterpress and gravure methods in that the printing areas (ink receptive, water repellent) and non-printing areas (water receptive, ink repellent) of the plate are on the same level. The offset press, in which the inked image is first transferred on to an intermediate, rotary rubber blanket, and then on to the paper has now almost entirely superseded the 'direct' printing press, and is used to print large pictorial areas of colour, usually needed in poster, map and showcard printing. Offset printing utilizes a thin metal plate which needs no makeready and will print good results on almost any type of paper. Offset-lithography has its disadvantages. Type matter must be on a film base; after the plate-making stage alterations and corrections are virtually impossible. The printed results

Operation of the planographic printing process:
1. Forme
2. Damping
3. Inking
4. Laying on the paper
5. Printing
6. Removing the finished print

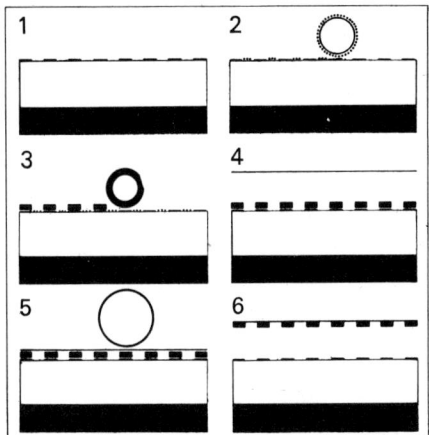

in black and white tend to have lower and shorter tonal ranges than reproductions by letterpress or gravure because the ink film is 'split' twice, but with improved inks the quality gap is gradually closing.

4. Screen Process Printing from master frames having stencil (cut-out) images formed by the Photo-stencil process

Screen process printing has grown out of the craft printing process known as silk-screen printing. Screen process is capable of quite long runs on all manner of surfaces and is a small but expanding section of the printing industry. A fine mesh of silk, nylon or steel is stretched over a wooden frame. A stencil with the image areas revealed as openings is stuck to the underneath of the mesh. A stencil may be cut out manually or produced photographically. A tacky ink is placed in the top of the wooden frame and a sheet of paper under-

The birth of photography

neath. The ink is drawn over the mesh with a squeegee, and this forces the ink through the image areas of the stencil on to the paper underneath. The ink weight printed is considerable and each sheet needs to be dried separately.

This process is ideal for short runs of large pictorial designs. All grades of paper and materials such as glass, plastic, wood, metal and textiles with flat or curved surfaces can be printed. Screen process is used extensively in the production of printed circuits. Printed results have a depth and brilliance of colour no other process can match; the technique is so simple in its basic principle that anybody can produce amateur printed copies of suitable pictures at a low price without any great degree of skill.

All graphic reproduction processes commence with an initial photographic stage which is of extreme importance, entailing the production of photographic negatives and positives which must faithfully copy the original picture's tonal gradation, visual balance and its component colours in a separated condition. Before this is possible the photographer needs to be competent and fully aware of the photographic reproduction process and its many variants, which become only too apparent during the process. Above all he must be capable of visually dissecting the original into basic colours, tonal balance and negative—positive images.

Elements of the screen process:
1. Uncoated screen
2. Coated screen with masks
3. Positioning the paper
4. Spreading the colour with a squeegee
5. Removing the screen

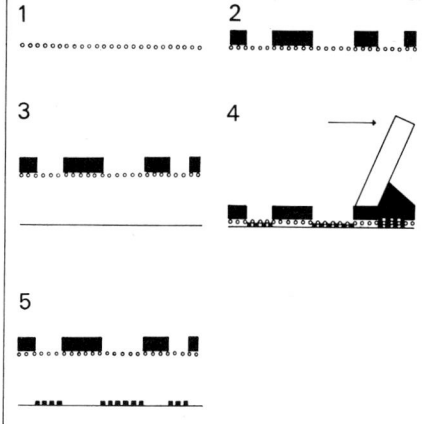

2. ORIGINAL PICTURE

Historical introduction

From the beginnings of civilization, pictures have accompanied the written word. Indeed, one of the earliest cultures, that of the Egyptians, developed a form of writing known as the 'ideogram'—a representation of experience, feeling and event in picture form; a more intimate relationship between text and illustration would be difficult to achieve.

A precise date for the invention of printing is difficult to give, but if we look upon the year 1450 as the start of illustrated books from type and blocks, then before this date illuminated initials, marginal illustrations and miniatures (small pictures explaining the text) were executed individually with great dexterity by artists of the period. Nearly all the books were devotional or educational—Bibles, psalters, bestiaries and herbals. These works of dedication set high standards of quality for the early printers to match. They endeavoured to do so but were handicapped by crude presses and illustrations reproduced from wood-cut printing blocks, made by cutting away the wood from the non-image areas. Broad, coarse lines were needed to give adequate strength and durability to the printing surface.

Owing to limitations imposed by the wood-cut process, illustrative work tended to be restricted to decorative borders, and these proliferated during the 16th century with the development of copper engravings, which reproduced a finer line than wood-cuts. The seventeenth and eighteenth centuries were periods of great typographic development rather than of new reproduction processes, although improvements were made in the skill and techniques of engraving an intaglio printing image in a metal surface.

Looking at the majority of books published by the printers of this period it would seem as though they had completely discarded the wood-cut in favour of intaglio metal-engraving, in spite of the handicap of printing illustrations separately from the type. Suddenly, in the latter part of the eighteenth century, wooden blocks reappeared, but the similarity to the early wood-cuts was only superficial—these printing blocks were wood-engravings rather than wood cuts, a technique developed by Thomas Bewick, an English engraver.

Original picture

An early wood-cut—Saint Christopher

In the early wood-cuts a sharp knife and gouge were employed to cut away the white, non-image areas leaving the black image areas standing in relief. The printing image was entirely in line—no tones were possible. Bewick found that by using the end-grain of box-wood instead of the plank he could achieve more control and engrave fine single lines, broken parallel lines, and reversed white lines out of black. By astute use of printing pressure he united the whole illustration into a picture of delicate lines and tones.

Colour reproductions, right up to the end of the eighteenth century, were generally carried out individually by hand using stipple engraving and drawn lithography, or achieved by early mezzotint and aquatint methods. The mezzotint was an intaglio process which began with a copper plate with one of its surfaces roughened all over. If this matt surface

Steel engraving—beggars at the door

Historical introduction

Bewick: wood engraving—the Starling

was inked the rough grain would retain the ink and print a solid black impression. The craftsman obtained his tones by smoothing off the roughness to varying depths working from black to white. Aquatint relied upon a fine powdered resin being carefully sprinkled over a copper surface, then heated to fuse it to the surface. The result was a fine grain structure on to which the design was drawn. Delicate etching followed, producing tones in accordance with the depth achieved. The fine grain structure remained, but modified according to the depth of etch, producing a final printed copy which was a combination of ink weights and grain patterns.

Until the middle of the nineteenth century the original was nearly always the printing block; an artist or engraver would create the original picture on the printing surface by artistic skill, employing one of the processes previously mentioned.

An Aquatint—the huntsman's "bag"

Original picture

Master paintings, pictorial illustrations and contemporary living scenes had to be translated and redrawn into originals suitable for the printing processes, rather than the printing processes being made suitable for reproducing the picture in its original form. This was all drastically changed in 1835 when photography was born and the whole application of illustrative printing began anew.

Through the eyes of a camera all things became possible and gradually, as the nineteenth century receded, the lens replaced the skilled hand of the engraver. The first book to be illustrated by photographs was Fox Talbot's *Pencil of Nature* in which photographic prints on paper were pasted into the centres of pre-printed borders. This book illustrated vividly the potential use of photography in printing. It broke the restrictive dam imposed by the early printing processes and let through a sea of creative thought.

Fox Talbot: photographic books "The Pencil of Nature"

It was not long before photographer and artist were merging their skills to produce original pictures embracing most human experiences. The camera also became an integral part of the printing process, supplying an intermediate medium by which an original picture could be transferred to a metal printing surface. Today, photographic colour transparency and colour print processes are capable of capturing a visual experience almost perfectly, and in turn, because of the application of photography, the early methods of graphic reproduction have led to the development of photoengraving, photogravure, photolithography and photostencil which serve each of the basic printing processes.

Visual interpretation

The original is usually provided by the customer and many diverse kinds are encountered. Their creation may be due to the skill of commercial artists and graphic designers producing pencil sketches, line drawings, wash impressions and coloured paintings, or to professional photographers taking black and white pictures, coloured transparencies and colour prints. Let us commence by trying to appreciate the motivation of the artist and photographer while they are creating the original.

A viewer's interpretation of a picture should be exactly the same as the original concept. The picture's message should be uncomplicated and obvious at the first glance.

An artist or photographer, usually under the direction of a graphic designer, is given a creative brief; that is, precise instructions based on the original concept, listing the reasons

A model of a standard creative brief: note the items listed

Creative Brief

Client FOCAL PRESS LIMITED.			Brief prepared by JAMES WALTER BURDEN.	
Media PRINTED PAGE : WHITE BACKGROUND.			Date	1 . 1 . 1970.
Size (flat) 51 x 24 cms	(folded) 24 x 16 cms	(trimmed) 23 x 15.5cms	Description	BOOK JACKET.
Paper CAXTON ART.	No. of workings TWO.		Layout required by	7 . 1 . 1970.
Process LETTERPRESS.	Quantity 10,000.		Artwork required by	1 . 2 . 1970.
Inks ACME Primary Red ac 2487 Opaque Black ac 63.			Type-mark required by	1 . 3 . 1970.

Produce a lay-out and finished artwork for a book jacket which illustrates the title

"PHOTOGRAPHY FOR GRAPHIC REPRODUCTION"

PRODUCTION PROGRESS	Number 6 5 6 5.	Date 1 . 4 . 1970.	Quantity 10,000.
Description COLOUR BOOK JACKET : LINE REPRODUCTION : BLACK AND RED PRINTINGS ON WHITE.			
Composing	Letterpress Printing	Litho Printing	
TYPE SET AND CORRECTED.	PRINTING CONCLUDED.	NOT REQUIRED ON THIS SHEET.	
7 . 4 . 1970.	14 . 4 . 1970.	14 . 4 . 1970.	
Process Engraving	Photolithography	Warehouse	
BLOCKS MADE AND PROOFED.	NOT REQUIRED ON THIS SHEET.	COVERS TRIMMED AND FOLDED.	
10 . 4 . 1970.	14 . 4 . 1970	16 . 4 . 1970.	

for its conception and indicating the desired message or expression.

After study of the creative brief the artist or photographer will create an original which in his view is the simplest subject, composed as an organized whole, that will transmit the correct pictorial message. Technique can reinforce the

Original picture

message, e.g. boldness can be created in a black line drawing or a stark photograph containing extreme contrasts; subtleness can be seen in pencil sketches, wash impressions and soft photographs possessing muted tones; gaiety can be avidly expressed in vivid colours, using paints, pieces of textile, dyes of transparent colour, or print processes.

The success of an original depends upon whether the mental ability and imagination of the viewer is the same as the designer. A viewer's perception is conditioned by his past experiences and store of knowledge; his mental ability governs the conception he constructs from visual images.

The graphic reproduction photographer is responsible not only for the translation of an original into negative and positive images, but also for the faithful duplication of the original's message. This is achieved by retaining the original's tonal appearance, texture, sharpness and colour balance.

A good photographer, before commencing any photographic work, will study the original, becoming acquainted with its highlight, middle tone and shadow areas, observing the fine detail, depth of colour and isolating the original's focal point. Only after this period of familiarization will it be possible with efficient photography to produce negative blackening in proportion to the original's luminosity, positive blackening as the exact opposite of the negative densities and a final printed copy possessing the same visual sensations as the original.

Mechanism of vision

Light energy can be seen in many guises; at any moment it may be uniform and harsh, or diminishing into a soft flicker. Visual stimuli are mixtures fused together in colour and intensity. These sensations are captured by the human eye which can function in dim moonlight or be quite comfortable bathed in brilliant sunlight.

Colour and tonal sensations conceived in our minds are the result of light stimuli apparent in objects surrounding us, exciting photosensitive structures within our eyes which arouse the corresponding sensation.

Our visual apparatus is separated into two sections.
1. An optical system forming an image upon the retina, controlled by muscles which position the eye.
2. The retina, consisting of photo-sensitive structures which record the image and produce impulses along the optic nerve fibres, arousing visual sensations within the central nervous system.

The lens of the eye, by decreasing and increasing its

Mechanism of vision

The human eye:
1. Lens
2. Iris
3. Retina
4. Eye muscles
5. Optic nerve

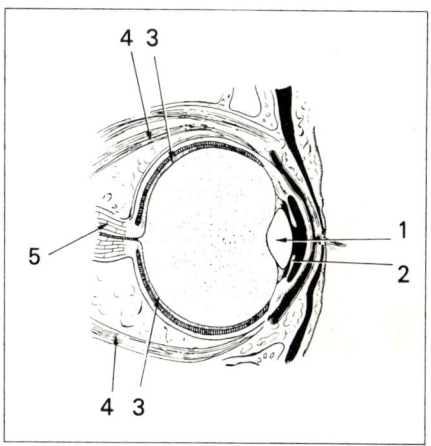

outer curvature, can focus on objects positioned at various distances from it. The elasticity of the lens is controlled by eye muscles; when relaxed the lens is focussed for distant vision.

The iris is a variable diaphragm positioned in front of the lens, controlling the amount of light entering the eye's aperture—the pupil. Its function can be understood by noting what happens when very intense light confronts the eye. When this occurs the iris constricts, cutting out the greater proportion of the light which would completely saturate or blind the light-receptor cells. Similarly, in weak light, the iris diaphragm dilates, allowing all the available light to fall on the light-receptor cells.

The retina is the rear lining of the eye comprising millions of photo-sensitive structures, long funnel-like tubes constructed in banks, their tips facing the lens of the eye. The retina's function is to detect light stimuli and produce nerve impulses, which will activate the brain, creating sensations attributed to the source of the stimulus.

These light-receptor cells are divided into two types, termed *rods* and *cones*. The resolving properties of an eye are greater at the centre of the retina—the *fovea*, where only cones are situated. Moving outwards from the fovea, rods and cones become interspaced, the number of cones being gradually replaced with rods, until at the periphery only rods are left. Rods contain a light-sensitive pigment, *visual purple*, which becomes bleached when struck by light causing an immediate impulse. Their light-sensitivity is extremely high, but this is achieved at the cost of being colour-blind.

Cones work in rather the reverse manner, distinguishing colour, but possessing low sensitivity. Rods and cones

Original picture

reinforce each other—the rods react to low intensities handing over the reception to the cones if the light becomes too intense. This is why objects appear colourless in dim moonlight, and indicates the pertinence of a quote by John Heywood, in the sixteenth century, 'When all candles bee out, all cats bee gray.'

Viewing tone and colour

Line drawings in a dense black ink on a white background represent the two most positive visual forces. White areas reflecting light are giving out energy or stimuli; black regions, because of their light absorbency, portray the absence of light. When we view these two extremes our visual senses record a pictorial message of density and contrast—white highlight areas give out a feeling of spaciousness and lightness, while shadow areas create a sensation of oppressive heaviness.

Fine line drawing

Heavy line drawing

Viewing tone and colour

With photographic prints the same visual feelings are felt when looking at the highlight and shadow ends. A great deal of play can be made on these sensations by altering the contrast range, that is, the degree of tonal movement between the shadow and highlight regions. Our visual interpretation of an individual tone is accentuated or diminished by the other tones surrounding it. The eye sees highlight areas first because of energy reflecting from them. Sparkle and brilliance may be portrayed by setting these highlights against a dark, solemn background. Conversely, a picture having predominant shadows and dark areas transmits a message of seriousness. When viewing the following examples you will probably feel that the shadow print depresses you, while the high key print is more stimulating.

Interpretations are also drastically affected by sharpness.

High key print

Heavy shadow print

Original picture

Out-of-focus print
Tonal gradient—no apparent difference between tones

In focus print
Tonal gradient—distinct difference between tones

To consider the sharpness of a picture we must inspect its outlines, that is, the edge or ending of a tonal weight against another, usually a background of lighter or denser tone. A sharp photograph is one in which these edges have no width, no transitional tones, and the tonal movement is immediate. The degree of sharpness can be indicated by a *physical gradient*—a line representing the physical movement or rate of descent from a darker tone to a lighter one.

The apparent sharpness of a picture is increased by emphasizing the edges of the tonal movement. This can be employed with advantage when a pronouncement of the focal point of the picture is required. To reinforce this the background can also be placed slightly out of focus.

When we see colours we are really looking at isolated regions of the visible spectrum. This isolation of colour is

Viewing tone and colour

Focal point: bowl of flowers—in focus, background—out of focus

Tonal gradient—accentuated difference between tones

caused by a pigment or dye naturally or synthetically inherent in the object, selectively absorbing certain regions of the spectrum and reflecting or transmitting the remainder. The visible spectrum moves through a complete gamut of violet blues, greens, yellows, finally ending in deep red. A warm sensation is generally felt when the red section is viewed, while blue gives the reverse, a cold feeling. Colours in a painting, transparency or colour print are in juxtaposition and will have a harmonious or discordant relationship with their neighbour.

The visual senses always try to create as much difference as possible between colours in a complementary fashion. For example, if a blue pigment is surrounded by green it becomes redder to the eye, because red is complementary to green. Similarly, if the same area is surrounded by red, the blue pigment will appear greener. You can prove this yourself by encircling a neutral grey pigment with different colours. It will be seen that the neutral appears tinted with the surrounding pigment's complementary colour. Some people talk about the vividness or depth of colour and this again can be misleading. The same pigment will appear to have more vividness and depth if it is surrounded by its complementary colour or printed as a large, isolated area.

During the reproduction process the technicians concerned have, for short periods of time, to remember a particular colour, translate it into a density of black silver and produce a dot or ink cell which will carry the correct amount of printing ink to reproduce this colour faithfully. This operation is made doubly difficult by the rather unreliable colour-retaining properties of the mind. It is an immense help to be able to classify this colour with precise terms and measure-

Original picture

ments, because an original picture's image is transferred from one medium to another anything up to ten times before the final printed copy is produced, entailing a tonal shift with each transfer, and complicated by the need for everybody concerned to view the original and its derivatives under exactly the same lighting conditions. Colour can be scientifically described and for all the people concerned in graphic reproduction it is a good habit to use these precise terms.

Colour

Science will entertain facts, but is rather unhappy with stimulated feelings or speculation. Bearing this in mind let us view the basic facts that confront us about colour. The easiest way to begin is by dividing colour into two sections, *synthesis* —the putting together of colour, and *descriptions*—terms indicating a colour's qualities. Under each heading there are three facts to be learned.

Synthesis

1. The visible spectrum is a small section of the much larger electromagnetic spectrum. It is seen as white light when all the wavelengths between 400 mµ and 700 mµ are radiated in a uniform distribution. If any of the component wavelengths in the visible spectrum are radiated individually then we see colour. Although there is an infinite number of colours in the visible spectrum it is helpful to divided it into three major regions—*blue*, *green* and *red*.

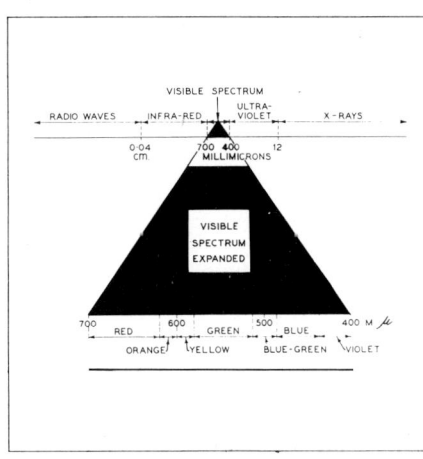

Visible spectrum expanded from the electromagnetic spectrum

VISIBLE SPECTRUM. Radiant energy which stimulates the eye producing the sensation of vision. Differences in spectral

Colour

distribution, e.g. the prominence of one wavelength over the others, result in a stimulus creating the sensation of colour.

2. If these three colours, blue, green and red, in the form of light beams are suitably mixed in various proportions then white light or any other colour can be visually matched. These three colours, being the first stage of colour vision, are termed the *primary* colours. Each of these primary colours has a *complementary* colour. The word complement means making up the whole, in this case the whole being the visible spectrum, so that when a primary colour and its complementary colour are mixed together the result is the spectrum. Therefore the complementary colour is produced by mixing the two remaining primaries together.

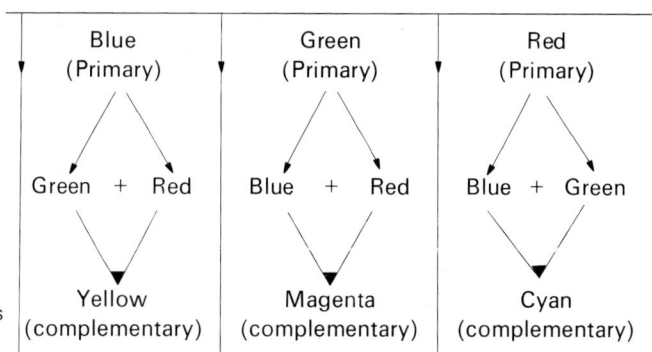

Visible spectrum

	Blue (Primary)	Green (Primary)	Red (Primary)
Primary colours	↓	↓	↓
	Green + Red	Blue + Red	Blue + Green
	↓	↓	↓
Complementary colours	Yellow (complementary)	Magenta (complementary)	Cyan (complementary)

PRIMARY COLOURS. Usually seen as monochromatic radiations which produce stimuli predominantly restricted to one of the three major regions of the visible spectrum, i.e. blue, green and red. When mixed in an additive manner can be made to match white or an infinite number of colours. Primary colours in pigment form transmit or reflect one third of the spectrum, absorbing the remaining two thirds.

COMPLEMENTARY COLOURS. These are sometimes rather confusingly referred to as the printing primaries, subtractive primaries or secondary colours. They are colours which carry out absorption in one third of the spectrum, transmitting or reflecting the remaining two thirds. Three in number and because of this selective absorption or subtractive feature are termed 'minus blue', 'minus green' and 'minus red' colours, i.e. yellow, magenta and cyan. When these colours combine in a subtractive manner a large number of colours may be matched.

Original picture

3. Coloured light is usually a mixture of wavelengths; it is rare to perceive a single colour or a monochromatic wavelength in nature. When we view colours it is difficult for our eye to determine if a colour is the result of two or more wavelengths emanating from a pigment, mixing in the eye to form a third colour, different from the two original wavelengths. Colour produced by this method is said to be an *additive* colour, the result of additive synthesis.

ADDITIVE SYNTHESIS. Light stimuli emitted in such a way that they mix an impinge on the retina either simultaneously, in rapid succession or in a mosaic pattern which the eye cannot resolve. If the primary colours of light are emitted simultaneously in equal proportions then white light will be the result.

The colour being viewed could also be the product of overlapping pigments producing a particular wavelength. In this case, because of selective absorption of the white incident light falling on the object the eye would view a colour already selected by the object's inherent pigment and the condition of the incident light. This colour is the balance of light left over after the object's pigment has subtracted portions of the visible spectrum. Colour viewed by this system is termed a *subtractive* colour, the result of subtractive synthesis.

SUBTRACTIVE SYNTHESIS. Absorbing media, such as pigments, dyes and filters, can be made to produce colours which are the result of simultaneous, successive or selective absorption of portions of the spectrum by each pigment present. If all three complementary colours were superimposed each one would absorb one third of the spectrum and black would be seen, because of the absence of the subtracted light.

Descriptions

1. When looking at pigments, dyes, etc., the first descriptive term which comes to our lips is the colour, or rather the basic colour, of the pigment. We distinguish them by saying that one is red, another blue or that one has a greenish look. Our retinas are affected by the most predominant wavelength contained in the pigment's stimuli. This underlying colour of the resultant visual sensation is known as the *hue*.

HUE. This is an attribute apparent in certain visual sensations which allows us to distinguish blue, green, red, yellow, brown, etc., from one another. This attribute is determined by the dominant wavelength of the spectral stimulus set up by the pigment.

Colour

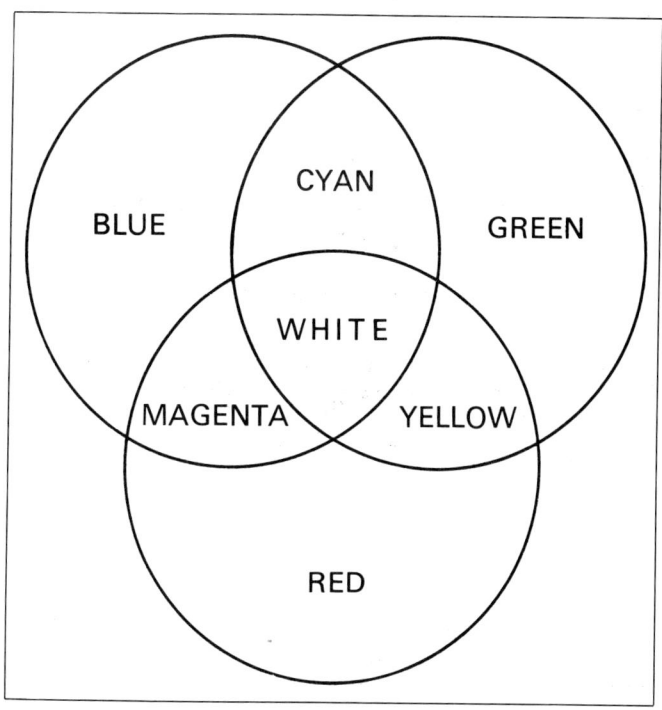

ADDITIVE SYNTHESIS
Blue, green and red circles of light projected onto a white screen give rise to two-band complementary colours where they overlap;
Red plus green gives yellow.
Red plus blue produces magenta.
Green plus blue results in cyan.
Where all three colours overlap, white light is obtained.
(It is recommended that the reader colours the diagram).

Original picture

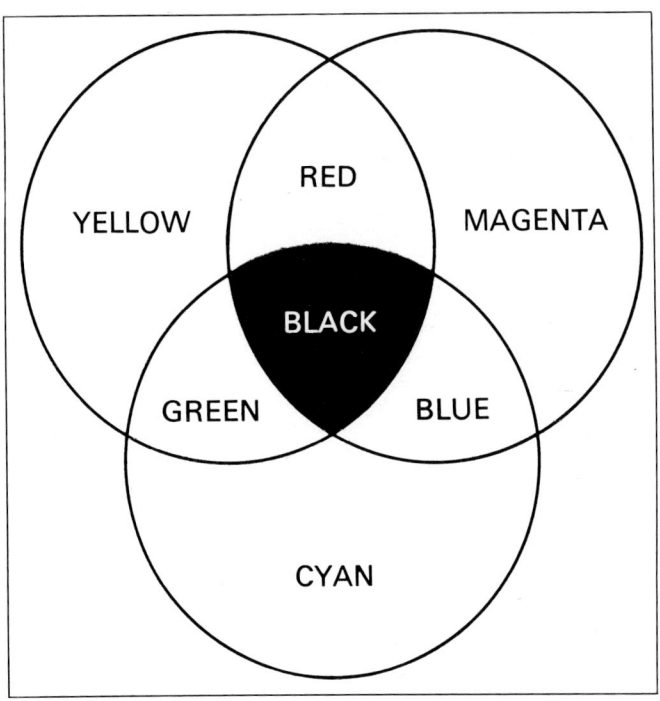

SUBTRACTIVE SYNTHESIS:
Yellow, magenta and cyan pigments, dyes or filters absorb the blue, green and red thirds of the spectrum respectively, but each passes the other two colours. Superimposing two such pigments, dyes or filters results in a primary colour being formed:
Yellow plus magenta gives red.
Yellow plus cyan produces green.
Magenta plus blue results in blue.
Where all three colours overlap there is an absence of light and black is observed.
(It is recommended that the reader colours the diagram).

Colour

2. After we have detected the hue, we become aware of the pigment's luminance—that is, the amount of light energy emanating from it. This strikes us as the brightness of the colour. Because of this sensation we are able to distinguish between a dark colour and a light colour of the same hue. A pigment with a red hue can be admixed with black to form dark red. The pigment has become degraded owing to the reduction of its luminance. Therefore when we speak of brightness we should be using the term *luminosity*, especially when searching for a subjective term to describe this attribute apparent in primary-coloured light sources.

LUMINOSITY. That attribute which permits a chromatic sensation to be classified by its luminous intensity, as a member of a range of colours which have the same hue but vary in luminance.

LIGHTNESS. This term must not be confused with luminosity, but is one that is readily used to describe complementary colours printed in a subtractive manner. *Lightness* is an attribute we describe by indicating the reflective powers of a pigment, the proportion of incident light it reflects back into our eyes.

3. On continuing a visual analysis of a colour the next subjective term usually employed is the purity of the colour, or how rich it looks. In fact the person is describing the amount of white admixed with the pigment. This has a drastic effect upon the colourfulness of the pigment. In the production of a water-colour wash impression, diluted paint is swept over a white surface leaving thin-looking colours containing a small amount of pigment. If the painting was repeated using undiluted poster paints then the white paper would not show through and the previous weak washes would take on a densely-pigmented condition producing a rich, colourful effect. This visual sensation is termed the *saturation* of the pigment.

SATURATION. That attribute of visual sensation which allows us to discriminate between the degrees of colourfulness apparent in a range of colour stimuli.

As mentioned previously colours can be measured scientifically. This is achieved via an instrument called a spectrophotometer which indicates the relative amounts of the three major spectrum colours apparent in the colour being analysed. A more detailed account of this colour measurement technique and its value may be found in the chapter on colour reproduction.

Original picture

Classification of originals

After viewing the original picture intelligently and describing its attributes in concise terms the next step is to classify it into one of these three major groups:

1. LINE ORIGINALS. Pictures delineated in solid black images on a white background, e.g. pen and ink drawings, type pulls, scraperboard illustrations, stipple and tint work.

2. TONE ORIGINALS. Pictures created in black and white, but with intermediate tones of varying density, e.g. photographic prints, chalk drawings, pencil sketches and wash drawings.

3. COLOUR ORIGINALS. Pictures presenting line and tone in colour, e.g. colour transparencies, prints, paintings and line drawings.

Preparation procedures follow, since many originals presented to the printer are below the required technical standard. In all cases certain general and specific operations have to be performed before the original is suitable for photographic reproduction.

Line original

Scaling the original

Generally speaking it is rare to receive originals which will print as a full page spread (usually appearing on right hand pages so that the illustration is viewed easily at the first glance). Most originals have to fit into a predetermined space which may be in between typematter or designed so that one or more edges of the picture 'bleed off' the page size when the printed sheets are trimmed. In a bled-off illustration due allowance must be made for trimming. The original must have its bleed edges between 3 and 5 mm ($\frac{1}{8}$ in. and $\frac{3}{16}$ in.) larger than the required size. Some originals are made, or happen to be the same size as the required printing dimension. This will mean a straightforward, same-size photographic copy, but the majority of originals have to be either enlarged or reduced to accommodate areas allotted to illustrations.

Before you can be sure an original is going to fit the space allowed for its printing, you must be sure that the picture area is in proportion to the final printing size. This may be easily checked by first drawing a light diagonal line with a soft pencil on the back of the picture. If you require an enlargement, gently glue the picture on to a large sheet of paper; then extend the diagonal line from the picture up and across the

Tone original

Original picture

paper. Now measure the required printing height or width out from the diagonal's left hand starting point. Construct vertical and horizontal lines to cut the diagonal. The lengths of these two lines from their base to the intersection will be the final printing size in proportion to the original. If this does not agree with the required size then the original must be masked to bring its width or height into proportion.

Scaling can also be calculated by a simple formula or with the help of a straight or circular slide rule. These methods will give more information than the diagonal method.

Proportional enlargement and reduction
1. Original format
2. Linear reduction
3. Linear enlargement

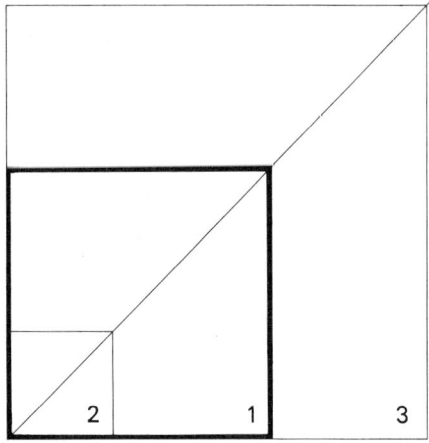

1. Finding the unknown final dimension

An illustration 400 mm × 320 mm needs to be reduced so that the long side becomes 100 mm. If the unknown side is termed X then its size may be calculated in the following manner.

$$400 : 320 = 100 : X$$

then $$X = \frac{320 \times 100}{400} = 80 \text{ mm}$$

The small side of the reduced picture will be 80 mm long.

An illustration 15 in. × 12 in. needs to be reduced so that the long side becomes 10 in. If the unknown side is termed **X** then its size may be calculated in the following manner.

$$15 : 12 = 10 : \mathbf{X}$$

then $$\mathbf{X} = \frac{12 \times 10}{15} = 8 \text{ in.}$$

The small side of the reduced picture will be 8 in. long.

Scaling the original

2. Calculating the magnification factor

Using the same example the original's longest side is 400 mm and must be reduced to an image size of 100 mm. The magnification factor is found by the following equation.

$$\text{magnification factor} = \frac{\text{image size}}{\text{original size}}$$

$$m = \frac{100}{400} = \frac{1}{4}$$

The magnification factor is $\frac{1}{4}$ or 25%.

Using the same example, the original's longest side is 15 in. and must be reduced to an image size of 10 in. The magnification factor is found by the following equation.

$$\text{magnification factor} = \frac{\text{image size}}{\text{original size}}$$

$$m = \frac{10}{15} = \frac{2}{3}$$

3. Working out the appropriate camera extension

An additional measurement which is helpful to the camera operator is the length of the camera extension at this magnification. This is worked out by the following formula once the focal length of the lens is known.

$$\text{Camera extension} = \text{focal length of the lens} \left\{ 1 + \text{magnification factor} \right\}$$

e.g. $f = 650$ $v = f(1 + m)$
$v = 650(1 + \frac{1}{4})$
$v = 812{\cdot}50$ mm

The camera extension at a magnification factor of $\frac{1}{4}$ using a lens of 650 mm focal length would be 812·50 mm.

An additional measurement which is helpful to the camera operator is the length of the camera extension at this magnification. This is worked out by the following formula once the focal length of the lens is known.

$$\text{camera extension} = \text{focal length of the lens} \left\{ 1 + \text{magnification factor} \right\}$$

e.g. $f = 25$ in. $v = f(1 + m)$
$v = 25(1 + \frac{2}{3})$
$v = 41{\cdot}66$ in.

The camera extension at a magnification factor of $\frac{2}{3}$ using a lens of 25 in. focal length would be 41·66 in.

Original picture

A worthwhile aid to precise working is a reproduction circular slide rule. The more complicated measurements can be calculated by rotating the two sets of numbers against one another. All the information arrived at previously by equation can be readily found. Usually the numbers are not marked off in any particular unit, so metric or typographical measurements can be used, but decimal calculations do seem to be the most popular. More advanced discs can be purchased for the photographer which also give exposure data.

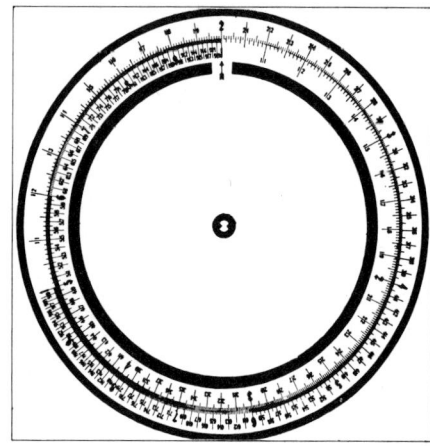

A circular calculator set at 2:1

Masking to proportion

If after calculation the required printing dimensions do not agree with the original's length and width, then its size must be brought into proportion by masking. Never cut off

Corner masks of white card for determining the picture size

Masking to proportion

unwanted areas of the picture, because originals are filed after reproduction to await use in a new production, usually in a new size.

To mask the original cut two L-shaped masks out of stiff card. Hinge a sheet of thin white paper over the original by folding approximately 25 mm (1 in.) of its top horizontal edge on the the back of the original, and lightly stick with glue. Now flap the white paper away and position the L-shaped masks so that the original's picture area is in proportion, gently flap back the white paper and make a pin-prick in each corner of the new picture area. Flap back and cut out the new shape making sure the edges are square. Remove the L-shaped masks, re-position the cut-out overlay and lightly glue it down. The surrounding white paper border makes an ideal background for the information lines which must be drawn to facilitate precise registration in the printing. These fine lines should be drawn with black indian ink, 0·2 mm in width, and are referred to as:

1. Central register marks ¢
2. Trim marks   ŧ
3. Fold marks   f

A colour transparency needs to be dropped into a sheet of fixed-out photographic film, preferably on a stable base. There should be at least 25 mm (1 in.) of clear film all the way round the transparency. The best way to do this is to lay the transparency on top of the film in the required position, trace round its shape, remove the transparency and cut out the shape with a sharp knife. Locate the transparency in the cut-out shape so that the emulsion side of the transparency is next to the

Information lines

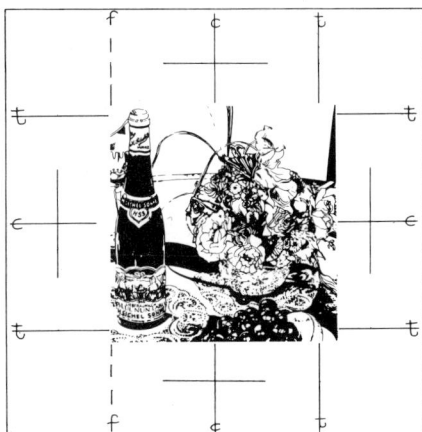

Original picture

emulsion side of the clear film. Glue it firmly to the film with clear tape, making sure the tape does not encroach upon the required image area of the transparency. Now carefully plot the centres of this image area and draw in with black indian ink (on the emulsion side of the film) the required information lines, e.g. central register marks, etc. It is a help to the photographer if the background to these central register marks is lightly dyed up with grey retouching dye until it has a density of approximately 1·0. This ensures that the register marks will not disappear because of excessive exposure when the transparency is being exposed. Finally, lightly attach a black paper mask over the transparency with the required image area cut out. This shows the photographer exactly to what area the final size and lay-out dimensions refer.

Instructions for graphic reproduction

Attach a sheet of tracing paper over the original so that it may be flapped back. On this sheet of tracing paper, over the appropriate areas of the original, indicate:

1. The original's number which corresponds to its space in the lay-out.
2. The size required.
3. The magnification factor.

On the same sheet of tracing paper indicate the following corrections:

4. Areas which need retouching.
5. Required background effect—as original, cut-out, vignetted or airbrushed.
6. Areas in which tints are required; the size of tint in percentage.

Essential written instructions can also be tabulated:

7. The reproduction process assigned to the printing. This determines whether the negative is to be reversed or not.
8. Stipulate the desired screen ruling (dots per inch or cm) for tone originals.

It is helpful to issue some sample sheets of the paper to be used for printing on. The reproduction technicians will be able to give an example of the result, which prevents disappointments and indicates the need for possible tonal corrections.

If these general preparatory operations are carefully carried out the originals will start along the appropriate reproduction flow-path correctly and without confusion. Occasionally

Line originals

unusual originals are encountered and these need individual treatment and extra instructions.

Line originals must be presented in only two tones—dense, opaque black and clean, reflective white. The line situation can be seen in many guises: type-pulls, pen and ink drawings, map-outlines with appropriate tint and stipple overlays are examples of just a few. If all these line presentations are going to reproduce well, that is as facsimiles, then the following requirements should be closely adhered to:

1. All image areas must be completely black and firmly drawn with sharp edges.
2. The background should be a smooth white surface of high reflectance.
3. Background materials need to be opaque and reasonably stiff or dry-mounted.
4. The original's size should be $1\frac{1}{2}$ times larger than the required printing size; the reduction helps to diminish any pictorial faults.

Good line originals have an air of spaciousness. The black lines provide the detail and contrast without becoming too thin (width not less than 0·075 mm (0·003 in.), while the white areas provide space and movement before becoming too small or isolated (width no less than 0·025 mm (0·001 in.). Black lines drawn closely together or minutely cross-hatched should be avoided.

Although these recommendations are well-known, a large proportion of line originals presented for reproduction do not conform and in many cases seem to have been drawn up with the reverse requirements in mind. Nevertheless the reproduction staff should, by using retouching methods, try to improve the original's photographic suitability.

The retouching methods are best learned as a number of steps.

1. Check and redraw if necessary any fine black lines which are too thin, broken or grey, with process black ink.
2. View and assess any minute white areas enclosed by black lines; increase their size if necessary while still maintaining their shape, using process white ink—do not employ chinese white as it absorbs untra-violet (u.v.) light and photographs as grey.
3. Any badly drawn or smudged image areas can be masked out by a piece of white paper and more suitable black lines redrawn on top. Intermediate image areas can be improved

Original picture

Unretouched line original

Retouched line original

by the inclusion of a suitable mechanical line tint (hatched lines of varying width).

4. If the original is too small to improve, it is advisable to have an enlarged photographic bromide print made. This print will contain larger areas which are easier to redraw and retouch.

Scrapboard technique

For the execution of line illustrations portrayed in fine lines with intricate detail the scrapboard technique, if well performed, is ideal. Scrapboard is a sheet of stiff board containing a layer of white chalk topped with a thin coating of black ink. Black lines appear as the top coating is carefully cut away exposing the white chalk underneath. Embossed scraper-

Line originals

boards are also made which leave patterns, tints, stipples and textures behind after cutting. With the facility to draw on top of the black coating with process white ink scraperboard is a line medium capable of giving the designer considerable artistic freedom.

Technical line illustrations

Black line drawings, such as technical lay-outs, illustrated machine parts drawn on a transparent material, e.g. Kodatrace or tracing paper, need to appear opaque in transmitted light, because if a same-size reproduction is required the original can be used as a positive, with films or printing plates being exposed directly from it. An alteration in size means a camera stage, so a sheet of white paper is used to back the transparent material rendering it opaque and reflective. Unwanted background grids should be drawn in a light blue colour, while requiring scaling or perspective graticules need to be in red.

Typematter

Many line drawings, especially technical illustrations, are accompanied by explanatory captions. These can be set directly on to film via a phototypesetting machine, but are usually letterpress type-pulls pasted on to the illustration. Type-pulls taken for reproduction purposes need to be on a 'coated' paper, printed in best quality black ink with the minimum of 'ink squash' effect. Type faces containing fine serifs do not photograph well and should be discouraged. Larger lines of type for headings or displays can be obtained by using one of the self-adhering pre-printed type sheets, e.g. Letraset. This system is satisfactory as long as due care is taken and the letters do not split on transfer.

Line and tint pictures

Line drawings of certain subjects tend to be visually thin and can be made more pictorially wholesome by the inclusion of mechanical tints—areas of dots or dot patterns of the same size, or stipples and textures.

The inclusion of these extra line media may be carried out on the original or if intricately shaped and minute areas have have to be covered, combined during the photographic process. These tint areas are best indicated on the original by applying a light blue wash. This wash will be lost during photography and replaced by a black dot tint of the size required, which must be clearly stipulated on the transparent overlay. If a reversed tint is needed, that is white dots on a black background, then the wash should be changed to red

Original picture

Line drawing

Line and tine picture

which will photograph as black leaving a clear area for the tint.

Colour line work

This could be a simple case of a black outline being filled with solid areas of colour, or a more complex extension of line and tint work so that the printed reproduction contains black outlines, solid colour areas and shades—colours plus a dot tint of black or tints—single or overlapping dot tints printed in colour. Once again the tint-laying and solid colour areas can be presented on the original as long as dimensionally stable material is used and precise register is maintained between the overlays and the base outline. Image areas, representing the colour portions, can be produced from the base outline during the photographic process, but in both systems any

Line originals

Line drawing plus positive tint

Line drawing plus negative tint

Colour line originals
1. Yellow 2. Magenta
3. Cyan 4. Black

Original picture

solid or tint area which abuts another solid or outline must slightly overlap to prevent the white paper showing through when minute register variations occur during the printing.

The designer must appreciate the colours obtainable by overprinting solids and tints so that maximum colour change is obtained from the basic inks used. A carefully planned colour line reproduction can be ruined by bad registration of the colours, so infinite care must be taken to ensure that each colour overlay, photographic copy and printing plate has register marks and image areas which will fit precisely once printing commences.

Colour plates overprinted
1. No overlap—white outlines
2. Overlaps—abut fit

Tone originals

The term 'tone original' embraces all pictures created in neutral tones or greys of varying strength ranging from white to solid black. They can be seen in the form of photographic prints, wash drawings or pencil sketches. When such originals are being created with subsequent printing in mind the modyfying effects of the reproduction process must be realised and taken into account. Generally speaking, the reproduction processes in printing lower the tonal movement between the lightest tones, and subdue the shadow tones. Each process has a maximum density capability. This is shown as a reflectance value after reading the darkest and lightest tones with a reflection densitometer. Tone originals presented with greater densities than this value will be printed with black tones flattened to dark greys or both highlight and shadow areas in a visually lower condition. The ideal presentation breaks down into four requirements.

Tone originals

1. The darkest tone, usually a black, should be immediately detectable from the other dark tones. Ideally the darkest tone should have a reflection density reading of the same value as the maximum density capable of being printed by the particular process, e.g. letterpress 1·6, gravure 1·8, lithography 1·4 and screen process 2·0. (N.B. These densities are single ink films on a good quality paper.)
2. The white areas need to be accentuated so that they 'jump' away from the nearest light tone. White areas are usually compared to the total reflectance of a block of magnesium oxide and ideally record 0·05. The next light tone should be at least 0·1 to give highlight accentuation.
3. The intermediate tones between the white highlight and black shadow areas need to be distinct movements. There should be no more than eight, so that including the highlight and shadow areas the printed result will contain ten tonal movements. Each step should be approximately 0·25 increase in reflection density.
4. Tone originals should be 1½ times larger than the required printing size and where a number are being used in the same production a common original size is extremely helpful.

Good tone originals have plenty of middle or medium tone areas. Most of the picture should be shown in soft, delicate greys, while dark and black tones are reserved for dramatic or important detail. Astute use of individual highlights and areas of white gives brilliance and a feeling of space to the picture. Graduation needs to be smooth but well denoted. If the tone original does not contain these attributes then they must be put in by skilful retouching. Immense improvements and alterations can be brought about when manual and air-

Tonal ranges capable of reproduction. (N.B. The ink film densities stated are average figures related to a single ink film printed on a coated paper)

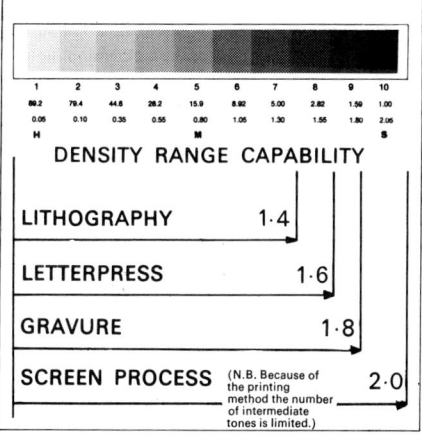

Original picture

Unretouched tone original

Tone original masked

brush techniques are used. A retoucher usually follows this routine when preparing a tone original:

1. Read instructions and mask-out, spray-down or vignette the background as required. This treatment emphasizes the picture's main subject.
2. View the middle and graduation tones; if necessary mask-out correct tones and spray the incorrect areas with an airbrush until the desired tonal weight is achieved.
3. Assess the detectability of the darkest tone; improve if necessary by overdrawing with a darker tone diluted from process black ink.
4. Accentuate the whitest areas by the addition of process white neat (for individual highlights) or diluted (for white spaces).

Tone originals

Airbrushing
Parts of the airbrush:
1. Needle valve
2. Trigger
3. Reservoir
4. Air Tube

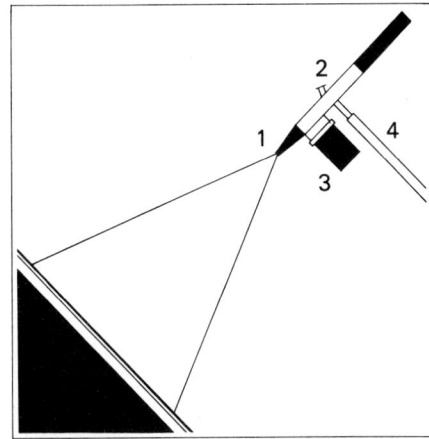

Retouched tone original

Wash drawings

A drawing of this type carried out in delicate washes of black water colour needs to be on a semi-matt white surface glued to a stable base. The surface can be moistened all over with clear water before drawing commences to lessen the chance of buckling. It is good practice to work to a limited stepwedge of tones which are suited to the required visual interpretation and printing process. On no account should lighter tones be obtained by adding white pigment; light tones should be a direct dilution of the solid pigment. Vignetted backgrounds and light tonal movement need to be shown with more contrast than is required in the final print.

Chalk and pencil sketches

When these media are used the problem of producing

Original picture

Tone original for visual appreciation:
negative curve-positive print

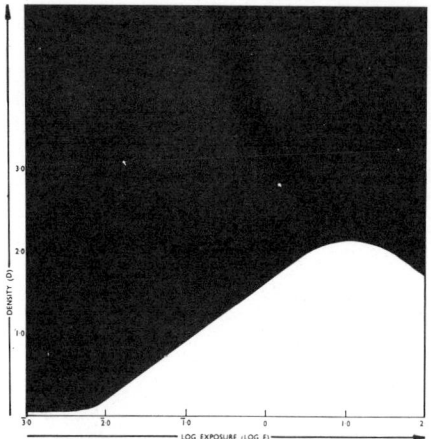

enough 'jump' between the white background and the lightest pencil stroke becomes evident. There are photographic emulsions prepared especially for this work in which increased contrast will be recorded on the white areas, but a lot of unnecessary work can be avoided if a little thought and skill are introduced at the conception of the original. The whitest possible background materials should be chosen on to which an 'open' sketch is drawn. Again an air of spaciousness is the keynote, along with the necessity to draw the lightest pencil strokes as firmly as possible.

Photographic prints

Photographs designed for reproduction should be made on smooth white, glossy promide paper at a larger size ($1\frac{1}{2}$ times)

Colour originals

Tone original for graphic reproduction
negative curve—positive print

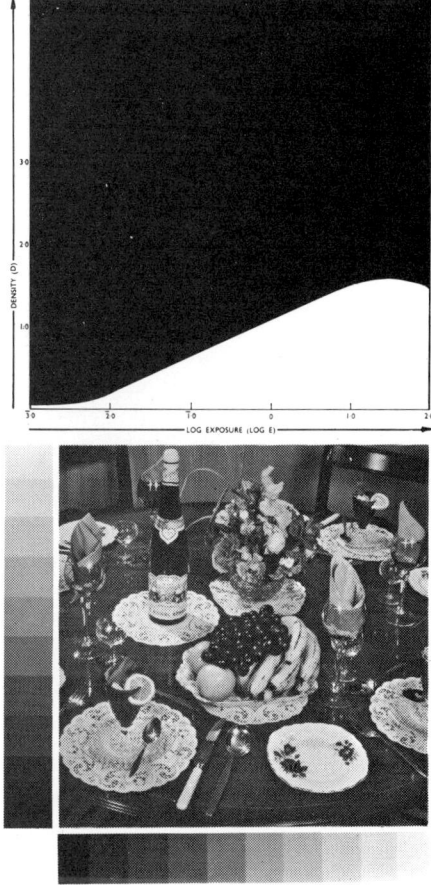

than the required printing size. If the prints do not require retouching or airbrushing then they can be glazed. All photographic prints need to be dry-mounted on to a sheet of stiff white board. Matt-surfaced prints should be avoided, but if there is no other alternative then white vaseline may be carefully applied to the surface, rendering it as smooth as possible. The overall density range needs to be lower than the one normally adopted for just visual appreciation. This can be be best explained by comparing the two, firstly as characteristic curves and secondly as visual comparisons.

Colour originals

Colour originals can be seen as line or tone interpretations. Assessment of these originals is based on the same attributes required for monochrome line and tone originals, but finalized

Original picture

on the suitability of the colours used. The colours may be the result of artistic skill presented in the form of designs, posters and paintings, displayed in bold, dominant brush-strokes or delicate, subtle washes. Photography is also widely employed as an instant supplier of colour originals, issuing forth a continuous flow of coloured prints and colour transparencies varying from explorations of the natural landscape to highly commercial fashion shots. All these diverse media exploit colour, tantalizing, stimulating and satisfying the human eye with controlled amounts of coloured light, e.g. the visible spectrum.

To reproduce a coloured picture the printer must transfer layers of ink on to white paper which will control exactly the same amount of coloured light as the original did. The human eye will then closely associate the printed copy with the original. The colour reproduction processes in printing are rather restrictive. For both practical and economic reasons tri-colour inks, e.g. *yellow, magenta* and *cyan* are normally used to control the reflection of the visible spectrum, e.g. *blue, green* and *red* light. Because of certain deficiences in the printing inks and individual processes a number of colours seen in nature cannot be satisfactorily reproduced. These may be broadly named as vivid blues, deep purples and dark greens, and need to be reproduced as the nearest tri-colour match or printed separately in a specially colour-matched ink via an extra plate. To avoid disappointment and possible unacceptability of the printed reproduction the coloured original created for printing should satisfy the following conditions:

1. All the colours contained in the original must be correlated to the printer's ink chart, which will show the various colour combinations obtainable with the inks and paper chosen for the reproduction.
2. Solid colours, dark shades of colour and light coloured tints should be as clean as possible with mixture colours, like brown, being individual pigments and not mixed from other paints. Weaker washes of colour must be direct dilutions of the solid colour pigment. Making light coloured tints by the addition of white pigment should be forbidden.
3. The colour original's overall density range and number of graduations should correspond with the density and tone capabilities of the particular printing process selected for its reproduction.
4. Autographic originals, such as paintings, impressions and washes, need to be executed on the same quality paper as the stock selected for printing. Colour photographs and trans-

parencies usually have extreme colour depth and vividness. Reasonably close colour reproductions can only be achieved on the best quality 'chromo art' papers using gloss inks.

Ideal colour originals for graphic reproduction have an aura of cleanliness with sharp, positive movements of colour from shadow to light tones. Dark colours and black are only used to give contrast and delineate important detail, so that the image on a black printing plate is reserved for adding contrast to the darker tones and producing sharper detail. This 'keys-up' the whole picture without printing in the lighter or clean colour areas. It is rewarding for everybody concerned if this type of original is always presented, but of course quite frequently originals with unsuitable characteristics are proffered for reproduction. Once again these originals can be improved by the skill of the retoucher carrying out the following remedial treatments:

1. Increase the cleanliness of the original by over-painting incorrect areas with more suitable paints, or in extreme cases mask-out with white paper and redraw and paint the incorrect area.
2. If necessary expand the overall density range to a value suitable for printing by adding lighter tones and process white to the highlights or process black and darker pigments to the shadow tones.
3. Undertake remedial work by introducing airbrush techniques; increase colour movement, subdue unimportant areas and accentuate the main subject detail.
4. If the original is photographic, e.g. a colour print or transparency, then the previous treatments are repeated with suitable photographic colour tints and dyes which fuse easily into the gelatine surfaces.

Colour transparencies and prints

In most modern coloured printing productions, e.g. mail order catalogues, holiday brochures and weekend supplements, extensive use is made of the integral layer transparency process. This process produces an instant photographic original consisting of three coloured dye layers, *yellow, magenta* and *cyan*. Extremely vivid, saturated colours can be achieved, but their visual colour and tonal quality varies drastically with differing light sources. This dilemma is solved when the customer, reproduction technician and printer view the transparency and its coloured reproductions under the same conditions, e.g. an illumination of 5000° Kelvin (see page 92.

Original picture

Whenever possible the coloured transparency should be made as large as possible; small films such as 57 mm ($2\frac{1}{4}$ in.) square and 35 mm sizes can be enlarged, but this increases the problem of retaining critical detail and colour balance. The end densities of a transparency created for colour reproduction need to be in the order of 0·2 in the lightest highlight tone and 2·0 in the darkest shadow, giving an overall density range of 1·8. The transparency is usually colour corrected by the addition of a film mask; this lowers the overall density range to one more appropriate for the printing processes. Transparencies which have a colour cast due to incorrect exposure or processing should be rejected, because even with the use of colour correction filters the reproduction is rarely up to standard. In all photographic work prevention is definitely better than cure.

Photographic prints on paper from coloured negatives are also popular originals and in many ways are more suitable for reproduction than transparencies as their density range is similar to those used in printing, complemented by the ease of handling and the fact that any colour or tonal correction can be quickly carried out by the retoucher.

3. LIGHT BEHAVIOUR

Historical introduction

Ever since the first cave-man squinted and shielded his eyes from the sun the nature and behaviour of light has intrigued men. Their attention was drawn to the visual excitement achieved while watching light rays form shadows, highlights, colours and eclipses. Progress in understanding the physical make-up of light was slow. One of the earliest speculations was that rays of light emanated from the eye and detected the subject to be viewed. This was later revised and the formation of light rays was attributed to luminous bodies emitting a stream of fast-moving corpuscles which stimulated the eye.

The corpuscular theory was vehemently supported throughout the early seventeenth century by the great physicist, Sir Isaac Newton. The only contemporary scientist to postulate a differing theory was Christian Huyghens who proposed that light was in the form of longitudinal pulses (not strictly waves). Starting with this early idea of light waves eminent scientists such as James Clerk Maxwell, Augustin Fresnel and Thomas Young carried out successful experiments to establish that visible light was a small part of a much larger electromagnetic spectrum and that light waves consist of a sequence of electrical disturbances which can be represented graphically as waves with crests and troughts. These crests and troughs represent +− or −+ polarity, and do not imply any geometric shape in the light ray.

By 1850 the wave formation of light was being generally accepted, although theories, no matter how logical and proven, are often confronted with phenomena which they cannot completely explain. Such phenomena appeared in the thoughts and experiments of Max Planck and Albert Einstein while working on photoelectric effects. From their observations the Quantum theory of light arose which proposes that light is emitted in discrete amounts or quanta, bursting forward like spasmodic machine-gun fire. Light rays convey energy in discrete minute units. These units are referred to as quanta or photons. The amount of energy in one quantum is proportional to the frequency—that is, inversely proportional to wavelength. Blue light having a short wavelength contains

Light behaviour

more energy than red light. This can be seen in the way blue light reacts more violently than red light with silver halide and di-chromated colloid emulsions. Today, scientists solve their problems and conduct experiments in both wave and quantum terms, respecting the occasional dualistic nature of light.

Visible spectrum

The photographer must concentrate his thoughts on two basic properties of light, firstly the *dominant colour* of the light and secondly the *luminous intensity*, because different colours stimulate the eye and expose photographic emulsions in varying ways, while the intensity controls the length of exposure. The wave-theory of light produces measurements and factors which can be readily applied to these two important properties. It is the wavelength that denotes the dominant colour while amplitude, the height of the wave, determines the luminous intensity.

While studying at Cambridge University in 1666, Sir Isaac Newton produced a distribution of colours from a single beam of sunlight. He allowed this beam to fall at an acute angle on an equilateral glass prism, and the emerging beam expanded into a range of colours from deep red to blue violet, each component colour merging imperceptibly into the next. This experiment became the classical way of producing a spectrum (visible colours) by dispersing light rays through a prism.

From 1800 it became apparent that the spectrum of the sun extended beyond the visible wavebands of red and violet. Light began to be accepted as an electromagnetic radiation extending over the visible spectrum, which was a small

Light-waveform
Wavelength—dominant colour
Amplitude—luminous intensity

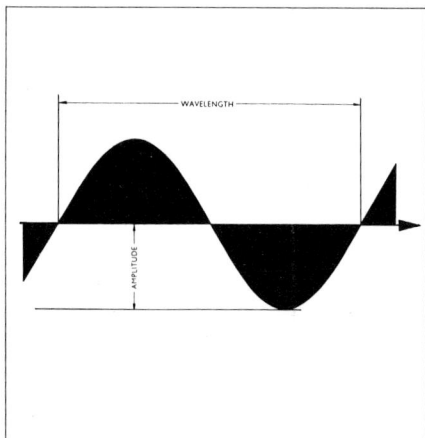

Dispersion of light

Visible spectrum

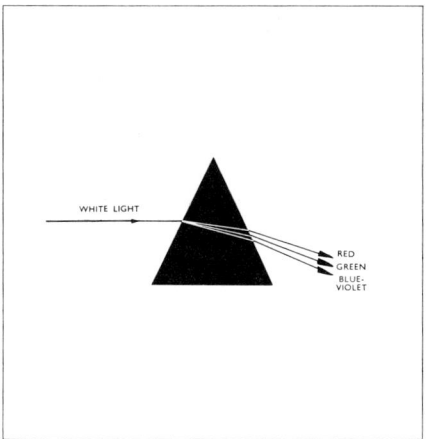

Visible spectrum

section of a much longer electromagnetic spectrum, most of which is invisible. We now realize that the full electromagnetic spectrum ranges in wavelength from long electrical wavebands of 10^{13} mμ to short gamma rays of 10^{-6} mμ. The component colours of the visible spectrum when viewed seem to be divided into three major regions, blue, green and red. This division is only carried out to provide a practical nomenclature. We must not lose sight of the fact that there is an infinite number of colours in the spectrum. Each colour can be isolated or recognized by its wavelength, which can be expressed in millimicrometers (mμ) or Ångström units (Å). With the advent of metrication these units will be superseded by the nanometre (nm), 1 nm equals 1 mμ. The limits of the visible spectrum are wavelengths of 700 nm deep red to violet wavelengths of 400 nm.

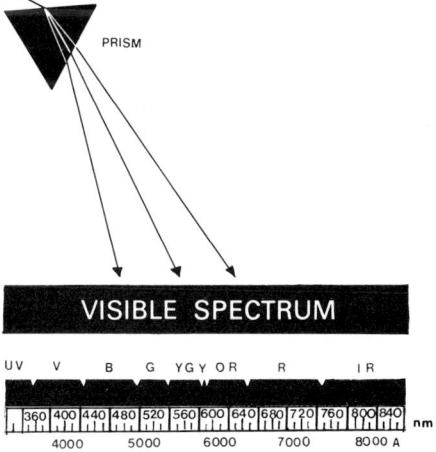

Light behaviour

Light terminology

In geometrical optics light is represented by a *ray*, i.e. a straight line indicating the direction of travel. A collection of rays from a near point source is called a *pencil* of light. A collection of pencils radiating from a source is called a *beam* of light. The study of light is based on three movements of light rays.

1. Rectilinear propagation

Light travels in straight lines. Light rays emitted from a point source may be shown as a number of straight lines radiating from the point. The straight line travel may be demonstrated by the use of a pinhole camera or blocking the path of a beam with an opaque object to produce a shadow and eclipse. In the case of the camera the fact that the image is real and inverted suggests that rays of light travelling in a single medium are propagated rectilinearly and the position and size of an image is found by drawing straight lines representing light rays passing and refracting through a lens.

Rectilinear propagation—the pin-hole camera

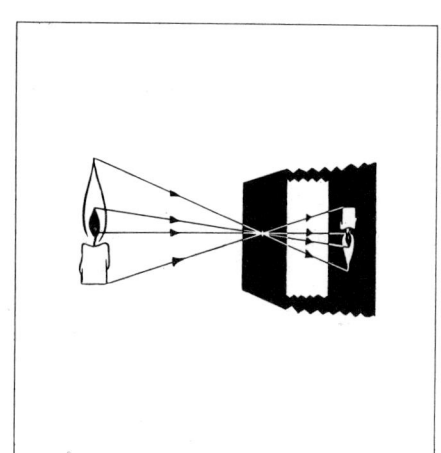

2. Reflection

Physical bodies are luminous or non-luminous. Luminous bodies emit their own light while non-luminous bodies become visible only by reflecting or throwing back light rays which emanate from a luminous body, e.g. the sun is a luminous body while the moon is only visible while reflecting the light rays from the sun. Light radiating from a source travels through space and impinges upon an object. According to the manner in which it is absorbed or reflected by the object's surface, a pattern of light, shadow and colour originates. This pattern

Light terminology

is visually interpreted by us into a three-dimensional picture.

A flat polished surface, such as a plane mirror, reflects most of the light that falls on it in a *regular* or *specular* manner, while a matt surface reflects incident light in an *irregular* or *diffuse* manner. All surfaces reflect light to some extent— a black body reflects very little, a white body reflects most of the light falling on it. The ratio of the reflected light from a surface to the total incident light reaching it is called the *reflectance* of that particular surface. In photography the angle and position of lamps should conform to the laws of reflection. The different visual tones of the final photograph are directly related to the varying intensity of the reflected rays from the original. The photograph is a representation of what a single eye sees. From the amount and intensity of light and shadow in the positive photographic print our visual mechanism arrives at an interpretation of the object in the picture. Therefore the reflectance of light informs us about the texture, size and shape of the object whether we can touch it or not.

Reflection:
∠ I = ∠ R
AB = BC

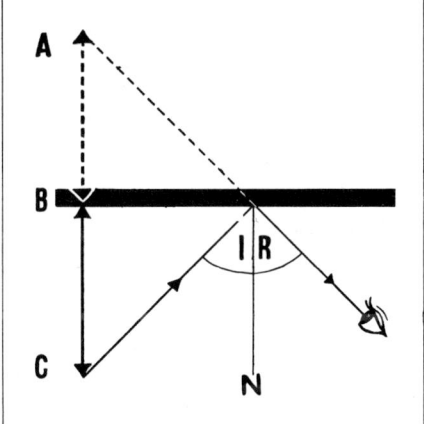

Laws of reflection

1. The incident ray, the normal, and the reflected ray lie in the same plane.
2. The angle of reflection is equal to the angle of incidence.

3. Refraction

If a light ray passes from one transparent medium to another its velocity is changed. If the incident ray is oblique it is refracted or bent, e.g. a light ray passing from air to water or air to glass. The amount of refraction depends on the

Light behaviour

optical density of the two media, the angle at which the light ray meets the surface, the shape of the surface and the light ray's relative speed in the two media. A certain proportion of the light ray is reflected by the second medium's surface, another portion is absorbed and the remainder refracted. The classic example of refraction is the bent appearance of a stick when partly immersed in water. When light passes from a rarer to an optically denser medium, e.g. from air to glass, it is bent towards the normal. Light passing from a denser to a rarer medium is refracted away from the normal. Light which falls perpendicularly will pass straight through although its speed will be retarded.

Refraction is caused by the retardation in the velocity of the ray as it crosses the boundary separating the two media. The ratio of the velocity of the ray in the first medium compared with its velocity in the second medium is termed the *refractive index* and is denoted by the sign μ. This figure indicates the extent of the deviation of the light from its original path. To calculate the refractive index for a particular substance using light of a predetermined wavelength the following equation may be used:

$$\text{Refractive index } \mu = \frac{V^1}{V^2} = \frac{\text{Velocity of light in air}}{\text{Velocity of light in the particular substance}};$$

e.g.
$$\mu = \frac{186{,}000 \text{ miles per sec}}{140{,}000 \text{ miles per sec}} = \frac{\text{Velocity of light in air}}{\text{Velocity of light in water}} = 1 \cdot 3,$$

$$\mu = \frac{186{,}000 \text{ miles per sec}}{124{,}000 \text{ miles per sec}} = \frac{\text{Velocity of light in air}}{\text{Velocity of light in crown glass}} = 1 \cdot 5,$$

$$\mu = \frac{186{,}000 \text{ miles per sec}}{110{,}000 \text{ miles per sec}} = \frac{\text{Velocity of light in air}}{\text{Velocity of light in flint glass}} = 1 \cdot 7,$$

Different glasses have different refractive indices and this point is taken advantage of by lens-makers for correcting defects from which the lens may be suffering. In 1620

Light terminology

$$\frac{\sin \angle I}{\sin \angle R} = \mu$$

1. Transmitted light
2. Refracted light

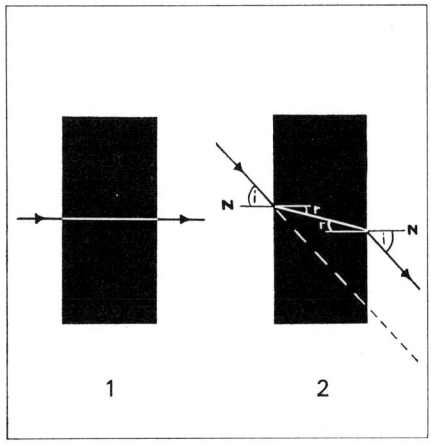

Willebrord Snellius, a Dutch astronomer, propounded two laws of refraction.

1. When a ray of light passes from one medium to another, the incident ray, the refracted ray and the normal at the point of incidence, are in one plane.
2. If *i* is the angle of incidence and *r* the angle of refraction, then for light of any one particular wavelength, the sine of the angle *i* divided by the sine of the angle *r* is a constant, called the refractive index of the second medium.

When a ray of light passes from a denser to a rarer medium, e.g. glass to air, the greater proportion of the ray is bent away from the normal, while a smaller portion is reflected back into the glass by the boundary. If we consider four indepen-

Total internal reflection:
1. Refracted light
2. Critical angle
3. Totally internally reflected

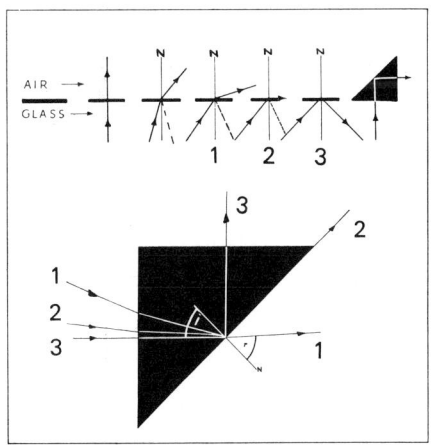

Light behaviour

dent light rays diverging from a point inside the glass, as the rays move into the air each ray, depending on its incident angle at the glass surface, will become more inclined towards the surface and the fraction of light internally reflected becomes greater, until at a specific angle (glass 41°) the refracted ray will skim the surface of the glass. This angle is termed the *critical* angle. Any incident ray meeting the surface at an angle greater than the critical angle will be totally internally reflected. If the critical angle for crown glass is approximately 41°, glass prisms with angles of 45° and 90° must totally and internally reflect any light ray entering one face and striking the hypotenuse face at a greater angle than the critical angle. If a prism of this nature was used in camera work to produce a reversed image, light rays meeting the face at a smaller angle than 41° would pass into air. Because of this the process prism has its long side silvered and so becomes virtually a mirror. In fact, on modern process cameras mirrors are used, being silvered on the face to prevent duplication of the reflected image.

Photometry

To produce constantly good quality photographic films a photographer must standardize the intensity and uniformity of his light source when illuminating the subject. In an endeavour to achieve this, applied photometry is an asset. For many years photometry was based on a unit source of light known as the standard candle, a candle of sperm wax burning at a rate of 120 grains per hour. Because the candle's intensity varied by approximately 20% this has been replaced by more scientifically reproducible standards.

Light units:
A. Luminous intensity—candela
B. Light source—one candela
C. Luminous flux—lumens
D. Illumination—lux
E. Reflectivity—reflection factor
F. Luminance—nit

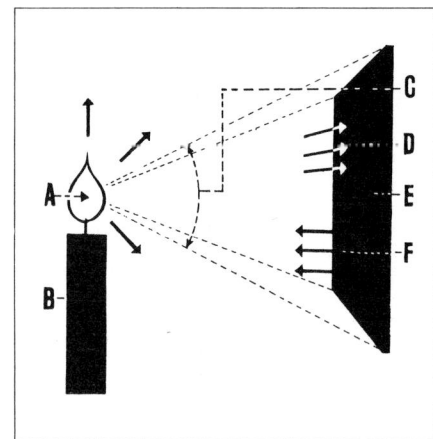

Photometry

Intensity of illumination

Imagine a 1 m (1 ft) radius sphere perfectly blackened on the inside surface, and the 1 candela (cd) source at its centre, emitting light equally in all directions. All the light from the source is absorbed. Now cut away 1 m³ (1 ft²) of the surface of the sphere. (This subtends unit solid angle 1 m² (1 ft²) at centre of the sphere.) The light coming out of the hole is 1 lumen (lm). If a 4 cd uniformly radiating source is placed at the centre of the sphere, 4 lm will pass through the hole. Therefore intensity of illumination may be found as follows:

Intensity of illumination

$$= \frac{\text{luminous flux, i.e. light coming out of the hole}}{\text{area over which this is distributed}}$$

= lumens/ft² (ft candles) (lux)

= lumens/m² (lux)

Before cutting away the piece of sphere, 1 lm was falling on this surface. The intensity of illumination was 1 lm per m². The whole surface was 1 m (1 ft) from a 1 cd source, so the intensity of illumination could also be expressed as 1 m–candela (1 ft candle) although this is rarely done.

Illumination control

In camera work for graphic reproduction the original is usually illuminated by four light sources. The amount of light passing through the lens and falling on the emulsion should be recorded through a light-integrating meter. This will smooth out any fluctuations occurring at the light source and provide constant exposures timed in light counts. The illumination of the original may be checked for uniformity with the aid of an exposure meter or photocell by dividing the original into nine sections and reading the intensity in each section. The four light sources may be moved in order to produce as near even illumination as possible.

When altering the distance of any light source the inverse square law may be applied. The law states, *the illumination of a surface is inversely proportional to the square of its distance from a point source*. If the distance from the source is doubled, the illumination is spread over the square of the distance, four times the previous area. If the distance is trebled, the illuminated area is increased nine times. Strictly speaking, this relationship is only valid for geometrical point sources.

Light behaviour

Inverse square law:
Distance 1m—coverage 1 square
Distance 2m—coverage 4 squares

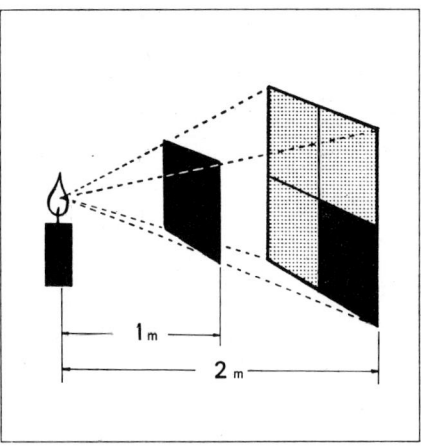

During the graphic reproduction process the inverse square law may be applied to the following:

1. The camera lamps, their distance and angle.
2. The camera extension using a fixed stop.
3. The lamp distance in a contact frame.
4. The lamp distance with the printing down frame.
5. The enlarger, lens to emulsion.

The law is used in the following manner:

Example. If, with a lamp at 300 mm, the exposure time is 15 sec, what would the exposure be with the lamp at 900 mm?

$$\frac{\text{New exposure time}}{1} = \left(\frac{\text{New distance}}{\text{Old distance}}\right)^2 \times \frac{\text{Old exposure time}}{1}$$

$$= \frac{900^2}{300} \times \frac{15}{1}$$

$$= \frac{9}{1} \times \frac{15}{1} = 135$$

The new exposure time, with a lamp distance of 900 mm, is 135 sec.

Example. If, with a lamp at 2 ft, the exposure time is 1 min, what would the exposure be with the lamp at 3 ft?

Photometry

$$\frac{\text{New exposure time}}{1} = \left(\frac{\text{New distance}}{\text{Old distance}}\right)^2 \times \frac{\text{Old exposure time}}{1}$$

$$= \frac{3^2}{2} \times \frac{1}{1}$$

$$= \frac{9}{4} \times \frac{1}{1} = 2\tfrac{1}{4}$$

The new exposure time, with a lamp distance of 3 ft, is $2\tfrac{1}{4}$ min.

Example. A table is illuminated by a lamp suspended 640 mm perpendicularly above it. Calculate the relative intensity of the illumination when the lamp is moved to 1280 mm above the table.

$$\frac{\text{Relative intensity}}{1} = \left(\frac{\text{Old distance}}{\text{New distance}}\right)^2 \times \frac{100\%}{1}$$

$$= \frac{640^2}{1280} \times \frac{100}{1}$$

$$= \frac{1}{4} \times \frac{100}{1} = 25\%$$

The relative intensity of the illumination at 1280 mm would be 25% of that at 640 mm.

Example. A table is illuminated by a lamp suspended 2 ft perpendicularly above it. Calculate the relative intensity of the illumination when the lamp is moved to 5 ft above the table.

$$\frac{\text{Relative intensity}}{1} = \left(\frac{\text{Old distance}}{\text{New distance}}\right)^2 \times \frac{100\%}{1}$$

$$= \frac{2^2}{5} \times \frac{100}{1}$$

$$= \frac{4}{25} \times \frac{100}{1} = 16\%$$

The relative intensity of the illumination at 5 ft would be 16% of that at 2 ft.

4. ILLUMINANTS

Historical introduction

The chronological order of artificial light sources in terms of usage are arc lamps, in which the emitted light is obtained from an arc discharge, in air, between positively and negatively charged carbon electrodes; incandescent filament lamps in which a carbon or tungsten filament is made to emit light in a free or inert gas atmosphere; and electrical discharge lamps in which the light radiates from the activity of electrons resulting from an electric charge through a gas or vapour.

It is only in recent years that advances in artificial illumination have been great enough to provide lighting suitable for almost any task undertaken. Early photographers laboured under the handicap of having only one light source, the fickle sun, because in the first half of the nineteenth century most people could only substitute for daylight either the oil lamp or candle. Gaslight was in its infancy and suffered from many deficiencies.

The arc lamp progressed from a glow discharge in an imperfect vacuum, supplied from a frictional electric machine. This was recorded in 1709. Later, in the year 1810, Sir Humphry Davy, the eminent scientist, demonstrated the electric arc between two carbon rods. He used voltaic batteries which only provided sufficient power to sustain the arc for a few minutes. Although arc lamps seemed to be the light source photographers had been longing for, and in fact were used in many leading studios for exposing portraitures on to daguerreotype plates, their efficiency and general acceptance was marred because their power was at this time supplied by Bunsen cells. However, the basic idea came to fruition in 1858 when Oliver Holmes invented an arc lamp plus a regulating mechanism which not only benefited photography, but also brought safety to the seafarer when its light blazed out across the dark, treacherous sea.

During the year 1848 this lack of artificial light prompted Joseph Wilson Swan to commence experiments that culminated in 1878 with the production of the first incandescent electric light bulb. Swan went on to escalate his experiments with the help of an improved vacuum pump from Sir William

Classification of illuminants

Crookes and the glass blowing skill of C. H. Stearn. This work produced a perfected lamp with a carbon filament in complete 'vacuo'. This practical and economical light source opened the way for the introduction of electricity into homes, workshops, offices and photographic studios.

As long ago as A.D. 980, the Japanese realized the luminescent properties of calcined (burnt to an ash form) oyster shells mixed with pigment. This knowledge fell into disuse until 1610 when a bootmaker and part-time alchemist, Vincentius Casiarolus of Bologna, calcined a local stone (barytes) in the search for silver, but formed barium sulphide instead, which to his amazement emitted an orange red glow in the dark for some time after it had been exposed to sunlight. Once again the fundamental knowledge of the fluorescent light was shelved until 1900 when the first practical electric discharge vapour lamps and early fluorescent lamps emerged as a result of the labours of two scientists, Peter Cooper Hewitt and John Stokes.

Classification of illuminants

There is a choice of seven main types of illumination for use in graphic reproduction. When considering them it is helpful to divide them into two main categories:

1. Solid body thermal radiators

The emission of light from this type of source is the result of heating a substance such as carbon or tungsten. As the current is increased the temperature of the substance increases. The light radiating from it becomes whiter and more intense. Illuminants in this group are enclosed arc lamps; open flame arc lamps; tungsten filament or gas-filled lamps; and quartz-halogen lamps.

2. Electrical discharge lamps

A lamp in this category generally consists of a glass tube containing an inert gas with an electrode at each end. An electric current is passed through the gas to produce light or ultra-violet rays which may be used directly or to excite phosphors coated on the inside of the glass tube. Examples are mercury vapour lamps; fluorescent tubes and lamps; and electronic flash (interval flashes and pulsed xenon).

Illumination requirements

When selecting a light source the following requirements should be borne in mind.

1. The nature of the emitted light

This is governed by two closely linked factors

Illuminants

(a) *Spectral distribution*

The light from the source in question may be dispersed by a prism or diffraction grating so that the white light emitted can be seen in its component colours. This coloured band is projected on to a grid indicating wavelength and relative energy. From this we can see which portions of the spectrum the lamp is rich in and if the spectrum formed is continuous; all colours present are adjacent and merge imperceptibly into one another. Or, it may be shown as narrow bands or lines present only in certain sections of the spectrum. In this case it would be termed a discontinuous or line spectrum. Monochrome graphic reproduction work requires lamps with emissions ranging from 350 to 580 nanometres to match the colour sensitivity range of ordinary and orthochromatic emulsions. Light sources having a discontinuous spectrum in these regions may be used. Colour reproductions require light sources producing continuous spectrums in which the chromatic composition is fairly uniform. Ultra violet rays are not required and are usually absorbed by the separation filters. In colour separation techniques the quantities of red, green and blue light in relationship with colour sensitivity of the emulsion and the colour temperature of the light source determine the exposure ratios between different filters. Lamps emitting only line spectra are unsuitable.

(b) *Colour temperature*

The colour temperature of any light source is based on a hypothetical concept—a black body radiator. This is a body which does not reflect light impinging on its surface, but will absorb all the radiation falling on it. This symbolic black body will also emit maximum radiation when heated. Colour temperature of a light source is defined as *the absolute temperature expressed in degrees Kelvin, which will cause a completely black body to emit light of exactly the same spectrum as the light in question.*

Degrees Kelvin ($°K$) is equal to the number of degrees Centigrade ($°C$) plus 273. Temperature is due to molecular movement; the faster the movement, the higher the temperature. The lowest possible temperature is that at which such motion stops. This is $-273°C$.

Colour temperature becomes more understandable if we replace the completely black body concept with a black metal poker. If the poker is heated in a fire, it begins to glow dull red and emit a faint light. The temperature of the poker could be recorded, i.e. $520°C$. Then the colour temperature of the faint light radiating from the poker would be $520+273=793°K$.

Illumination requirements

poker, without melting it, to approximately 1200°C it would become white hot and emit a strong light. The colour temperature of this light would be 1200°C + 273 = 1473°K. Now if we substitute the metal poker for more realistic substances such as tungsten wire and carbon rods it can be seen that strictly speaking colour temperature can only be quoted for light sources relying on heat. This group of light sources are termed 'thermal radiators' and produce continuous spectrums.

In lamps where light is emitted by an electrical discharge causing a gas or vapour to become luminous, e.g. mercury vapour lamps, the composition of the emitted light does not depend upon the temperature. Consequently it is impossible to quote the colour temperature of such a light source which always produces a line spectrum. However, an approximate colour temperature may be indicated solely on the visual colour impression of the light emitted, although this is of no value when considering the light's photographic effect.

As stated earlier the spectral distribution is closely related to colour temperature. This relationship can be seen from the following diagram containing three curves. A spectrum of light having a colour temperature of 3000°K contains more red light than blue and will appear yellowish when viewed, while light sources with a colour temperature of 5500°K have a fairly well balanced spectral distribution resulting in a fairly intense white light. Of course at the other extreme a clear northern sky may reach and exceed 20,000°K resulting in a spectral composition rich in ultra violet and blue light. When exposing colour films or integral layer masking films such as Kodak Tri-mask and Agfa-Gevaert Multi-mask, it is essential that light of the correct colour temperature is used.

Colour temperature:
1. Tungsten lamp
2. Daylight
3. Enclosed arc lamp

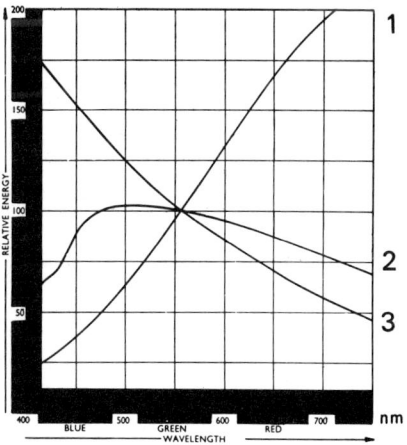

Illuminants

If white flame arc lamps are used to expose Tri-mask film it is desirable that the colour temperature of the arc lamp is brought down from 5000°K to 3200°K by the use of an 85B filter. Kodak and other manufacturers market a complete range of light balancing filters, each producing a precise increase or reduction in colour temperature.

Table 1—Colour temperature

Light source	Approximate colour temperature
Standard candle	1930°K
Vacuum tungsten lamp	2400°K
Gas-filled tungsten lamp (general service)	2760–2960°K
Warm-white fluorescent lamp	3000°K
'Photographic' lamp	3200°K
Photoflood lamp	3400°K
Plain carbon arc	3800°K
Daylight fluorescent lamp	4500°K
White-flame arc	5000°K
'Mean noon sunlight'	5400°K
Electronic flash tube	6000°K
Average daylight (sunlight and skylight combined)	6500°K
Colour matching fluorescent lamp	6500°K
Enclosed arc	10,000°K
Blue sky	12,000–18,000°K

2. The luminous intensity of the emitted light

The intensity and actinism (effectiveness of light of given intensity in exposing emulsions, dependent upon wavelength) must be high enough to give short but controllable exposure times with the range of emulsions used. Heat radiating from the light source should not be too intense. If it is intense, the heat rays should be filtered out with water and the lamp cooled by extractor fan, to prevent any distortion of the original during long exposure times. The lamps when positioned to the left and right of the original at an angle of 45° should produce a uniform illumination over the entire copy holder. Because fluctuations in the mains voltage supply will alter the luminous intensity, spectral distribution and colour temperature of the light, it would prove advantageous to stabilize the voltage supply and expose via a light integrating meter.

Illuminants

Arc lamps

The light from these lamps is a mixture of radiations from the

Illuminants

Spectral curve of an enclosed arc lamp

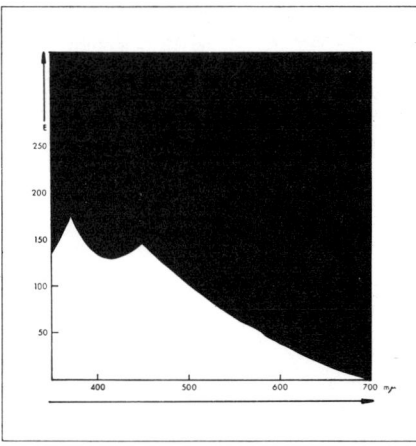

gas column and the red hot arc crater. Solid carbon rods are prone to 'arc wander' and burn irregularly. To prevent this a central core of softer carbon or mineral salts such as cerium fluoride is provided. This produces a crater, a depression in the positive carbon tip which stops arc wander and influences the spectral composition of the light radiating from the gas column. All arc lamps produce a mixed spectrum; line spectra mainly in the ultra-violet and blue region of the spectrum from the gas column and a continuous spectrum in the visible portions from the carbon extremities.

The enclosed arc lamp, as the name suggests, has the carbon rods enclosed in a glass cylinder and they burn in an atmosphere of carbon dioxide and nitrogen. This permits a long gas column which emits a light rich in ultra violet and

Spectral curve of open arc lamp

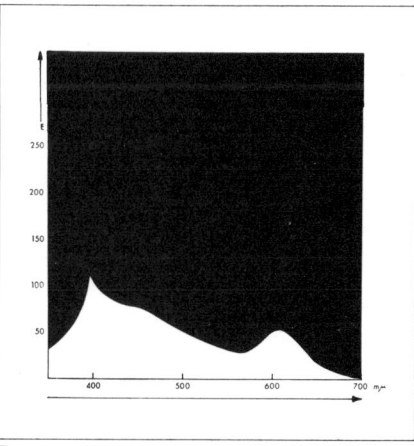

Illuminants

blue rays. Under these conditions the life of the carbon rods is considerably longer than that of the open arc lamp. The enclosed arc lamp is widely used when exposing and hardening light-sensitive colloids.

Open arc lamps operate with a short gas column in a free atmosphere. This produces high luminous intensity, but requires a high amperage which means rapid consumption of carbon rods and expensive running costs. The colour of the light depends on the core of the carbon rods and the length of the gas column. 'White flame' open arc lamps are the most suitable for colour reproduction because the rods have central cores containing mineral salts (fluorides). The colour temperature is 5000°K which is approximately the same as daylight. Because of the physical movement of the gas column due to the consumption of the rods, fluctuations occur. After switching on, the lamps should be allowed to settle down for a short time in order to stabilize the arc gap. Intensity variations can be smoothed out by the use of a light integrating meter. Spectral composition changes with impurities occurring in the mineral salts, the length of gas column and variations in electrical supply. Electrical imput can be reasonably standardized with a variable resistance. An attempt to produce a constant length of gas column can be seen in motorized arc lamps. Two or three copper covered (increases conductivity) carbon rods are used, each one being slowly moved in as the carbon tips become consumed.

Arc lamps should be used in well ventilated rooms or have fume extractors built into the reflectors, because the fumes over a period can be harmful to the operator. Open arc lamps produce light of high actinism and intensity resulting in relatively short exposure times. For many years they have been widely used on process cameras and printing frames. In practice certain disadvantages have been encountered. The lamps occasionally emit clouds of smoke and dust. Long exposure times are accompanied by a relatively high output of heat which sometimes shrinks or warps reflection originals and buckles or distorts colour transparencies.

Incandescent lamps

This is the commonest artificial light source. Each lamp consists of a tungsten filament suspended in a glass bulb filled with the inert gas 'argon' (sometimes with iodine vapour added). The filament is heated by an electric current until it reached maximum temperature and becomes incandescent. The light remains fairly constant except for fluctuations in luminous intensity and spectral distribution because of

Illuminants

variations in the electrical supply and a gradual decline in performance as the lamp ages.

For photographic work two types of incandescent lamp have emerged. The first is termed the 'photographic' lamp and is generally used in small enlargers and printing frames. This 500 W lamp emits light of high luminous intensity, but only at the cost of a reduction in its working life (approximately 100 hours). The colour temperature is approximately 3000°K. From this it can be seen that the spectral composition of the light is rich in the red portion, but very low in the blue region. Because of this, if this lamp is employed in colour separation work the exposure through the blue filter is extremely long.

The second incandescent light source is the 'photoflood' or overrun lamp, in which a higher voltage than the recommended value (which produces a long life) is being fed into

Spectral curve of incandescent lamp

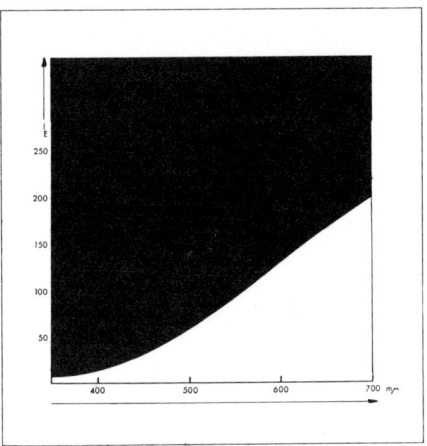

the filament. As a result the filament is running at a higher temperature and radiates an abnormally intense light, but this increase in luminous intensity is accompanied by a very short working life. The No. 1 photoflood has a 3 hour life while the larger No. 2 photoglood lives for 8 hours. Both have characteristic continuous spectrums with a marked drop in the blue region complemented with a colour temperature of 3400°K, 200°K below the melting point of the tungsten filament, a contributive factor to the short life. In practice, in addition to the increase in exposure times when compared to arc lamps, incandescent lamps generate a great deal of heat which can cause dimensional instability in originals.

Illuminants

Tungsten halogen lamps

This light source, generally termed the quartz-iodine or tungsten iodine lamp, is basically an incandescent photographic lamp with an approximate life span of 1000 hours. Its tungsten filament is not housed in an evacuated glass envelope, but surrounded by a quartz (crystalline silica) tube. This enables the tungsten filament to work at a higher temperature. The conventional tungsten incandescent lamp darkens and loses its efficiency with age because of the filament evaporating and depositing on to the inside of the glass envelope.

To combat this the particles of the tungsten filament which normally evaporate and darken the glass combine with the iodine vapour in the quartz-iodine lamp and enter into a continuous cycle of combination and dissociation, finally

Spectral curve of tungsten halogen lamp

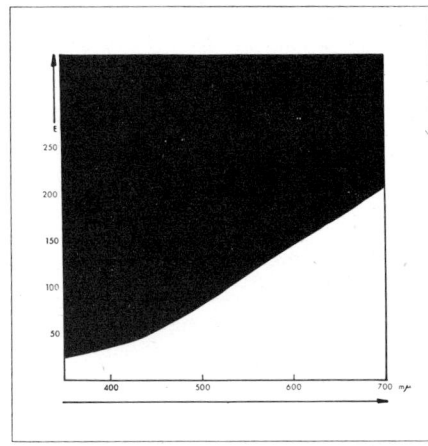

redepositing the tungsten filament. Although the emitted light has the same characteristics as the conventional incandescent lamp, it has several important advantages. Firstly, the lamp's internal action helps to eliminate the blackening dilemma and in doing so produces an improved emission with regard to colour temperature (3000°K) and luminous intensity. Secondly, for monochrome reproduction, it is proving to be an efficient, fumeless and fairly economical light source.

Mercury vapour lamps

This light source, as far as graphic reproduction is concerned, is the basis of electrical discharge lighting. At the present moment the high pressure mercury vapour lamp is preferred. This consists of a glass tube filled with mercury

Illuminants

Spectral curve of mercury vapour lamp

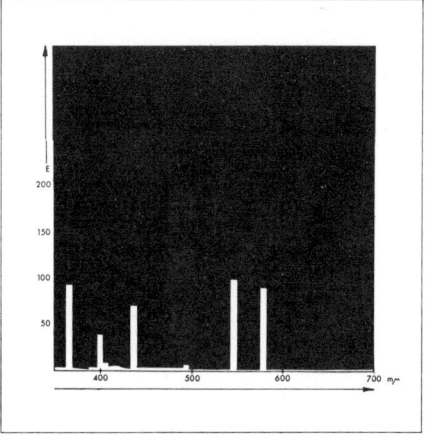

and with an electrode positioned at each end. When the lamp is switched on an electrical current passes from one electrode to the other and vapourizes the mercury. This vapour emits light, the spectrum of which is a classic example of line spectra with lines of light energy in the u.v., blue and green regions. This distribution makes a red filter exposure impossible. This spectral emission makes mercury vapour lamps only suitable for monochrome reproduction work. Their luminous intensity is low, but the illumination from the source is extremely uniform and constant over a relatively large area, which accounts for their popularity as illuminants for large sized photographic enlargers and transparency holders. One practical disadvantage is that the exposure times tend to be rather long when compared with more recently developed light sources, e.g. pulsed xenon arc lamps. They also need a build-up period before reaching maximum efficiency, combined with the fact that the lamp cannot be restarted until it has completely cooled down.

Fluorescent lamps

These lamps are an advancement on the mercuty vapour lamps, but rely basically on the same discharge principle. The glass tube contains argon, an inert gas, and a small amount of mercury vapour. When the lamp is switched on the vapour emits u.v. radiations which excite phosphors coated on the inside surface of the glass tube. The phosphors 'fluoresce' and in doing so convert the invisible shortwave u.v. radiations into visible light, its spectral composition being determined by the type of phosphors used. This

Illuminants

Spectral curve of fluorescent lamp

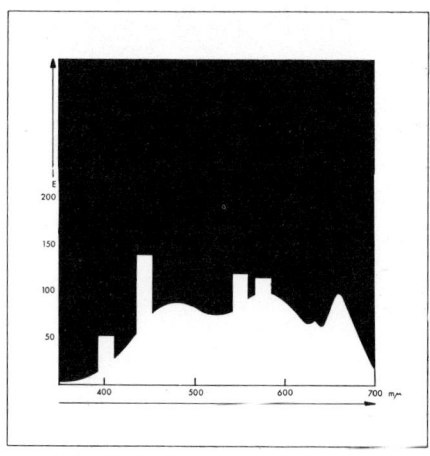

provides a choice of spectrums, e.g. 'daylight' (6000°K) fluorescent tubes, 'white' (4000°K) and 'warm-white' (3000°K).

The phosphors emit a continuous spectrum which becomes the background to the line spectra emanating from the mercury vapour. This combination can produce light which has a colour temperature figure attributed to its visual appearance which may be alien to its spectral composition. Therefore the Kelvin value stated may not always be reliable in terms of photographic effect with light-sensitive emulsions. Coloured fluorescent tubes in rotating turrets have been used for colour separation work. The lamps are constructed so that on rotating the turret only the bank of tubes facing the original light up. The argument for a filtered light source in place of the conventional filter is strongly supported. The advantage of this system is the uniform illumination achieved, but this is offset by long exposure times, especially with the red filter separation.

For graphic reproduction photography white fluorescent tubes are more useful when arranged into a battery illuminating a transparency holder. They also have a relatively long working life complemented by a uniform, diffuse illumination without harmful heat effects. For general illumination of workrooms and light tables, as long as the spectral composition of the emitted light is constant throughout the factory, there is no better illuminant at the present time.

Electronic flash lamps

No matter whether the final function of the lamp is a single flash source or a pulsed xenon arc, electronic light

Illuminants

sources work on the following basic principle. A condenser stores up electricity ready to discharge through the gas. In order to make the gas conduct electricity a higher voltage pulse is applied. This disrupts the molecules of gas, producing ions which carry a current through the gas and discharge the condenser. Much of the electrical energy so used up is converted into light, and since the condenser discharges rapidly, the light is concentrated into a very short period of time. The flash is thus both brief and intense.

The amount of electrical energy contained in an electronic flash unit is usually expressed in joules (J) (watt-seconds). The amount of light produced by this electrical energy will, however, depend upon the luminous efficiency, which may vary from one Xenon lamp to another.

Light-sensitive silver halide grains react violently to short, intense electronic light flashes. This is a reaction between the light's quanta (amount of energy contained in the exposing light) and the formation of the latent image. The shorter the wavelength of light, the more energy per quantum. Silver halide grains, as in ordinary emulsions, require a minimum amount of energy to produce the electron shift which gives rise to a latent image. This energy can only be supplied by quanta of blue or shorter wavelength radiation. For example, when a silver halide grain is exposed to tungsten light the incident ray breaks down the structure of the grain, liberating bromine ions and attracting silver ions to the sensitivity centres formed during the manufacture of the emulsion, while the grains were being ripened with heat.

With the short, intense exposures (approximately $\frac{1}{3000}$ sec) obtained with the electronic flash lamp a great deal of bromine ions are liberated, but because of the short exposure time, many of the free silver ions do not have sufficient time to reach the sensitivity centres. This means less silver forming at the sensitivity centres and provides an explanation for the lower density ranges on negatives produced with electronic flash, when compared with films exposed to tungsten light.

Electronic flash—single flash source

This provides the photographer with a light source capable of producing sufficient light intensity to penetrate a colour transparency plus mask, with only a few, short flashes; or a constant and efficient light for exposing the image of continuous tone negatives through a halftone screen. The intensity of the emitted light may be increased by steps, or if the maximum intensity is too weak, a number of flashes may be given. The electronic flash lamp produces improved highlight

Illuminants

Spectral curve of electronic flash

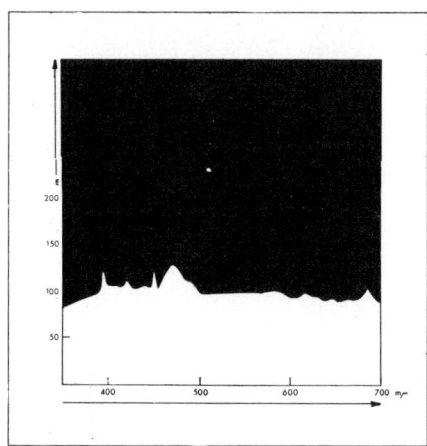

detail when compared with tungsten and arc lamps. Practical results have indicated that a superior tonal separation is achieved when only one flash is given and declines with the increase in the number of flashes.

It is difficult at times to determine the correct exposure with this light source. This problem may be eased if the following procedures are considered.

1. Use the largest lens aperture possible as long as optimum sharpness is retained.
2. Adjust the intensity so that the least number of flashes are used.
3. Make sure the size of the flash head corresponds to the transparency size. Having a too large flash head is just as wrong as employing a too small head.
4. Use a standard grey scale as a test strip to check the end densities readings and tonal separation.

This light source combines extremely high luminous intensity, with low current consumption. The colour temperature is 6300°K producing a spectral composition extremely suitable for colour separation work. One head may be utilized in the enlarger, contact frame and process camera. The emitted light is devoid of fluctuations, fumes and heat effects.

Pulsed xenon arc lamps

A quartz tube containing xenon gas has an electrical charge pulsed through it. The energy produces a series of flashes that occur so rapidly (100 flashes per sec) that the light output appears as a continuous emission. This light source has a very high luminous intensity, but becomes very hot. This is rectified by a cooling system using air draughts or water.

Lens flare

The spectral composition of the emitted light compares favourably with that of sunlight which makes it an ideal light source for colour separation work, and for those who can afford the relatively high installation cost plus expensive tube replacements this light source is a popular substitute for open arc lamps. This light has a typical colour temperature of 5200°K and emits a line spectrum, but the lines are so closely grouped that the energy output appears continuous.

Lens flare

After setting the lamps on a process camera for angle and distance, the photographer should position himself so that his eyes are in the same plane as the lens. This enables him to view the original and correct the lighting for any undesirable

Lens flare:
1. Original on black background—negative curve
2. Original on white background—negative curve

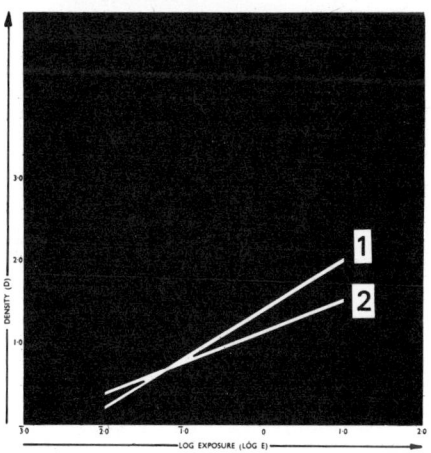

Illuminants

direct reflections that may be occurring from the glass of the copyholder.

When a number of lamps are employed to produce a uniform emission, not only the original is illuminated, but most of the surrounding surfaces will have incident light reflecting from them. This stray light in many cases rebounds into the lens, causing lens flare, which results in a detrimental effect on the negative comparable to a uniform fog over the entire film.

To minimize stray light the copyboard, the ceiling, and the walls adjacent to the original should be painted matt black. During exposure switch off all extraneous light so that the only light reaching the lens is the reflected rays from the camera lamps. Care should be taken in the preparation of the original. All white reflecting surfaces, especially surrounding the original, must be kept to a minimum. After clamping the original in the copyholder the perimeter of coverglass not being used should be covered with light-absorbing black card.

The next item to control is the lens itself. In the beginning it is advisable to invest in a coated lens (its outside surfaces are bloomed or coated so that reflected rays entering at very oblique angles are absorbed) fitted with a lens hood. At least once a day the lens and reflecting mirrors or prisms should be cleaned, because even a microscopic layer of dust (inevitable when open arc lamps are used) will result in stray light. Dirt and scratches must be removed by cleaning or professional polishing and if gelatine filters are being placed in the centre of the lens, they should be perfectly flat, free from scratches, kinks and distortions. Following the path of the reflected light the next surface which needs intermittent inspection and cleaning is the interior of the camera bellows. On some large process cameras it has been found necessary to fit light-absorbing baffle plates close to the light-sensitive emulsion.

To summarize the movement of light during exposure the following steps may be considered:
1. The emitted light is selectively reflected and absorbed by the original and surrounding surfaces.
2. The reflected light passes through the compound lens undergoing many inter-reflections among the surface boundaries.
3. The outer extremities of the emerging light are absorbed by the black camera bellows.
4. Finally, the reflected image plus any stray light exposes the light-sensitive emulsion.

Calculating the flare factor

No matter how complete precautions may be, it is wise to measure intermittently the amount of stray light reaching the emulsion by the following method.

Calculating the flare factor
Apparatus

1. One sheet of grey card with a reflectance of 50%. Size: the maximum working area of the camera.
2. A continuous tone stepwedge. Size: 120 mm × 20 mm approximately.
3. A strip of black paper of high absorption. A strip of white high reflectance. Size: both twice the size of the continuous tone stepwedge.

Procedure

Place the white strip of paper in the centre of the grey card and position the black strip next to it.

Focus the camera to half size and load a piece of continuous tone film with a stepwedge (X) in its centre.

Position the image of the white paper so that it falls on the stepwedge.

Expose and develop to produce a density of 0·4 on the image of the black paper strip.

Look at the photographic result and find a step on the negative stepwedge which corresponds to the density of the black paper image 0·4.

If the step that produced the 0·4 image has, for example, a density of 1·7 and a corresponding transmission of 2·00%, then this step on the continuous tone stepwedge has transmitted 2·00% of the light reflecting from the white paper

Flare factor: Apparatus
1. Grey card, white and black strips
2. Continuous tone stepwedge
3. Stepwedge and white paper image next to black paper image

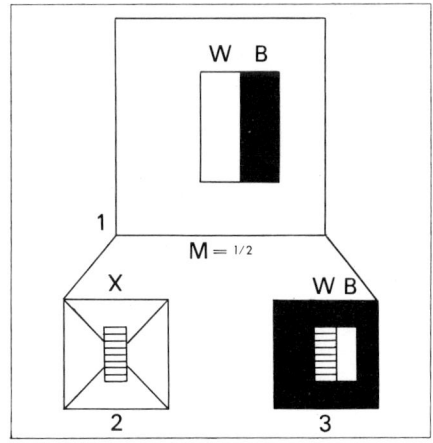

Illuminants

Density reductions due to flare

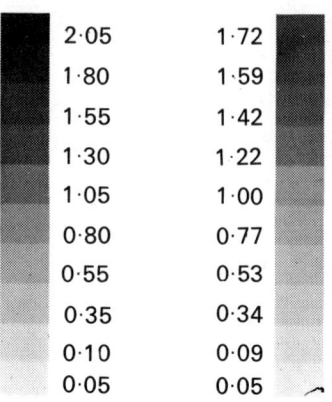

2·05	1·72
1·80	1·59
1·55	1·42
1·30	1·22
1·05	1·00
0·80	0·77
0·55	0·53
0·35	0·34
0·10	0·09
0·05	0·05

strip. The black paper strip will also have reflected 2·00% of the incident light.

Now measure the reflection density of the black paper strip, e.g. Reflection density = 2·00; this density should only reflect 1·00% of the incident light.

The flare factor is therefore found by deducting 1·00% from 2·00 = 1·00%. The flare factor in this example is 1·00%, the normal flare factor encountered on graphic reproduction cameras.

As the image is being exposed on to the photographic emulsion the stray light or flare is degrading its contrast range, so that each density on the original is reduced by the flare. This density reduction is more noticeable in the dark shadow tones than the light highlight tones. With a flare factor of 1% the above density reductions between original and photograph would take place.

Viewing conditions

In graphic reproduction the correct assessment of colour is vital; because the coloured appearance of dyes and pigments varies under differing light sources a standard illumination for transmission and reflection viewing must be maintained.

A colour temperature of 5000°K has been chosen as this standard, in preference to higher or lower values, in order to give definite distinctions in the red and blue sections of the spectrum. The level of illumination must be uniform and constant, but no specification can be indicated for the light itself because (i) the light may be modified by an apparatus,

Viewing conditions

and (ii) to obtain the required quality of lighting a mixture of light sources may be employed.

Viewing transparencies

Coloured transparencies should be illuminated by a diffuse transmitted light and any illuminated area surrounding the transparency should be blacked out. The light source may be a mixture of tungsten filament light (2400°K) and light blue fluorescent light of much higher colour temperature.

Assessing coloured proofs or reflection originals

The light should be housed in a booth, so that any external lighting, direct or indirect, will not interfere with the standard illumination. The illuminants should be so situated that specular reflection in the direction of the observer is minimized to reduce the effects of:

1. forming an image of the light source,
2. excessive glare, and
3. desaturation of colours.

In practice it has been found that most difficulties are overcome when the light fittings are positioned vertically each side of an almost vertical viewing surface. All surfaces not being viewed should be covered with a mid-grey medium. When a side-by-side comparison of a reproduction made from a combination original (reflection original plus a transparency) is being carried out, all the coloured images should be contained in the booth and controlled to the same level of mean luminance.

A colour booth

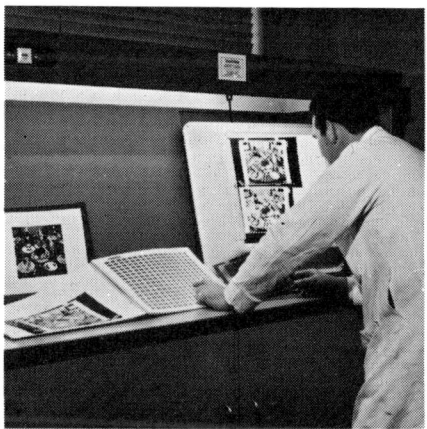

5. THE CAMERA

Historical introduction

Today, cameras are being manufactured in all shapes and sizes, designed to perform many varied and specialized tasks. They are all variations on the same basic theme—the camera obscura, a darkroom with a minute hole in one of its walls through which the scene outside is projected to produce an inverted image on the opposite wall. The camera obscura's early employment was mainly restricted to observations of solar eclipses and perspective tracings on which artists based their pictures.

Many people have had their names linked with the invention of the camera obscura, but no one man can be ascribed with the invention.

This image formation is the product of light rays travelling in a certain manner and like the action of a glass lens was probably seen in the first instance as a natural phenomenon. Then after many distorted explanations by bemused observers it slowly developed into a plausible reality.

As early as 1039 an Arabian scholar, Alhazen, described its basic principles in such a manner that it seems to have been well understood amongst Arab scholars of this period. The pinhole produced a fairly sharp image, but owing to its small size only a small proportion of the light reflecting from the original scene passed through. This resulted in a dimly-lit image. Any efforts to increase the size of the hole resulted in an unsharp image. This dilemma was overcome in 1550 by Girolamo Cardano, a Milanese physician, when he enlarged the aperture in the wall and positioned a bi-convex spectacle lens at the centre of the hole. From this moment on the camera took on many varying guises, but always with the intention of reducing its size and increasing its portability. The camera at this stage was not a photographic camera in the true sense of the word, although a number of far-sighted men did realize its potential and laboured in vain trying to capture its image by chemical means. The first successful and permanent photograph was produced on a metal plate, nearly 300 years after the introduction of a lens, by the Frenchman, Nicéphore Niépce in the year 1826. During a lifetime devoted to photo-

Historical introduction

Early lenses by
1. Cardano
2. Wollaston
3. Chevalier

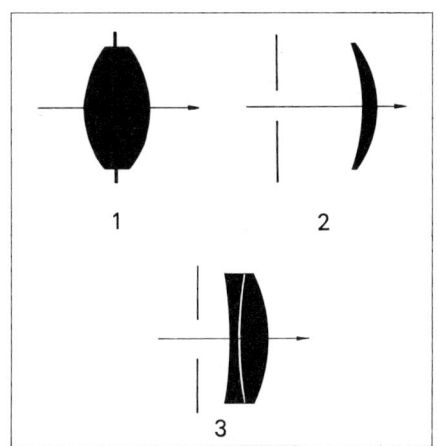

graphy and printing Niépce made a number of cameras with the then revolutionary, but now familiar refinements, such as expanding bellows, variable iris diaphragms and interchangeable lenses.

Cameras would not have developed so rapidly after Niépce's success without the diligent work and research carried out by the optical designers of the day. Cardano's single bi-convex lens made of crown glass suffered severely from numerous aberrations. This type of lens persisted for 250 years until the year 1812 when an English physician, William Hyde Wollaston, produced a lens capable of reducing certain aberrations. He achieved this by positioning the stop in front of a meniscus-shaped lens. This lens design improved definition, but suffered from chromatic aberration which was subsequently overcome by the use of a convergent lens of crown glass combined with a divergent lens of lower dispersion made from flint glass. This was introduced by Charles Louis Chevalier, the famous French optician, in 1829.

The nineteenth century fostered the growth of photography; as its application widened the demand for more sensitive emulsions and faster lenses increased. A lens capable of transmitting a great deal of light through a large aperture was sorely needed. This need was met by a new approach to lens design propounded by Josef Max Petzval, a Hungarian mathematician. Astigmatism and field curvature aberrations were still obvious, but its rapidity in portraiture made this lens design an important milestone in lens development.

Gradually the optical designers were reducing the lens aberrations and with the advancement in optical glass

The camera

Compound lenses by
1. Petzval
2. Taylor (triplet)
3. Taylor (quadruplet)

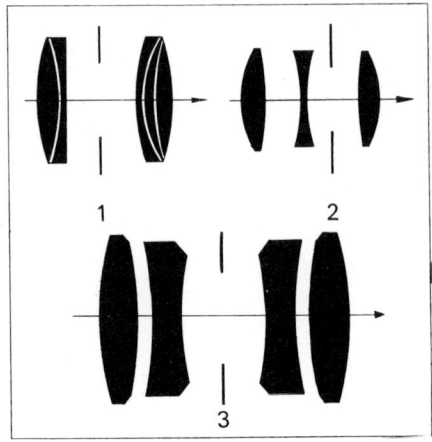

manufacture made by Abbe and Schott in Germany during the year 1888 great strides were taken. After the lead of Wollaston, Chevalier and Petzval in original conceptions it was left to Harold Dennis Taylor in 1893, at the turn of the century, to produce a triplet lens of simple design which proved to be remedial for most lens aberrations.

The pinhole camera

A pin-hole may be employed to produce an inverted image. The size and perspective is in relationship to the distance between the diaphragm and the light-sensitive emulsion. The image quality is determined by the size and sharpness of the pin-hole. The quality improves as the hole is reduced in size until a limit is reached when the image

Pinhole camera:
object-pinhole-image

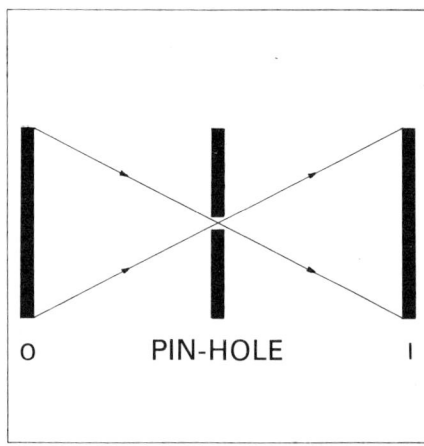

Convex lens:
Object-lens-image

The simple lens

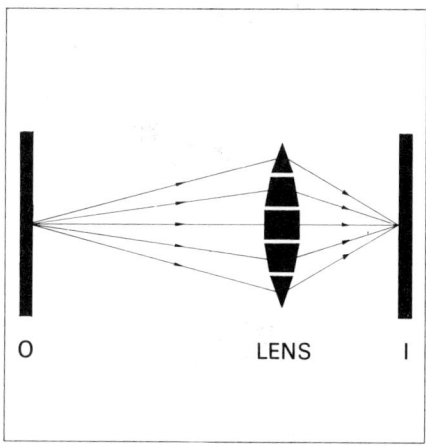

O   LENS   I

deteriorates through diffraction effects. Therefore an optimum pin-hole diameter has been calculated. This is approximately $\frac{1}{125}$ of the square root of the distance in inches from the pin-hole to the light-sensitive material.

Pinhole cameras may be made cheaply and easily and are capable of reproducing reasonable images, with the added advantages of no focusing, distortion-free images and wide angles with infinite depth of field. Conversely, even with a correct pin-hole diameter, final photographs have an overall softness and lack of definition. In addition exposures tend to be too long. To overcome this two alterations are needed. Firstly, to let in more light rays, and secondly to converge them so that a sharp, real and inverted image is formed. This need is satisfied by the insertion of a simple convex lens.

The simple lens

A lens is a transparent medium (glass or plastic) with a refractive index different from that of the surrounding medium. Its boundaries are smooth and curved symmetrically around a centre line known as the axis. The lens may be thicker or thinner at the centre than at the periphery. Lenses are usually made from blocks of optical glass. They have their boundaries ground and polished into curved surfaces which are specific portions of spheres (a plane surface being considered as a sphere of infinite radius), the spheres' radii being termed the curvature radii of the surfaces.

The function of a lens depends on the fact that parallel light rays reflecting from an object may, owing to refraction at the curved boundaries of the lens, be made to converge and meet at a point on the axis.

The camera

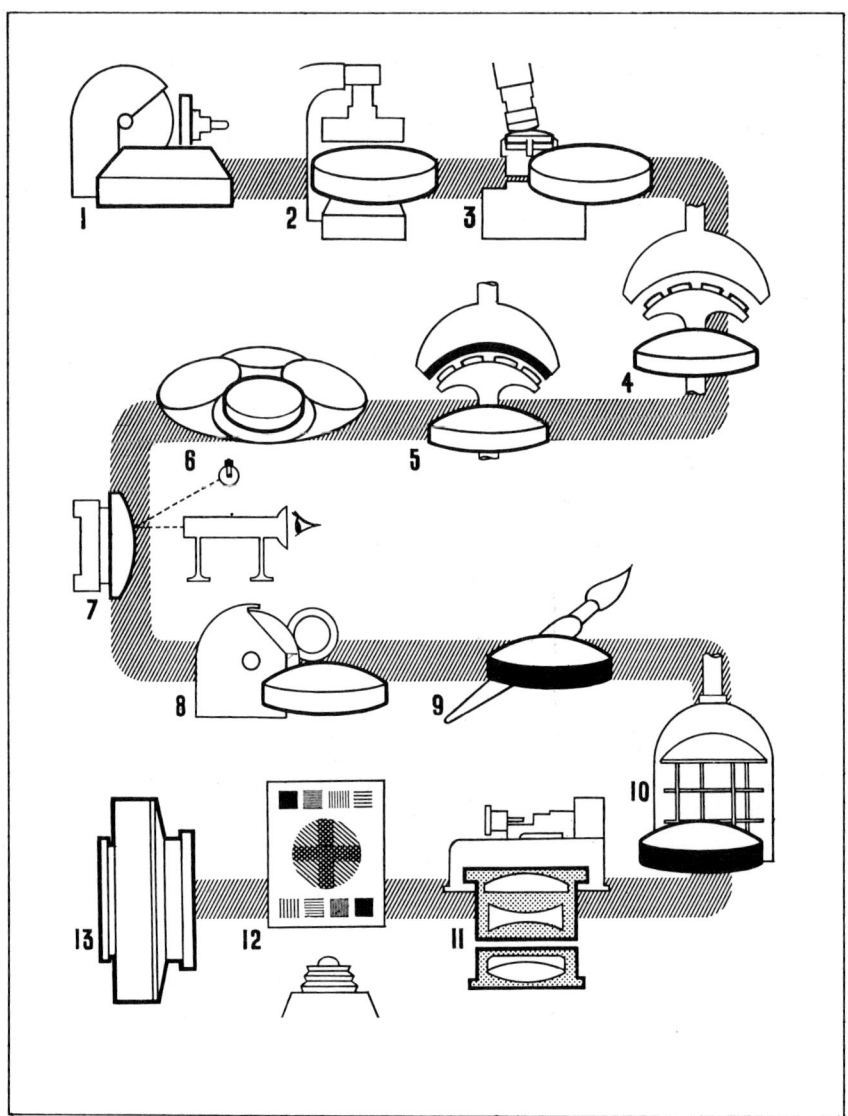

Lens manufacture:
1. Cutting raw glass into blanks
2. Moulding the blanks
3. Rough grinding the blanks
4. Fine grinding mounted blanks
5. Polishing
6. Checking curvature with test plate
7. Centring lens for edging
8. Edge grinding
9. Blackening of edges
10. Coating of surfaces
11. Positioning lens units in mount
12. Test for resolution
13. Finished lens with diaphragm shutter

The simple lens

The construction and function of a simple lens may be illustrated by the use of a basic prism shape. Lenses which are thicker in the centre than the periphery are termed *convex* or *convergent* lenses and are basically two prisms placed base to base. Light rays which pass through them are converged to a point and in doing so produce an inverted image. Simple lenses of this type have at least one convex shaped surface and are termed *double convex, plano-convex* or *convergent meniscus*.

Convergent lenses:
1. Object-basic prism shape-image
2. Double convex
3. Plano convex
4. Convergent meniscus

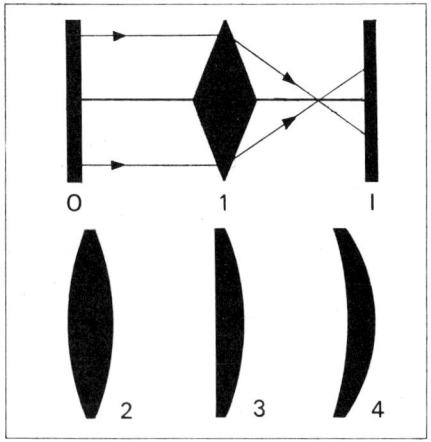

Lenses which are thinner in the centre are called *concave* or *divergent* lenses and are basically two prisms placed apex to apex. This type of lens causes light rays falling on it to diverge outwards and in doing so to produce a virtual image.

Divergent lenses:
1. Object-basic prism shape-image
2. Double concave
3. Plano concave
4. Divergent meniscus

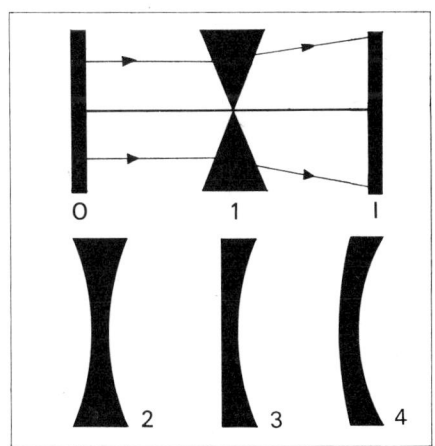

The camera

Simple lenses of this type have at least one concave shaped surface and are termed *double concave, plano-concave* or *divergent meniscus*.

The simple lens is not capable of producing an extended, distortion free image. A simple lens will have a number of defects or aberrations (these will be considered later) which are caused by the fact that in lens manufacture spherical surfaces are used. By combining simple lenses with differing refractive indices and varying spherical curvatures the optical designer can produce a compound or complex lens comprising of a number of glass *elements* which will produce satisfactory images over a limited area.

Focal length

The distance between a lens and a sharply focused image is determined by three factors.

1. The refractive index of the glass used to form the lens;
2. The curvatures of the lens surfaces; and
3. The distance from the lens to the object.

The combination of the first two factors determines the focal length of any particular lens. The focal length of a lens is the distance from the centre of the lens to the principal focus. This is the point on the axis where parallel light rays reflecting from a distant object converge to form a sharp image.

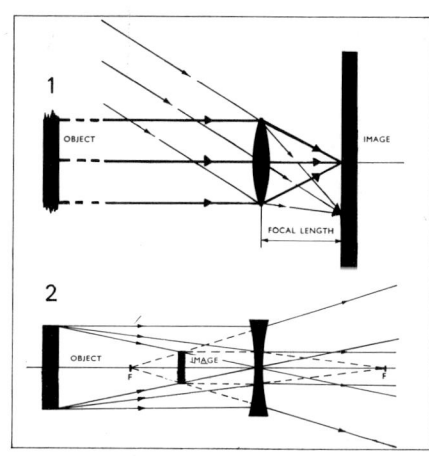

Focal length:
1. Convex lens—one focal length image
2. Concave—one focal length image

The focal length determines the working area (maximum image size) and the maximum enlargement and reduction

Focal length

capabilities of the lens. The focal length (*f*), object distance (*u*) or image distance (*v*) are related by the basic equation

$$\frac{1}{f} = \frac{1}{V} + \frac{1}{U}.$$

Lens equation

$$\frac{1}{f} = \frac{1}{v} + \frac{1}{u}$$

F = focal length
v = camera extension
u = object
I = image
M = magnification factor

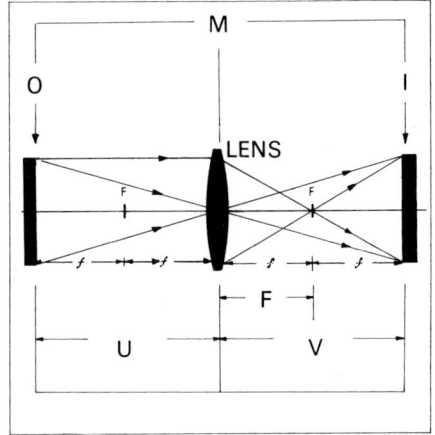

Example: Metric
The camera extension and object distances are 900 mm and 1800 mm respectively. Determine the focal length of the lens used.

$$\frac{1}{f} = \frac{1}{V} + \frac{1}{U}$$

$$\frac{1}{f} = \frac{1}{900} + \frac{1}{1800}$$

$$\frac{1}{f} = \frac{2+1}{1800}$$

$$\frac{1}{f} = \frac{3}{1800}$$

$$\frac{f}{1} = \frac{1800}{3}$$

$$f = 600 \text{ mm}$$

Example: Imperial
The camera extension and object distances are $37\frac{1}{2}$ in. and 75 in. respectively. Determine the focal length of the lens used.

$$\frac{1}{f} = \frac{1}{V} + \frac{1}{U}$$

$$\frac{1}{f} = \frac{1}{37 \cdot 5} + \frac{1}{75}$$

$$\frac{1}{f} = \frac{2+1}{75}$$

$$\frac{1}{f} = \frac{3}{75}$$

$$\frac{f}{1} = \frac{75}{3}$$

$$f = 25 \text{ in.}$$

The camera

Simple lenses are assumed to have no thickness and the object distance and image distance are measured from the centre of the lens. Compound lenses do have a thickness and the object distance is calculated from one point in the lens and the image distance is calculated from another. These specific points are termed the *nodal* points of a lens.

Nodal points:
A composite lens behaves as if it were a simple lens which receives the light at one point and discharges it at another. These two points are the nodal points, and the dotted lines passing through them are the principal planes.
O, object, I, image, N, front nodal point, N, rear nodal point, U, object distance, V, image distance.

Focal length of a compound lens

$$\frac{1}{f} = \frac{1}{f_1} + \frac{1}{f_2} - \frac{d}{f_1 f_2}$$

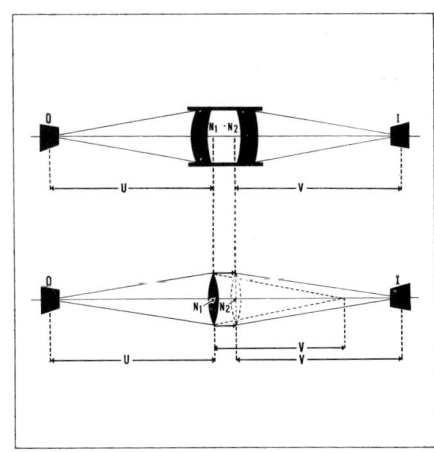

Example: Metric.
The focal length of the first element is 960 mm, while the second element has a length of 1440 mm. The two elements are separated by a distance of 96 mm. Find the combined focal length.

$$\frac{1}{f} = \frac{1}{f_1} + \frac{1}{f_2} - \frac{d}{f_1 f_2}$$

$$\frac{1}{f} = \frac{1}{960} + \frac{1}{1440} - \frac{96}{960 \times 1440}$$

$$\frac{1}{f} = \frac{1}{960} + \frac{1}{1440} - \frac{1}{14,400}$$

$$\frac{1}{f} = \frac{15 + 10 - 1}{14,400}$$

$$\frac{1}{f} = \frac{24}{14,400} \quad \frac{1}{f} = \frac{1}{600}$$

$f = 600$ mm

Example: Imperial
The focal length of the first element is 36 in., while the second element has a length of 56 in. The two elements are separated by a distance of 4 in. Find the combined focal length.

$$\frac{1}{f} = \frac{1}{f_1} + \frac{1}{f_2} - \frac{d}{f_1 f_2}$$

$$\frac{1}{f} = \frac{1}{36} + \frac{1}{56} - \frac{4}{36 \times 56}$$

$$\frac{1}{f} = \frac{1}{36} + \frac{1}{56} - \frac{1}{504}$$

$$\frac{1}{f} = \frac{12 + 9 - 1}{504}$$

$$\frac{1}{f} = \frac{20}{504} \quad \frac{1}{f} = \frac{1}{25 \cdot 2}$$

$f = 25 \cdot 2$ in.

Image formation

Two lines representing light rays reflecting from an object are plotted when calculating the position of a real, inverted image:

1. a ray parallel to the axis impinging on the convex lens,

Image formation:
1. Parallel ray
2. Central ray

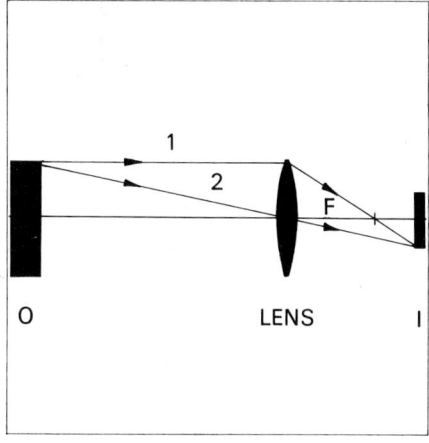

changing direction after refraction and emerging through the principal focus; and

2. a ray passing through the centre of the lens with its direction unchanged.

Magnification of the image

If the camera is focused on distant objects—*infinity*—the image will form on the principal focus in the focal plane. If the object is moved nearer the lens the position of a sharp, in-focus image will alter. In graphic reproduction photography object distances are never very great and may be generally illustrated by the following diagrams.

1. If the object is positioned two focal lengths away from the lens, the image formed will be real, inverted and *same size* as the object.
2. If the object is now moved more than two focal lengths away, the image will become real, inverted and smaller than the object, i.e. *reduced*.
3. As the object is brought nearer the lens between one and two focal lengths, the image will be real, inverted and larger than the object, i.e. *enlarged*.
4. Finally, if the object is positioned within one focal length of the lens, the image will not be real, but virtual, erect and

The camera

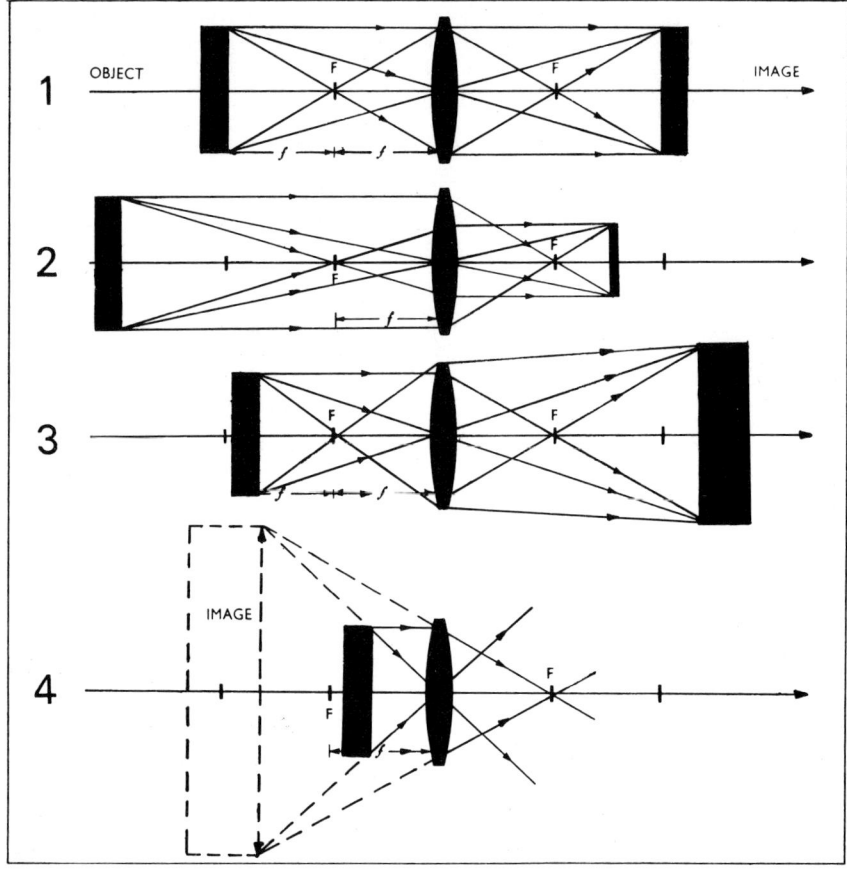

Magnification
1. Same size:
    u and V are equal, both $2f$
2. Reduction:
    u is long, $3f$. V is short, $1\frac{1}{2}f$
3. Enlargement:
    u is short, $1\frac{3}{4}f$. V is long, $2\frac{1}{4}f$
4. Magnified:
    u is less than $1f$. V is long, $1\frac{3}{4}f$

Magnification of the image

larger than the object, i.e. *magnified*. This, of course, is useless in a camera, but represents the condition under which a magnifying glass is used.

It can be seen from these diagrams that the object and image distances are linked. When one expands the other reduces. For every position of the image there is a corresponding position of the object. Because of this they are termed the *conjugate* (meaning 'yoked' or 'linked') *foci* (points or positions) distances. Focusing the camera so that the image becomes a particular size, relies on the adjustment of the conjugate foci distances in relationship to the focal length of the lens. Although most modern process cameras have automatic focusing a camera operator should be able to calculate the position and size of the image. From the basic focal length formula a number of helpful equations have evolved.

1. *Equations:*

$$\frac{1}{f} = \frac{1}{V} + \frac{1}{U}$$

$$M = \frac{I}{O} \text{ or } \frac{V}{U}$$

$$V = f(1 + M)$$

$$U = f\left(1 + \frac{1}{M}\right)$$

f = focal length of the lens

M = magnification factor

O = object size

I = image size

V = camera extension

U = object distance

2. *Example:* f = 480 mm, Original = 540 × 780 mm, Image = 180 × 260 mm

Calculate V and U

$$\text{Magnification} = \frac{I}{O}$$

$$M = \frac{180}{540}$$

$$M = \tfrac{1}{3}.$$

Camera extension

$$V = f(1 + M)$$

Object distance

$$U = f\left(1 + \frac{1}{M}\right)$$

The camera

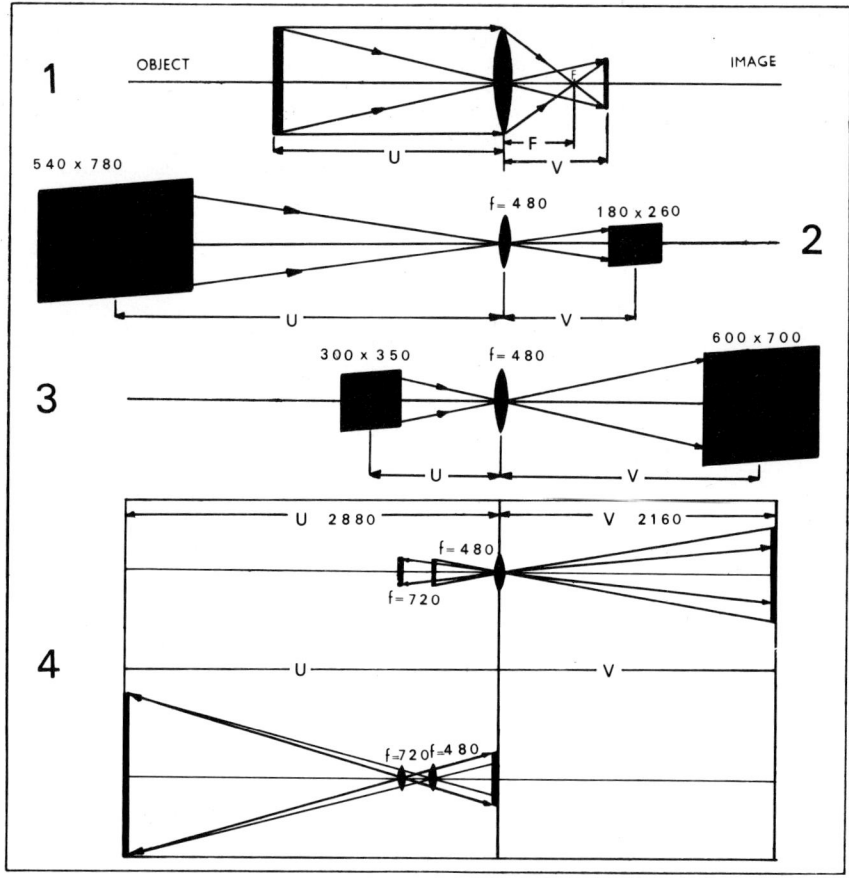

Conjugate foci
1. Camera information:
   Lengths of f, O, I, V and U
2. Reduction:
   The object is reduced to a third of its size.
3. Enlargement:
   The object is enlarged to twice its size
4. Maximum enlargement:
   f 480 mm = M $3\frac{1}{2}$
   f 720 mm = M 2
   Maximum reduction:
   f 480 mm = M $\frac{1}{5}$
   f 720 mm = M $\frac{1}{3}$

Magnification of image

$V = 480(1 + \frac{1}{3})$ $\qquad\qquad U = 480\left(1 + \frac{1}{\frac{1}{3}}\right)$

$V = 480(1\frac{1}{3})$ $\qquad\qquad U = 480(4)$

$V = 640$ mm $\qquad\qquad U = 1920$ mm

3. *Example:* $f = 480$ mm, Original = 300×350 mm, Image = 600×700 mm

Calculate V and U $\qquad$ Magnification $= \dfrac{I}{O}$

$$M = \frac{600}{300}$$

$$M = 2$$

Camera extension $\qquad\qquad$ Object distance

$V = f(1 + M)$ $\qquad\qquad U = f\left(1 + \dfrac{1}{M}\right)$

$V = 480(1 + 2)$ $\qquad\qquad U = 480(1 + \frac{1}{2})$

$V = 480(3)$ $\qquad\qquad U = 480(1\frac{1}{2})$

$V = 1440$ mm $\qquad\qquad U = 720$ mm

4. *Example:* maximum camera extension = 2160 mm.
   maximum object distance = 2880 mm.
Find the maximum enlargement and reduction capabilities of the following lenses, $f = 480$ mm and $f = 720$ mm. Remember the camera extension determines the enlargement, while the object distance controls reductions.

(A) $\quad f = 480$ mm, $V = 2160$ mm and $U = 2880$ mm.

Camera extension = enlargement $\qquad$ Object distance = reduction

$V = f(1 + M)$ $\qquad\qquad U = f\left(1 + \dfrac{1}{M}\right)$

by equation, $\qquad\qquad$ by equation,

$\dfrac{V}{f} - 1 = M$ $\qquad\qquad M = \dfrac{f}{U - f}$

$\dfrac{2160}{480} - 1 = M$ $\qquad\qquad M = \dfrac{480}{2880 - 480}$

The camera

$$4\cdot 5 - 1 = M \qquad M = \frac{480}{2400}$$

$$3\cdot 5 = M \qquad M = \tfrac{1}{5}$$

(B) $f = 720$ mm, $V = 2160$ mm and $U = 2880$ mm

Camera extension = enlargement   Object distance = reduction

$$\frac{V}{f} - 1 = M \qquad M = \frac{f}{U - f}$$

$$\frac{2160}{720} - 1 = M \qquad M = \frac{720}{2880 - 720}$$

$$3 - 1 = M \qquad M = \frac{720}{2160}$$

$$2 = M \qquad M = \tfrac{1}{3}$$

The smaller focal length lens will give greater enlargements and reductions, but its smaller object and image covering area must be taken into account.

Lens aberrations

The aim of graphic reproduction photography is to provide photographic results which have captured the original so accurately that numerous facsimile copies may be printed from them. Therefore the lenses used for this work must possess the least amount of aberrations possible. For this reason the camera operator must be able to recognize and identify all the aberrations, no matter how slight they may be, in a modern compound lens.

Lens aberrations:
Object-plate camera-image

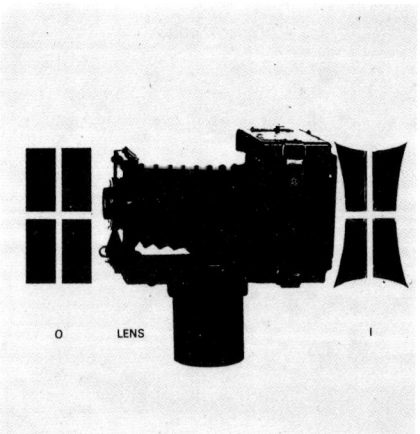

Spherical aberration

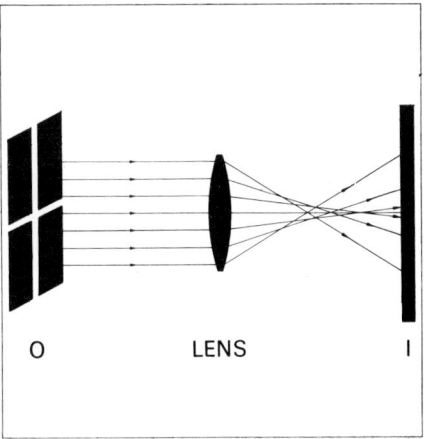

Magnification of image

O   LENS   I

The most interesting and worthwhile way of identifying the different lens aberrations is to create them. The keen student should obtain a small simple bi-convex lens and position it in the front of an old plate camera or a strong cardboard box with an aperture cut in one end for the lens and the open end provided with a ground glass screen and a film holder.

Point the camera at an object containing opaque lines in the form of a cross and square—a window frame of this shape is ideal. Now, adjusting the conjugate distances, focus the image as sharply as possible.

It will be noticed that the sharp lines of the object will take on a soft, out-of-focus appearance. This is because of *spherical* aberration. Rays entering the centre region of the lens are converged to a focus farther away from the lens than the rays converging from the periphery. Consequently, both groups of rays cannot be focused sharply at the same point. To reduce this aberration a double convex lens is used with a 1–6 radii proportion. In fact a plano-convex, with the flat surface facing the image plane, would reduce the defect nearly as well. A stop in front of the lens also reduces spherical aberration.

When light rays fall on a prism-shaped block of glass, the light not only refracts, but disperses into its different wavelengths forming a spectrum. Each coloured component forms its own image, and the different images do not coincide at the same point. This is *chromatic* aberration.

A simple double convex lens is really a number of prisms producing an image on the focusing screen which will be soft with a surrounding coloured fringe. If the camera is focused on the violet-blue rays (short wavelengths) nearer the lens there will be a red fringe, while if the focus is shifted

The camera

Chromatic aberration

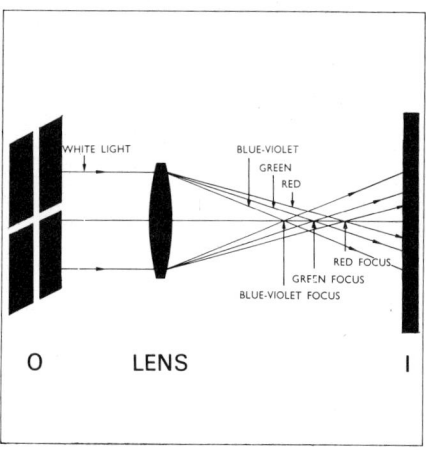

to the red rays (long wavelengths) the coloured fringe will be violet-blue. This aberration is corrected by the construction of a simple compound lens (a doublet) termed an achromatic lens.

Another aberration which to a lesser degree is common in some modern compound lenses is *astigmatism*. This means, 'not meeting in a point' and is evident in off-axis images. The cross in the window-frame object would be split into two images. The vertical line would focus nearer the lens, and the horizontal line further from the lens.

When the focusing screen is midway (the circle of least confusion) between these two focused lines, the image is small, circular and in 'the position of best focus'. Astigmatism may be slightly reduced by stopping-down the lens.

Astigmatism

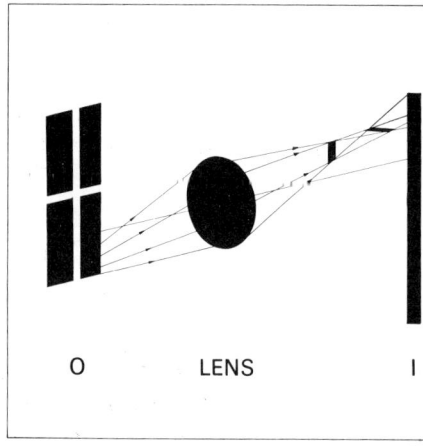

Lens aberations

Coma

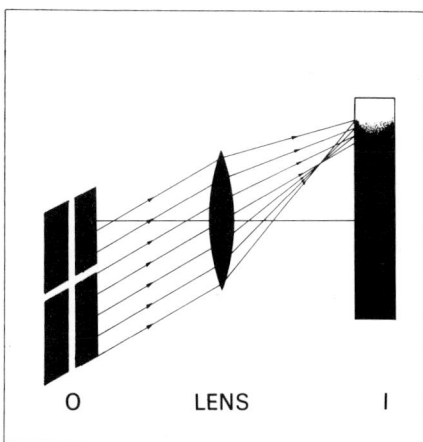

The next aberration to consider is *coma* which is spherical aberration of oblique rays, caused by concentric ring divisions (zones) of the lens not being of equal magnifying power.

The image assumes a comet-like appearance increasing in its distortion in direct proportion to the distance of the object from the axis. Coma is well corrected by drastic stopping-down, but this is detrimental to large image areas requiring good definition.

With a simple lens, if the image of a point on the object (each corner of our window-frame) is recorded at its point of best focus, the results would not be a flat plane, but a curve caused by the aberration—*curvature of field*.

This is spherical and concave towards the lens and object. In graphic reproduction photography, where a flat field is

Curvature of field

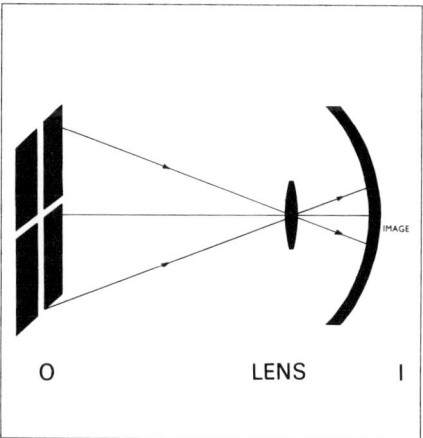

The camera

Distortion:
1. Barrel distortion
2. Pincushion distortion

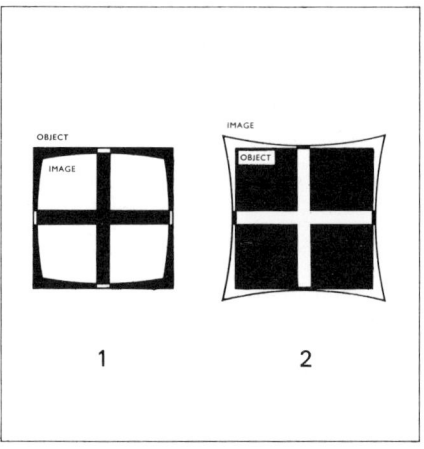

essential, curvature of a field is intolerable. This aberration is corrected by opposing the curvature of one lens component with a contrary curvature in another.

Even if the field surface is flat and the image sharp the lens will still be unsuitable for graphic reproduction photography if the straight lines in the object are not reproduced as straight lines in the image. With a simple lens a separate stop, whether positioned in front or at the rear of the lens, while reducing one aberration, introduces another—*curvilinear distortion*.

The stop positioned in the front displaces the light rays towards the centre of the field producing *barrel distortion*.

The stop positioned at the rear displaces the rays away from the centre of the field producing *pincushion distortion*.

Curvilinear distortion occurs near the edges of the screen

Compound lenses:
Rapid Rectilinear

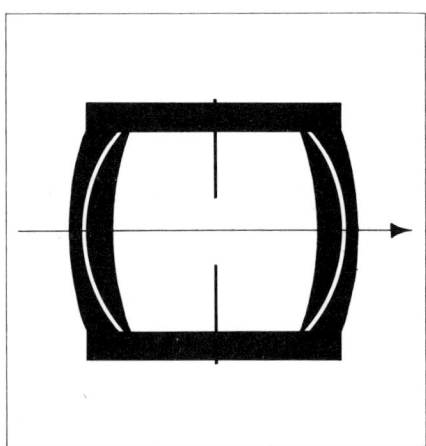

Compound lenses

and may be encountered when exposing at the maximum working area of a lens. This aberration and astigmatism may be removed to a great extent by placing two compound lenses (each a plano-concave and bi-convex together) symmetrically together with the stop between them.

In this compound lens structure—a *Rapid Rectilinear* lens—we have the beginnings of an aberration free lens.

Compound lenses

Lens aberrations are reduced or corrected by:

1. selecting the appropriate radius for the curvature of the spherical surfaces;
2. using glass of different refractive indices and dispersion; and
3. the position of the stop and the distance between component elements.

One of the earliest distortion free lenses was the symmetrical doublet known as the Rapid Rectilinear lens. This consisted of two achromatic lens elements. The first element of this lens would produce 'pincushion' distortion, and the second 'barrel' distortion, but when used in conjunction they eliminated distortion and coma.

Lens design was retarded for some time by the quality of the optical glass available. Rapid Rectilinear lenses permitted a high degree of astigmatism and curvature of field. To correct these aberrations and produce flatness of field the glass would need two desirable properties, which at that time seemed diverse requirements—high refractive index combined with low dispersion. The crown and flint glass available produced high refractive indices with high dispersion powers. However, in 1888 Abbe and Schott resolved this dichotomy and shortly afterwards the first *Anastigmat* lens, the Ross Concentric, was developed.

This apparent advancement was marred by the fact that the new combination was the opposite of the structure employed to reduce spherical aberration. Therefore, anastigmats enabled flat fields to be reproduced, but large marginal spherical aberration became apparent and needed to be restricted by the use of a small stop-size.

In 1893 Dennis Taylor took a convergent (convex) and divergent (concave) lens of equal powers (thus reducing curvature of field, separated them by a distance, which produced a total positive lens. Distortion was overcome by dividing the convergent element into two and positioning one on each side of the divergent element. Dennis Taylor found that by the

The camera

correct choice of glass the whole compound became achromatic. Astigmatism was corrected by positioning the stop in the centre near to the divergent element. Spherical and coma aberrations were eliminated by selecting correct curvatures. This immense step in lens design became available in the form of the Cooke *Triplet* lens.

This triplet was re-designed once again by dividing the divergent lens into two elements and these were placed either side of a central stop. The design proved to be capable of larger angular fields, an essential feature for graphic reproduction photography. A modern design based on this construction is the Taylor–Hobson *Apochromatic* Process lens. The post-war era has seen no dramatic change in lens design, but steady improvements have been made to existing designs.

Reproduction work imposes stringent resolution requirements on the process lens employed. The lens is usually large, e.g. 25 in. (650 mm) focal length, to accommodate large originals, but must be capable of producing aberration-

Compound lenses:
1. Ross Concentric
2. Cooke Triplet
3. Taylor Hobson Aprochromat

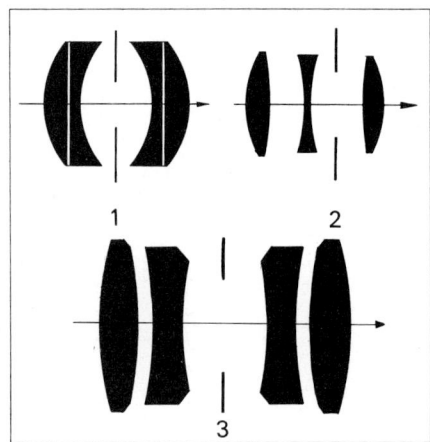

free negatives when used to reduce or enlarge the original.

Most printed matter these days incorporates coloured originals which must be separated into three or four negatives —one for each printing colour. This is carried out by exposing through tricolour (blue, green and red) filters. Their individual, separated coloured images must converge at the same focus point so that identical size is maintained. This only happens when the lens is highly corrected and termed an *apochromatic* lens.

Lens blooming

When incident light strikes air-glass boundaries approxi-

Lens blooming

Lens blooming:
1. Ray of light
2. Reflection from coating
3. Reflection from the glass

The two out of phase reflections cancel each other out

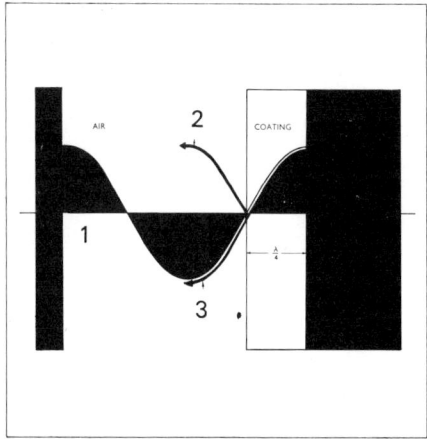

mately 5% of the light is reflected. Therefore it can be seen that in a compound lens containing a large number of boundaries, up to 45% of the light may be reflected backwards, finally emerging after internal reflection at unwanted angles, reducing the image contrast and degrading tonal gradations. (See 'Flare' in Chapter 4.)

Dennis Taylor first noticed in 1892 that when a lens surface becomes bloomed by atmospheric conditions, destructive interference takes place in the bloomed film covering the glass surface, and in doing so reduces the amount of reflected light (flare). Therefore it could be seen that to increase the transmission of a lens a bloom may be deposited on to the glass surfaces.

The substance must have a low refractive index—equal to the square root of the refractive index of the glass used. Magnesium fluoride has such an index and is used on most modern compound lenses, being coated to a controlled thickness (exactly a quarter of a wavelength thick—approx. $0{\cdot}127$ μm ($\frac{1}{200000}$ in.) by evaporation in a high vacuum. This bloom will increase the total transmission of the lens by approximately 25%.

A lens hood is a funnel-shaped shield, blackened on the inside, occasionally adjustable in length, which is fitted on the front ring of a lens to prevent light rays from outside the object area from impinging on the lens. Its function is to reduce internal reflections (flare) which degrade the image. The forward aperture of the hood, which may be rectangular, is, in fact, an extra diaphragm and should be of the same proportions as the maximum image area, but as small as the focal length of the lens will allow.

The camera

Lens aperture

The lens aperture, stop or diaphragm has three main functions:
1. to limit the amount of light passing through the lens;
2. in doing so, to provide a means of variation, so that a constant, correct exposure time may be maintained; and
3. to keep illumination at a constant intensity on the image plane. In poor quality lenses small apertures improve the definition.

In graphic reproduction photography the camera operator has a choice of diaphragms, as most process lenses provide facilities for Waterhouse diaphragms or slip-in stops, which come in sets providing a selection of circular, square and shaped apertures.

The iris diaphragm is usually housed in the centre of a lens in which a number of thin overlapping blades are rotated by turning a ring encircling the outside of the lens barrel. Each minute turn of the ring produces a circular aperture of varying diameter. Every lens has a maximum and minimum iris diaphragm aperture. Their diameters are determined by the focal length of the lens.

The speed of a lens is determined by the amount of light it can transmit on to the image plane. This will depend on the number of glass components and the maximum working size of the lens aperture. Most lenses which have an iris diaphragm are scaled-off in f numbers. This number indicates the size of the aperture which is formed when the iris diaphragm is turned to predetermined positions. The f number should be strictly associated to the focal length of the lens. The diameter of a parallel beam of light which completely fills the entrance

Waterhouse diaphragms:
Slip-in square and "dog-eared" stops

Lens aperture

Iris diaphragm:
1. f/16
2. f/32
3. f/64

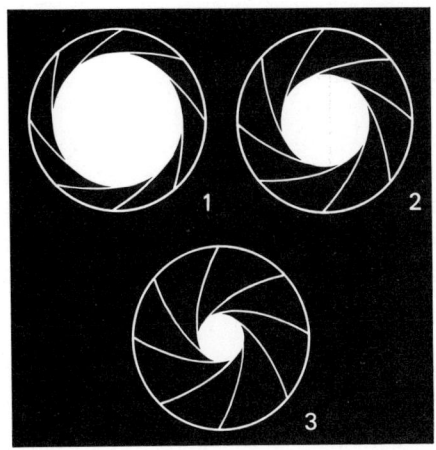

of a lens is termed the *effective aperture*. This beam of light passes on until it reaches the iris diaphragm or *actual aperture*. Because of the condensing action of the front element the diameter of the beam passing through the iris diaphragm, even opened to maximum size, is considerably smaller than the effective aperture. The *f* number is a general term for the *relative aperture*. This is the size of the actual aperture in relation to the focal length of the lens.

It will be seen that the majority of process lenses are marked off in the following *f*/numbers:

2·8, 4, 5·6, 8, 11, 16, 22, 32, 45, 64, 90 and 128.

These numbers are aperture positions which will produce

Relative aperture:
1. Effective aperture
2. Actual aperture
3. Relative aperture

$$\text{Relative aperture} = \frac{\text{focal length}}{\text{Diameter of the Effective aperture}}$$

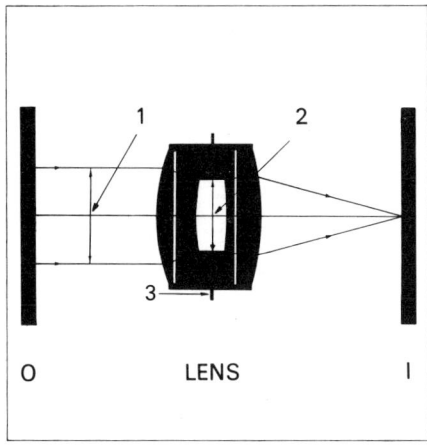

The camera

actual diameters of 1/2·8, 1/4, 1/5·6, etc., of the focal length of the lens. In most cases the diameter of each actual aperture is exactly half that of the preceding one, so that it passes only half the amount of light. This is achieved by multiplying each number by approximately 1·4 ($= \sqrt{2}$) to obtain the next aperture number. This system is extremely useful as a change to the next smaller aperture number will necessitate halving the exposure time, while an increase to the next higher aperture number will mean a doubling of exposure time.

Reversal systems

The need for a reversal system is well understood by experienced camera operators, but to students and others it is usually a confusing subject. Let us begin, therefore, by simply defining the terms.

LATERALLY REVERSED. When a gallery camera equipped with a lens is positioned straight on to the original, the lens produces an image on the emulsion side of the film in which right and left sides have changed positions when compared with the original. The image on the emulsion side is a mirror image, that is, laterally reversed.

LATERALLY CORRECT. When a laterally correct image is required, i.e. the image appears the same as the original on the emulsion-side of the film, a prism or reversing mirror must be used.

The need for reversal is dependent upon the printing process used (this will be tabulated later) and the type of original supplied. The need for reversal or non-reversal must be

Non-reversed negative:
1. Object
2. Lens
3. Image

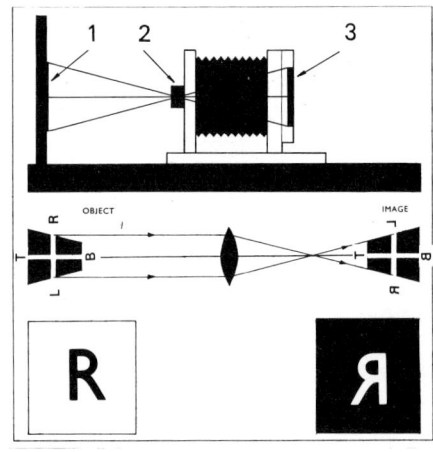

Reversal systems

Reversed negative:
1. Object
2. Prism and lens
3. Image

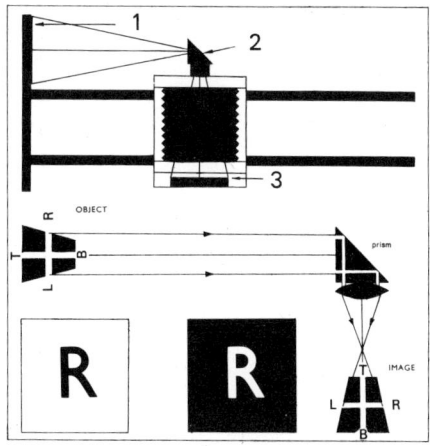

stipulated before work commences. A negative produced by a lens with the camera facing the original has a correct-reading image on the back of the film when the emulsion side is away from the observer (this negative is termed a non-reversed negative). A contact positive made from this negative, emulsion to emulsion, will read correctly when the emulsion faces the observer.

There are many methods of producing a reversed or laterally reversed image, but in this instance we will concentrate on optical systems used in the camera. These may be listed as:

1. prisms and mirrors fitted on cameras which have been turned at right-angles to the copyboard;
2. Straight-line reversing systems for horizontal darkroom cameras;
3. reversing systems designed for vertical darkroom cameras; and
4. projection methods through the transparency holder of the camera or enlarger.

In the case of gallery cameras a right-angled prism is attached to the lens and the camera body turned at right-angles to the copyboard so that the prism faces the copy-holder which is centred (on the focusing screen) by moving it to the side.

A negative produced by a prism and a lens has a correct-reading image on the emulsion side of the film (this negative is termed a *reversed negative*). A contact positive made from this negative, emulsion to emulsion, will read correctly when the back of the film faces the observer.

The camera

Straight-line reversing systems

With the introduction of the darkroom camera another reversal system had to be developed because the darkroom camera, being permanently fixed in the wall, could not be turned through 90°. To produce a reversed image in a straight line three reflections plus a lens are needed.

Early straight-line systems obtained this with two prisms and a mirror, all in front of the lens, but this suffered badly from flare and long exposure times because of considerable light-loss occurring on the many reflecting surfaces.

In recent years the perfection of aluminizing and quartz-coating optical glass has produced mirrors of high reflectance. This has led to the present system which use a vertical lens positioned between three angled mirrors. The third mirror receives the image and deflects it on to the surface of the

Straight line reversal—three mirror system:
A. Non-reversed negative
 1. First mirror
 2. Second mirror
B. Reversed negative
 1. First mirror
 2. Second mirror
 3. Third mirror

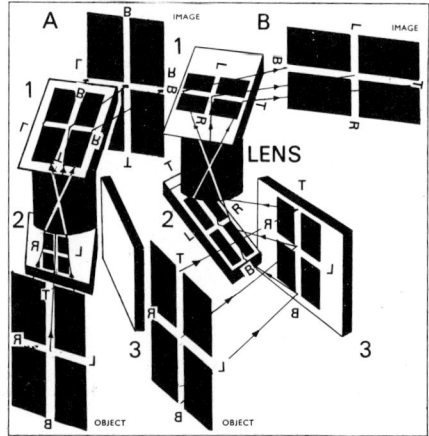

second mirror which reflects the image up through the lens to the first mirror. This in turn collects the image and turns it along the camera extension to the light-sensitive emulsion. To produce non-reversed negatives the third mirror is bypassed by turning the second mirror round to face the copyholder.

In a number of straight-line systems a supplementary lens is moved into a normal position for non-reversed negatives. This minimizes the reflecting surfaces, and therefore reduces flare and exposure times.

A different approach to straight-line reversal may be found on most Klimsch horizontal darkroom cameras. This is the utilization of the roof-mirror principle. Two silver-surfaced mirrors meet at right angles, with their centre join at 45° to the lens axis. The two mirrors are positioned in front of

Straight-line reversing systems

Straight-line reversal—Roof-mirror system:
A. Non-reversed negative
  2. Single 45° mirror
B. Reversed negative
  1. Roof mirror

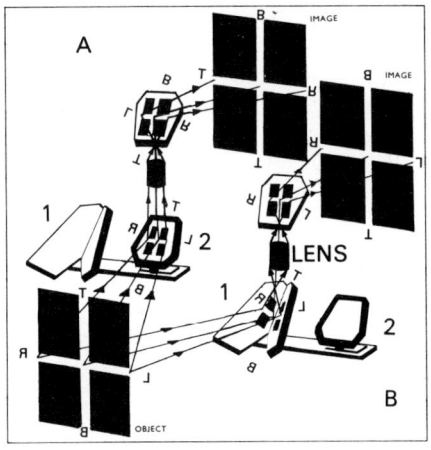

the lens and produce two reflections; the third reflection, directed towards the emulsion, is caused by a third mirror behind the lens. For non-reversed negatives the roof-mirror is slid to one side and replaced by a single 45° mirror which faces the original and collects the image, passing it on through the vertical lens to the rear mirror and finally to the emulsion.

It may be noted that on most vertical darkroom cameras, which have a lens in a horizontal position, the opposite results are achieved. The roof-mirror produces a non-reversed image and a single mirror is used for reversed negatives.

A recent innovation is the Wray reversal unit housed on a rotating turret.

If the indirect method of working is employed it is normal

Straight-line-reversal—Rotating turret system:
1. Single 45° mirror
2. Roof-mirror—reversed negatives
3. Single 45° mirror—non-reversed negatives
4. Flashing lamp

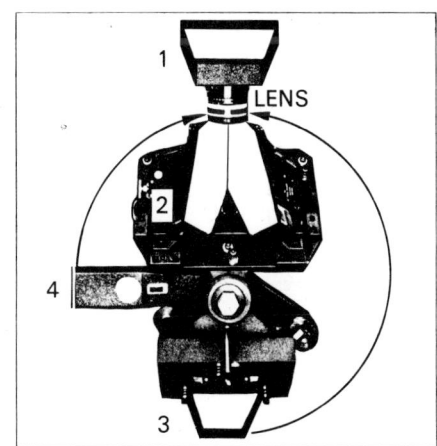

The camera

procedure to place the continuous tone negatives in the transparency holder of the camera. By positioning the lamps behind the negative it may be projected through the lens on to the emulsion.

Reversal of the original is effected by making sure the correct-reading side of the continuous tone negative is facing the lens. Similarly, with enlarger-type cameras, when reversal is required the continuous tone negatives are placed in the transparency holder with the correct-reading side facing downwards, towards the vacuum baseboard.

Graphic reproduction cameras

The following descriptions deal only with the basic principles of graphic reproduction cameras. More detailed information may be obtained from manufacturers.

Graphic reproduction cameras are made to produce negatives and positives for line, halftone, continuous tone and colour separated reproductions, copying from reflection and transmission originals. If the capital is available it is always wise to buy specialized photographic equipment, because in the final analysis a result of higher quality is produced in a shorter space of time.

Gallery cameras

This type of camera is robust and sufficient for most straightforward reproduction work up to 500 mm × 400 mm (20 in × 16 in.). The length of rail and bellows extension usually permits reductions to at least 20% of the original and enlargements to 175%. The focal length of the lens in this case would be approximately the diagonal of the maximum

Gallery camera: Plan view
1. Object
2. Lens
3. Image

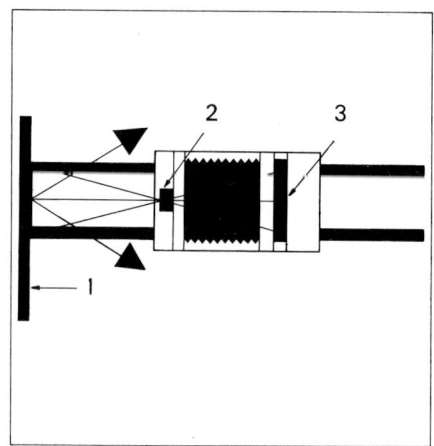

Graphic reproduction cameras

image size 650 mm (25 in.). In the more recent darkroom camera design a lens turret has been incorporated into the front body so that by merely rotating this a lens of different focal length may be brought into alignment with the original. In doing so the operator is provided with a new scale of enlargements and reductions with every lens.

The bed of a gallery camera is positioned on a turntable which, when set at right angles to the copyboard, facilitates image reversal with a prism screwed on to the front of the lens. The gallery camera tends to take up a great deal of floor space and productivity is hampered by the necessity to walk to and fro with the darkslide. Because of these factors gallery cameras, although excellent for basic training, are rapidly being replaced with relatively inexpensive darkroom cameras built to basic designs without sophisticated refinements.

Vertical darkroom cameras

The front lens body is permanently housed in the darkroom wall. Alteration in the size of the image is the result of moving the camera back along its bed and the copyholder up and down rails fixed to the outside wall. After placing the original in the copyholder and setting the lamps the camera operator completes all his operations within the darkroom. The whole idea of the darkroom design is to reduce the camera's floor area, increase production and provide a versatile piece of photographic equipment. These cameras contain many refinements including:

1. automatic focusing;
2. lens turret containing a selection of lenses;

Vertical darkroom camera: side view
1. Object
2. Mirror and Lens
3. Image
4. Darkroom wall

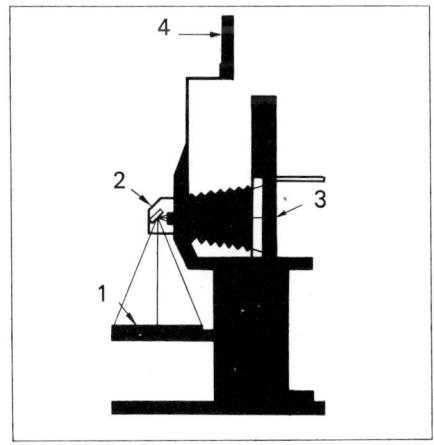

The camera

3. vacuum pressure for holding films, contact screens and originals;
4. screen cassette, facilitating the removal or introduction of the glass halftone screen at any time; and
5. pin and punch, three point lay registration systems.

Horizontal darkroom cameras

The camera back is permanently housed in the darkroom wall. Alteration of the image size is produced by moving the front lens body and the copyholder along a low, rigid camera bed. This type of camera incorporates all the advantages and refinements of the vertical darkroom camera while producing extra large films without any alteration or fluctuation in their squareness or size.

Horizontal darkroom camera: Plan view
1. Object
2. Lens and straight-line-reversing system
3. Image
4. Darkroom wall

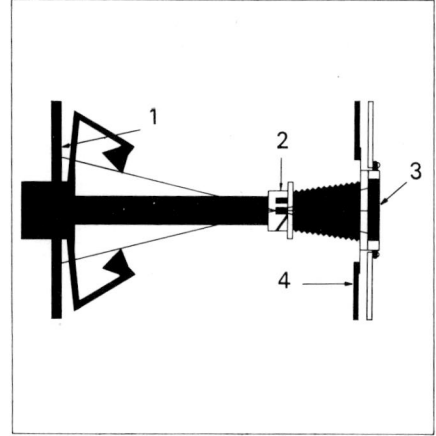

Vertical enlarger-type cameras

This camera stands completely inside the darkroom. Its manufacture arose from the increasing use of colour transparencies as originals. Most colour transparencies, being small, in the first instance need an intense, concentrated light source combined with a selection of lenses to produce the high degree of enlargement usually required. This type of camera is extromoly versatile. It may be used as a conventional camera, because with suitable lamps attached to sides of the vacuum baseboard the whole construction is similar to a gallery camera placed upright on its railends. In this case, the original is positioned on the vacuum baseboard and the film placed in the transparency holder. However, as stated before, it is better to use the equipment for its own specialized function, which is the reproduction of colour transparencies and screen positives from continuous tone negatives.

Graphic reproduction cameras

Vertical enlarger-type camera: Front view
1. Object image
2. Lens
3. Object or image

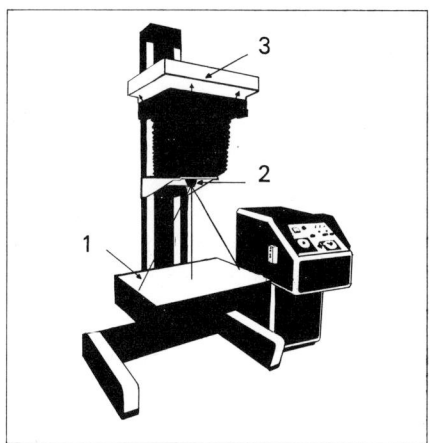

This approach to camera design is becoming increasingly popular because of its flexibility and choice of refinements, which include:

1. a selection of lenses housed in a turret;
2. an exhaustive choice of light sources;
3. scale focusing on a motor drive and re-setting devices;
4. registration control throughout with the pin and punch method; and
5. filtration at the light source.

Roll-film horizontal darkroom cameras

This camera design is essentially the same as any camera containing a roll-film which is wound-on after each photograph is taken. It specializes in monochrome reproduction work where there is a great deal of straight-forward line and halftone originals. Because speed of operation is necessary there is no visual focusing. The copyholder is designed to allow precise centring of the original. The scaling system producing enlargements and reductions is operated from outside the darkroom on the bed of the camera. The front lens body and copyholder are motor driven, their relative positions being calculated from the magnification factor and located very accurately on percentage scales.

As the camera operator is working alongside the camera bed on the outside of the darkroom the control panel with screen distance movement, etc., is positioned on the front side of the camera back. Three film rolls of varying width are loaded above the vacuum film holder. Usually in one mechanical operation a length of the particular width required is wound-down and squeegeed on to the vacuum holder. After

The camera

Roll-film horizontal darkroom camera: side view
1. Object
2. Lens
3. Image

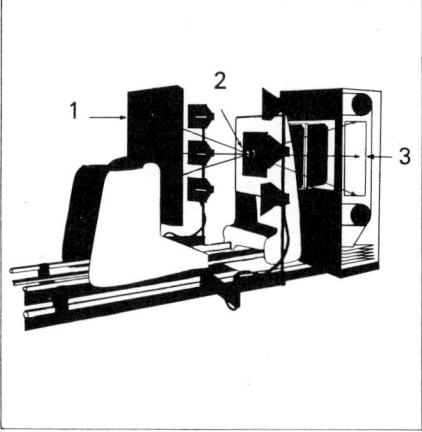

exposure it is released, wound-on and then cut off, finally slipping into a box-carrier to await processing.

Contact frames

Most reproductive work entails a great deal of negative to positive exposure. This may be producing contact positives from master negatives or halftone positives using a contact screen positioned over the continuous tone negative. Whatever the result an efficient contact frame is essential. Basically there are two main types. The first is a contact box with the light source underneath and the vacuum pressure in the lid.

This design is ideal for straightforward contact work, line

Contact box: side view
1. Light source
2. Glass
3. Vacuum pressure

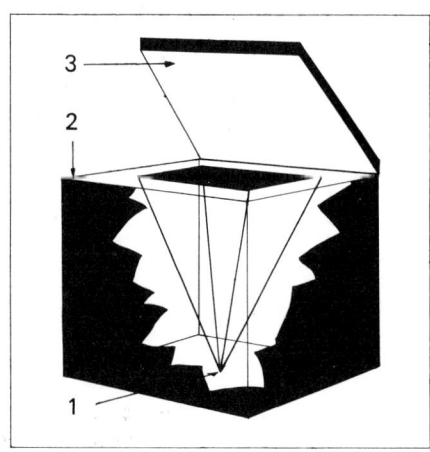

Graphic reproduction cameras

Contact frame: Side view
1. Light source
2. Glass
3. Vacuum pressure

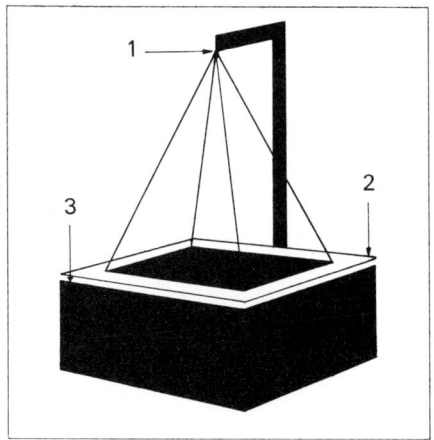

and tone combination techniques, and, with the inclusion of an electronic flash light source, colour separation from colour transparencies, but is rather restricted for the production of screen positives. For this work the upright frame, with an overhead, adjustable height light source, is much more versatile.

Multiple image work is best carried out on a vertical step and repeat machine which is usually employed in the plate-making department. Where an image is very small it is more practicable to step it up on a sheet of film first and then print down this film, containing a block of multiple images, only three or four times.

In the next section all the previously mentioned equipment has been planned into a reproduction camera studio.

Vertical step-and-repeat machine: side view
1. Light source
2. Glass
3. Vacuum pressure
4. Light-sensitive plate

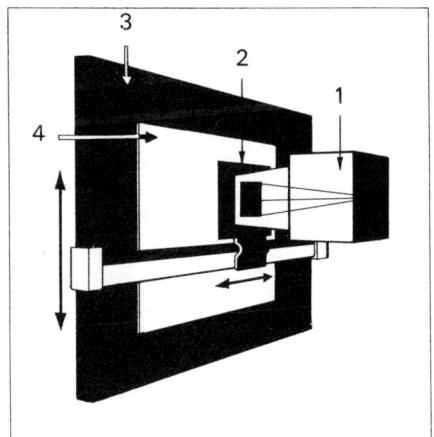

The camera

Camera studio design

Efficient designing of a printing or platemaking plant is of utmost importance, because once operational the chance of alteration is slight. Although we are mainly concerned with the camera studio layout this is not isolated but an integral part of the whole design. Therefore it will be advisable to look at complete designs first and then to concentrate on the studio plans. These designs are general in their outlook, but contain essential factors. All final plans must include the individual needs of the particular company.

A multi-storey building is not perhaps ideal for printing, but in most city centres it is the only construction available. In this case the following design would be practicable.

If a single-storey building is to be built for a company moving into a new growth area, then planning will become enjoyable and satisfying. Extremely efficient layouts can be drawn up after these factors have been fully considered:

1. general construction of the building;
2. future productivity figures for each section;
3. movement of product and personnel;
4. internal environment, health and safety conditions;
5. communications, services and waste disposal; and
6. fire precautions.

A tentative plan would probably view as illustrated in Design B.

A company faced with a chance of completely reorganizing their factory and staff should seize the opportunity to introduce new ideas, especially into the working conditions of employees. As cleanliness is so important in good printing it would seem advisable to install shower units so that employees may wash before and after work. Ideally, entire working clothes suitable for the particular section should be supplied clean each day. For example, a camera technician would require a light pair of trousers and tee shirt in white nylon, a white nylon coat and a pair of house shoes.

Studio requirements

When planning a camera studio, no matter what the size or capital outlay, it is worthwhile trying to include the following requirements:

1. the overall floor plan to be divided into wet and dry areas;
2. a logical series of movements and direction while producing photographic results from the original;
3. ensure that every technician will be able to work independently without the inconvenience of sharing or waiting for equipment;

Camera studio design

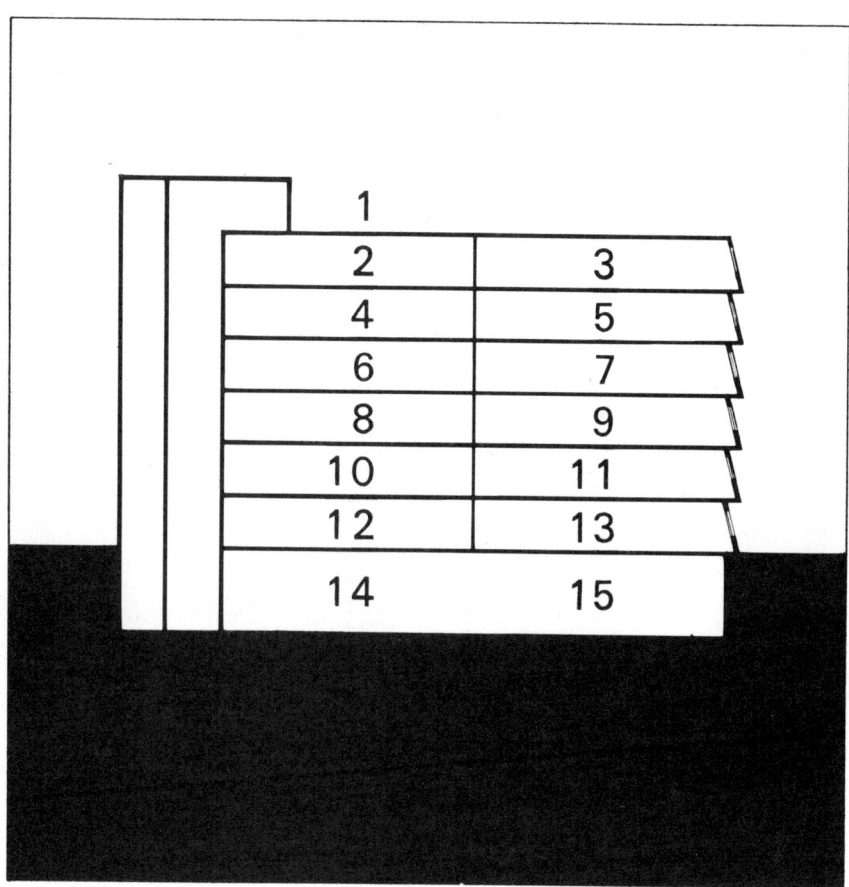

Design A: Multi-storey design: side view
1. Roof studio
2. Commercial photography
3. Graphic design
4. Reproduction photography and electronic scanning
5. Retouching and planning
6. Plate-making
7. Proofing
8. Service and maintenance
9. Light-machine room
10. Quality control dept
11. Dining hall
12. Finishing processes and dispatch
13. Offices and reception
14. Paper warehouse
15. Heavy-machine room

The camera

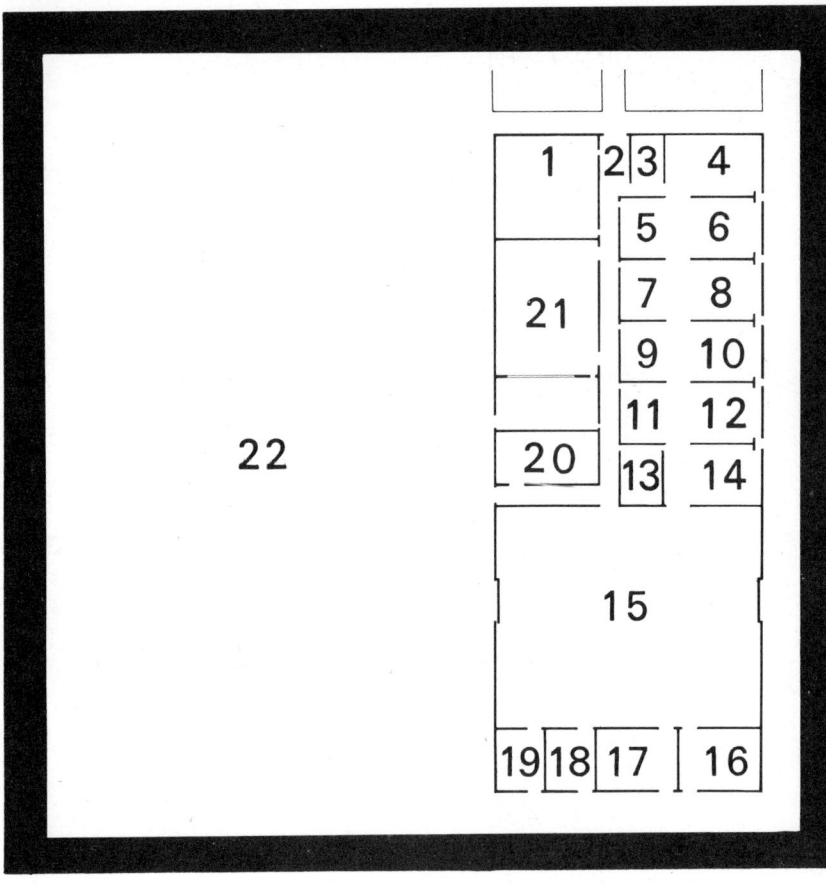

Design B: Single storey design: Plan view
1. Offices—sales
2. Reception
3. Print viewing
4. Offices—works
5. Graphic design
6. Commercial photography
7. Electronic scanning
8. Reproduction photography
9. Planning
10. Retouching
11. Step and repeat
12. Plate-making
13. Quality control dept.
14. Proofing
15. Printing machine area
16. Finishing and dispatch
17. Paper testing and warehouse
18. Ink store and mixing
19. Maintenance
20. Cloaks, toilets, showers and services
21. Dining hall
22. Car park and service road.

Camera studio design

4. centralize all stores and chemicals with the maximum use of pre-packed liquid solutions;
5. light-trap entrances to all darkrooms;
6. the use of automatic processing machines, housed in one major processing room;
7. a line of electrical sockets over the cameras so that different types of illuminants on wheeled stands become fully-interchangeable between the cameras; and
8. the overall appearance of the studio should be one of spaciousness with two obvious, unobstructed walk-ways throughout the length of the work area, giving easy access and egress to the studio.

Studio layout

We will now look at an example design of a reproduction camera studio on the fourth floor of a multi-storey building. Generally the first consideration would be the specialized equipment and labour force needed to reproduce the type of work encountered. Listed below is equipment, compiled from a number of camera studio layouts of this type.

1. Roll-film high productivity camera for monochrome reproduction, rapid production of line and halftone negatives.
2. Horizontal darkroom camera for monochrome and colour reproduction, reflection copying in line, halftone and continuous tone, mostly in large sizes.
3. Vertical enlarger-type camera for colour transparency reproduction, enlargements and contact screen positives.
4. Two automatic processing machines for general processing, one for lith-type emulsion and one for full-range and mask-range continuous tone emulsion.
5. Contact frames for production of positives; a number in individual booths where retouchers may produce their own screen positives.
6. Ancillary apparatus, densitometers, etc. for quality control, density readings and information to check and maintain the desired quality.

In this department four camera technicians would be needed to operate the three cameras and electronic drum scanner, working on a monthly turn-around system so that each man is familiar with all the equipment. The camera operators would produce negatives, masks and combination positives for the eight retouchers and two planners. The retouchers would be responsible for making their own straightforward contact and contact screen positives in the individual contact booths.

The camera

Studio lay-out: Plan view
1. Enlarger-type camera
2. Mixing-room
3. Horizontal dark-room camera
4. Automatic processing room
5. Wet area, drying cabinets, etc.
6. Roll-film dark-room camera
7. Contacting room individual frames and contact boxes
8. Walk-in light-trap
9. Vario Klischograph electronic scanner
10. Drum-type electronic scanner
11. Retouching booths
12. Large planning tables

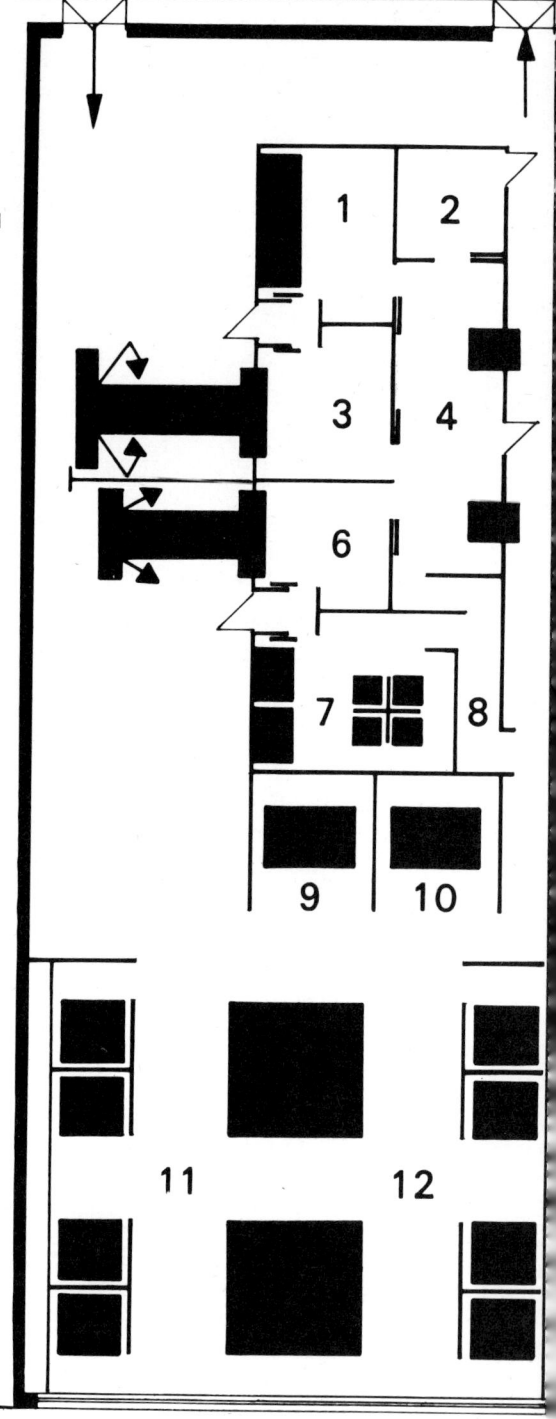

Camera studio design

This layout, labour force and choice of equipment would be capable of tackling most graphic reproduction work from a small monochrome letter-heading to multi-coloured mail-order catalogues.

6. LIGHT SENSITIVE EMULSIONS

Historical introduction

Very few discoveries of any importance are made instantly by one person; rather the reverse, a painfully slow development takes place as one fact, often discovered by accident, reinforces another. Hundreds of years before photography was thought of intelligent observers noted that light was capable of producing reactions in pigments, foliage and human skin. As knowledge increased and experimental investigation began, the isolation of chemicals and their properties became possible.

At the start of the eighteenth century a German professor, Johann Heinrich Schulze, while researching into phosphorescence, exposed a bottle of silver chloride mixed with chalk sediment to light. To his amazement the mixture darkened. Schulze realized this was because of the reaction of light and repeated this experiment, isolating the light's action through stencils—paper with cut-out shapes and letters. After suitable exposure the images appeared as blackened silver against the surrounding white chalk sediment. Schulze proudly exhibited these results to his friends who, when trying to diagnose the cause, came up with many fanciful ideas, such as that the inside of the bottle was a mirror reflecting images which Schulze had hidden on his person.

Fifty years later Carl Wilhelm Scheele, a Swedish chemist, proved, by using ammonia to divide the unexposed silver chloride from the exposed areas, that Schulze's blackened images were indeed grains of metallic silver. The next person to take up the cudgel in the early eighteen hundreds was Thomas Wedgwood, who is usually remembered as a member of the family manufacturing the famous pottery, rather than the first person to realize the potential of projecting the camera obscura's image on to a material coated with light-sensitive salts. Unfortunately he never fulfilled this particular inspiration and had to be satisfied with producing many profiles.

Wedgwood laid various objects containing interesting textures, such as fern leaves and insect wings, on to sheets of white leather saturated with a mixture of silver nitrate and silver chloride. After suitable exposure to light these objects

Historical introduction

were removed, leaving their silhouettes in blackened silver. Once again the limitations of his knowledge prevented Wedgwood from permanently fixing his images, and selected friends had to view his handiwork in dim candle light. Perhaps his premature death at the age of 34 retarded the possibility of an early photographic process.

The first permanent photograph is attributed to Joseph Nicéphore Niépce in 1826. Although of utmost importance, this scene was captured and recorded on a pewter plate by photo-mechanical means rather than a photographic process. The true photographic processes began in 1835 when Louis Jacques Mandé Daguerre, a Frenchman and Niépce's earlier partner, introduced his Daguerreotype Process. This began with a polished copper plate being coated with silver. The prepared base was carried into a darkroom and made light sensitive by applying iodine vapour to the silver coating. The copper plate holding the silver iodide emulsion was positioned in the camera and exposed for several minutes. Development of the latent image commenced on the application of mercury vapour which adhered to the exposed areas. Fixation was carried out in a solution of common salt, until the introduction of superior hyposulphite of soda solution. The direct, laterally-reversed positive result on its copper base was attractively toned in gold chloride.

Daguerre's process had an almost hysterical acceptance throughout Europe, but because of its limited potential, difficult image duplication and prohibitive cost the Daguerreotype Process, with all its accompanying ceremony and glamour, gradually gave way to the English Calotype process invented by William Henry Fox Talbot in 1841.

Louis Jacques Mande Daguerre
(1787–1857)

Light sensitive emulsions

William Henry Fox Talbot
(1800–77)

The Calotype process was based on a versatile negative to positive print system. A sheet of paper was immersed in baths of potassium iodide and silver nitrate producing a light sensitive emulsion of silver iodide. The moist, light sensitive paper was placed in a camera and given an effective exposure.

Development was carried out in a solution of gallic acid and silver nitrate. The developer reacted physically, depositing more silver on to the exposed areas rather than chemically (reducing the exposed areas to a metallic silver state). At the time, because of its improvement of the Calotype process, this developer became the centre of a great deal of dissension, as its invention was attributed to another photographic pioneer, the Rev. John Bancroft Reade. After development the paper negative was fixed in hyposulphite of soda.

However, the next stage in the process proved to be a more important contribution. Numerous positive copies were produced by repeating the process after the prepared paper had been exposed in contact with the negative which was now translucent after being soaked in oil. The final positive print on paper lacked the fine definition and jewel-like appearance of the Daguerreotype plates, but this lack of critical definition was counteracted by the inexpensive simplicity of the process.

1851 marked the beginning of the 'wet plate' era, when Frederick Scott Archer, an English sculptor, presented the patent-free Wet Collodion process to all those interested in photography. This process was capable of producing excellent results, far superior to any other contemporary process. A glass plate was cleaned and coated with an emulsion layer formed from mixing soluble iodide in a solution of nitro-cellulose (collodion). This emulsion was sensitized in a

Historical introduction

darkroom by immersing it in a silver bath containing silver nitrate. As in the Calotype process the plate was exposed while still in a moist state and physical development followed in a pyrogallol or ferrous (iron) sulphate developer. Fixation was once again carried out in a solution of hyposulphite of soda.

Although the manipulation of the process was unwieldy and a travelling photographer took everything with him, including the kitchen sink, many new applications of photography began with use of the Wet Plate process. Pictures of current events, and snap-shooting began, breaking away from the formal portraiture and group photographs. For example, news pictures from the front lines during the Crimean and American Civil wars, captured by the moist silver iodide plates, were the first to show the horrors of warfare. Graphic reproduction, or photo-mechanical printing as it was known then, owes its establishment to the Wet Plate process, and the process continued to be used for this type of photography for nearly 100 years.

During the closing months of the year 1871, Richard Leach Maddox, an English physician, substituted gelatine for collodion and silver bromide for silver iodide. The result of this modification was the first gelatine dry plate. This new innovation contained many advantages. Sensitivity was increased, preparation of the light sensitive emulsion could now be carried out in advance of exposure, and a more straightforward processing system could be completed almost at one's leisure. The travelling photographer's equipment was now drastically reduced and a whole new range of subjects became possible. From this moment on there was

Frederick Scott Archer
(1813–57)

Light sensitive emulsions

Structure of matter

MATTER

ELEMENTS

An element is a substance which cannot be divided or reduced to a simpler or more basic substance.

e.g. Oxygen
Bromine
Silver
Carbon
Sulphur

COMPOUNDS

A compound is a substance with its own properties, but may be divided into its constituent elements.

e.g. Silver bromide

MIXTURES

A mixture is a variable combination of elements or compounds which is easily separated into its constituents.

e.g. A typical developer

SMALLEST PARTICLES OF SUBSTANCES

ATOMS

An atom is a minute structure containing electrically charged particles, generally seen as a positively charged nucleus surrounded by orbiting negatively charged electrons. The nucleus determines the atom's mass or weight and chemical identity, the outer ring of electrons its volume and chemical reactions.

IONS

An atom can be made to gain or lose electrons and this produces a more positively or negatively charged particle which is then termed an ion. When producing compounds some elements combine more readily than others; this is because their atoms are more willing to gain or lose electrons. Ions can also be formed from groups of atoms.

MOLECULES

A molecule is a small group of atoms, chemically joined together in a particular manner or pattern. When a compound is formed from two different elements this is the result of the atoms of these elements sharing or interchanging their electrons.

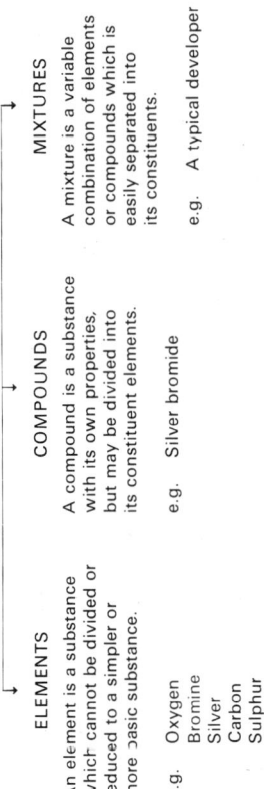

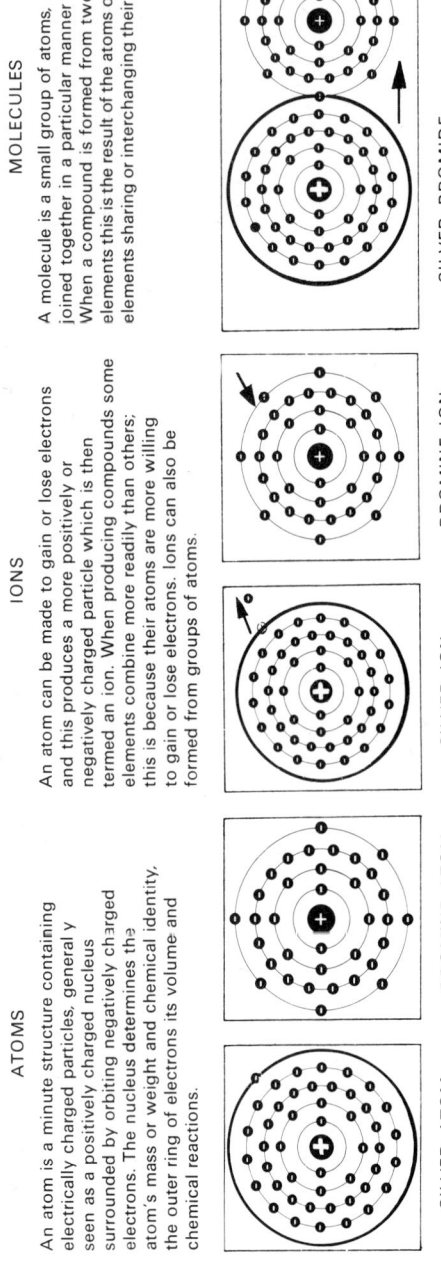

SILVER ATOM BROMINE ATOM SILVER ION BROMINE ION SILVER BROMIDE

Chemical introduction

Measurements and reactions

MEASUREMENTS AND WEIGHTS

To understand the chemical combinations and reactions in photography we need to know the quantities of the chemicals in question. The information is more applicable when the relative numbers of atoms and molecules taking part in the reaction are known. Weights of individual atoms and molecules are termed atomic and molecular weights.

Atomic weight—The basic unit of measurement is the atom of oxygen and is given a relative weight of 16. All the other atoms have their own atomic weights expressed in similar numbers. One sixteenth of the value of oxygen becomes the basic unit of these numbers.

e.g. Ag = 108 Br = 80

Molecular weight—This is itself just a number. The proportionate quantity in grams is called gram-molecular weight.

e.g. AgBr = 188

IONIC AND NON-IONIC COMPOUNDS

Chemical compounds form themselves into two classifications—ionic and non-ionic compounds. Ionic compounds are held together by opposite charges on ions combining into complementary pairs. Non-ionic compounds are bound together in a different way.

e.g. ionic—Silver bromide

non-ionic—Metol

Some substances do not normally exist as single molecules, but as crystals built up from ions.

e.g. Silver bromide

Crystal lattice

○ Bromine ions   ● Silver ions

CHEMICAL FORMULAE

Each atom or group of atoms is capable of combining with a certain number of atoms of another element. An atom of hydrogen combines with only one atom of another element since it is capable of forming only one bond. The number of bonds an atom or group is capable of forming is called its valency.

Hydrogen, bromine, silver and the nitrate group have valency 1. Hence formulae of typical compounds are:

H—Br   Ag—Br
H—NO$_3$ Ag—NO$_3$

Oxygen, magnesium and the sulphate group have valency 2. Hence:

H\O/H   H\SO$_4$/H

Br\Mg/Br Br\Mg—SO$_4$

We do not normally put the bonds in chemical formulae in this way unless it is desired to show position of atoms relative to one another.

CHEMICAL REACTION

A chemical reaction between two compounds takes place between the molecules or ions of which the compounds consist. It may involve combination between molecules or transfer or exchange of atoms or ions between one compound and another.

ADSORPTION

This is not a true chemical reaction—it is a surface reaction. Solids are held in a permanent state by a cohesive force holding its molecules together. All along the surface boundary this cohesive force is prevalent, but unsatisfied. Molecules of a solution, liquid or gas, being in a mobile state, are capable of moving up to the solid's surface. The cohesive force attracts and finally captures these molecules holding them tightly against the surface. This adsorption process becomes apparent during certain stages of the photographic process, especially in the addition of optical sensitizing dyes to the light sensitive emulsion.

139

Light sensitive emulsions

no need for the photographer to prepare and process his own emulsion prior to, and after exposure. This led to the formation of numerous photographic dry plate manufacturers, such as the Britannia Dry Plate Company which eventually became Ilford Limited.

Dr. Richard Leach Maddox
(1816–1902)

Two years later Hermann Wilhelm Vogel, a German chemist, discovered that by the inclusion of certain dyes the emulsion's sensitivity (emulsions at this time were only sensitive to u.v. and blue light) could be extended into the green section of the spectrum. Vogel's work was the foundation of ortho and panchromatism, which was achieved later when improved dyes, such as erythrosin (*red*), absorbing green light for orthochromatic emulsions, and pinacyanol (*blue*) for panchromatic emulsions were produced.

Photography was now being widely used in all manner of applications. Improvements were coming fast and furiously: flexible film bases; improved emulsions; whiter printed papers; cinematography; and finally, in 1910, the first emulsions capable of colour photography.

Chemical introduction

The camera's image is captured and rendered visible by chemical means. To appreciate the reactions involved we must turn to basic chemistry. The make-up and interactions of many substances may be explained once the terms and rules of chemistry are realized. To begin with all matter may be classified and sub-divided into constituent parts.

Action of light

When explaining movement of light rays such as reflection, refraction, etc., it is easier to think of light in terms of wave-formation. The photographic image is the result of chemical reaction or change and this is due to the light-sensitive gaining, losing or transferring energy. Because this chemical change is initiated in the first instance by incident light rays it is more feasible to look upon light in its new role as energy or quanta. We are now entering the field of *Photochemistry*—the action of light upon light-sensitive emulsions. This work revolves round two basic laws which are:

1. *The Grotthuss–Draper Law.* Chemical change is produced by absorbed light—not transmitted light.

2. *The Bunsen–Roscoe law.* Chemical change is proportional to the exposure when I (intensity) is multiplied by T (time).

To explain the interaction of light upon light-sensitive emulsions let us use an analogy. Imagine the light-sensitive emulsion as a bank of life-size targets and the light source a machine-gun. When the gun fires spasmodic bursts of bullets (quanta) some of the targets fall, others stand firm. From this, law number one evolved—if the bullets pass through unhindered the targets stand, but if the bullet hits and embeds (absorbs) itself into the target then a reaction takes place—the target falls.

Further investigation carried out by Robert Bunsen, a German Professor of Chemistry, and Henry Roscoe, an English chemist, during the middle of the nineteenth century, proved that in most cases the number of fallen targets was directly proportional to the number of bullets fired—law number two.

Of course, the interaction of light and light-sensitive grain structures is not as simple as this analogy, but to begin with it helps to explain a very complex process.

Photochemical requirements

A photochemical system capable of permanently recording the image of a camera must satisfy the following requirements:

1. high-sensitivity response to weak reflected light rays;
2. the chemical change must be proportional to intensity and time of exposure in order to produce images made up of tonal gradations;
3. The final chemical compositions in the image and non-image areas should be capable of permanent fixation; and
4. the photochemical system should be as straighforward

Light sensitive emulsions

as possible, withstanding robust manipulation during processing.

It may be realized from this list that few systems foot the bill completely and from the few that do, one is more suitable than the others. This is the light-sensitive system made from the basic element of *silver*.

Emulsion manufacture

Silver nitrate, alkali halides, gelatine and distilled water are basic ingredients of a silver-halide emulsion. They fulfil three basic functions.

1. Silver nitrate + Alkali halide
   = production of light-sensitive silver halides
   ↓
2. Gelatine
   = an active binder holding the finely divided silver halide grains apart
   ↓
3. Distilled water
   = solvent to carry above substances
   ↓
   Light-sensitive emulsion

Silver halides

A partially insoluble silver halide is produced when silver nitrate is mixed with an alkali metal such as potassium bromide, potassium iodide or sodium chloride. In these examples the resultant silver halide is:

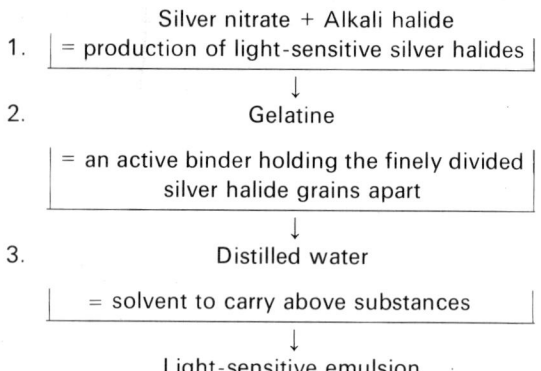

	silver nitrate	+	alkali metal	=	insoluble silver halide	+	soluble by-product
1.	$AgNO_3$	+	KBr potassium bromide	=	AgBr silver bromide	+	KNO_3 potassium nitrate
2.	$AgNO_3$	+	KI potassium iodide	=	AgI silver iodide	+	KNO_3 potassium nitrate
3.	$AgNO_3$	+	NaCl sodium chloride	=	AgCl silver chloride	+	$NaNO_3$ sodium nitrate

If the silver halides, silver bromide, silver chloride and

Emulsion manufacture

Silver halides in solution:
A. Silver halides in water
B. Silver halides in gelatine

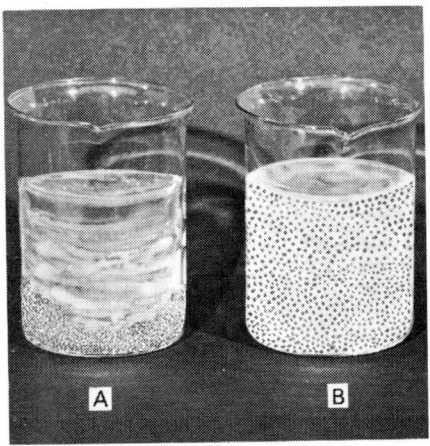

silver iodide were used individually to produce an emulsion, its sensitivity would not be particularly high, but a combination of the three basic silver halides will produce an increase in the degree of sensitivity.

Gelatine

This natural substance is obtained from animal bones and hides. In emulsion manufacture it becomes a virtually ideal binding agent because of the following properties:

1. After the precipitation of silver nitrate and an alkali halide, the insoluble silver halide grains are held by the gelatine in a separated state.
2. With the application of heat, gelatine will liquefy and flow, and when cooled it solidifies into a firm jelly. It will, as an emulsion binder, swell when immersed in solutions, so that processing chemicals may easily percolate through to the silver halide grains.
3. Most gelatins contain sulphur compounds which have a contributive effect on the ultimate sensitivity of the emulsion and this is why gelatine is referred to as an 'active' binder.
4. During exposure to light, *bromine electrons* shift away from the silver nucleus. After exposure the bromine electrons have a tendency to reunite with the silver, but the gelatine prevents this by accepting and holding the bromine electrons.

Water

Because purity is essential throughout the manufacturing process, distilled water is used. Ammonia is added in the production of high-speed emulsions when large silver

Light sensitive emulsions

halide grains are required. As soon as the silver halide grains are formed the manufacturing must continue under safelight conditions. Cleanliness and purity are the passwords in this process. Preparation and mixing of solutions is carried out in controlled atmospheres and chemically-inert containers.

Manufacturing process

The manufacturing process of photographic emulsions may be classified into six major stages.

1. *Precipitation.* Silver halide crystals (grains) are formed in gelatine.
2. *Physical ripening.* The silver halide crystals grow in size and distribution.
3. *Cooling and washing.* Harmful by-products are removed.
4. *Digestive ripening.* Sensitivity centres are formed on the silver halide grains.
5. *Doctoring.* To achieve the required sensitometric qualities and sensitivity properties different emulsion types are mixed with various products.
6. *Coating.* The completed emulsion is coated on to a suitable base material.

All these individual stages contain many operations which may be varied almost indefinitely to produce a wide range of emulsion types varying extensively in sensitivity (speed), contrast (gamma) and graininess (grain size).

We will now take a slightly closer look at each stage.

Precipitation

Gelatine is added to an alkali halide solution forming a base mixture. The silver nitrate solution is then added to this base mixture. This produces a chemical reaction which results in the precipitation of silver halide crystals (grains) which are held separate by the gelatine. The manner in which the silver nitrate solution is added has a profound effect upon the size of the silver halide crystals formed. Rapid addition produces crystals of approximately equal size. This results in an emulsion of high contrast, but low sensitivity. Conversely, slow addition complemented by continuous and vigorous stirring results in the production of large-sized crystals, the basis of a high speed, low contrast emulsion. Other variables which act on the size and character of the silver halide crystals may be employed, such as intermittent addition, addition at the end of the physical ripening stage and alteration in the temperature of the alkali halide solution.

Manufacturing process

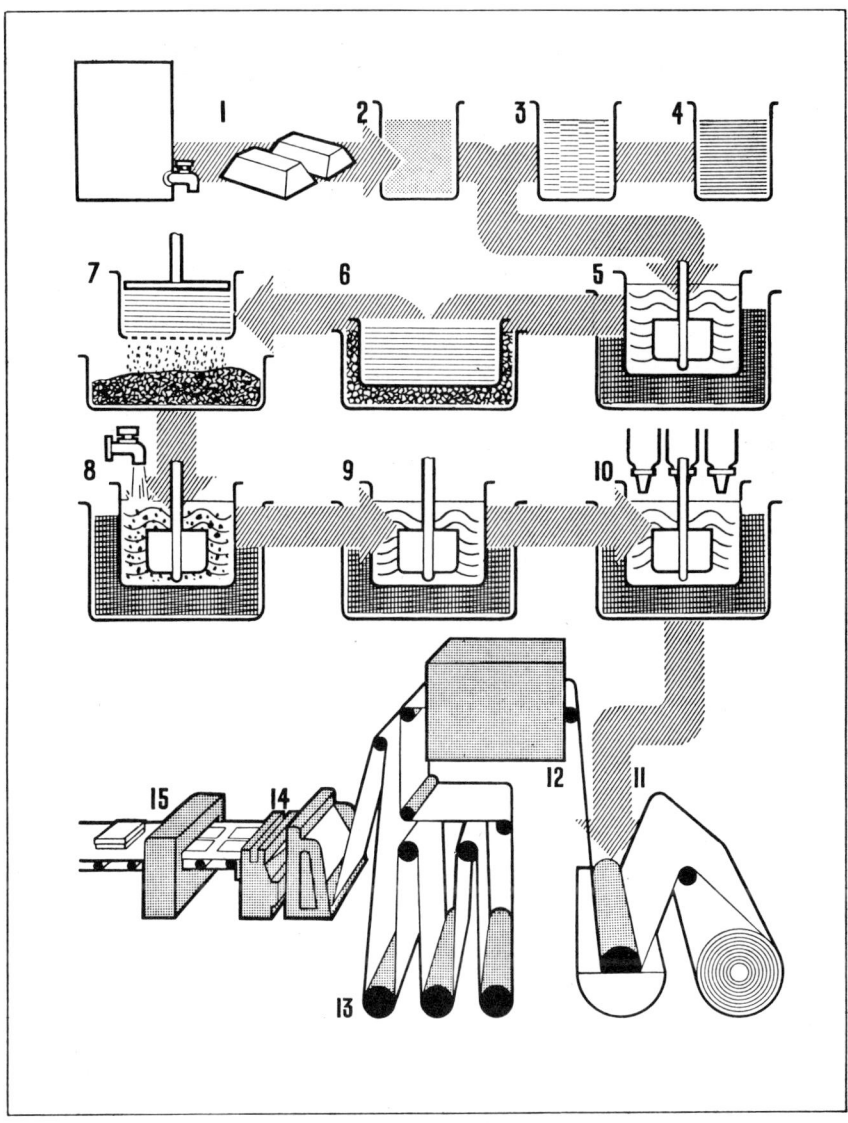

Manufacture of sensitized materials: this flow chart gives a simplified idea of the stages by which sheet film is produced. 1. Silver and nitric acid. 2. Silver nitrate, 3 and 4. Gelatin and sodium or potassium bromide and iodide. 5. Mixing of the emulsion. 6. Cooling to jellify the emulsion. 7. Shredding the jelly. 8. Washing the jelly to remove the unwanted salts resulting from the formation of silver halide. 9. Melting the jelly and digestion. 10. Addition of colour sensitizers, plasticizers, hardeners, spreading agents, etc. 11. Coating on the film base. 12. Chilling to set the emulsion. 13. Drying. 14. Cutting of the continuous roll of film into required sizes. 15. Final inspection and packing for dispatch.

Light sensitive emulsions

Photographic emulsion:
1. Silver halide grains shown diagramatically
2. Photomicrograph of an emulsion (× 2,000)

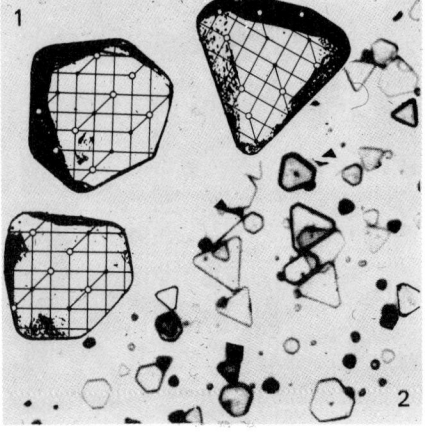

Physical ripening

When the silver halide grain is formed, the emulsion is stirred for a predetermined time at a higher temperature. This ripening or cooking increases the size of the grains and the sensitivity of the emulsion. This grain-growth is caused by two processes:

1. *Coalescence.* Two or three adjacent grains unite to form one single, larger grain.
2. *Ostwald ripening* (named after its propounder, Wilhelm Ostwald, a German professor and Nobel Prize-winner). The ripening or heating of the solution is carried out while a silver halide solvent is present. Various solvents are used to suit the particular silver halide; in the case of iodide-bromide emulsions an excess of potassium bromide or ammonia is needed. This solvent dissolves the weaker, smaller silver halide grains and transfers their substance to the larger grains, thus increasing their size.

The attainment of the required grain size and its distribution may also be affected by the following factors:

1. *The temperature of the emulsion.* The silver halide's solubility increases as the temperature rises.
2. *A surplus of ammonium halide.* The silver halide's solubility increases once again in the presence of ammonium bromide.
3. *The lowering of the gelatine content, plus* vigorous stirring helps to unite crystals.

The duration of physical ripening ends when the silver halide grains have reached the desired size.

Manufacturing process

Emulsion grains: micrographs (× 2,000) taken after
1. Precipitation, at the end of
2. Physical ripening

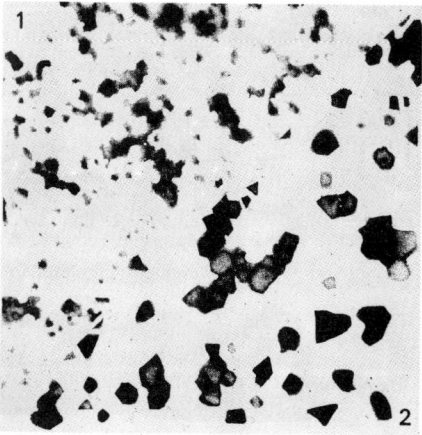

Cooling and washing

After the physical ripening extra gelatine is added so that when the emulsion cools it jellifies, but the emulsion still contains an unwanted excess of alkali halide and aikali nitrates. Therefore their removal is essential. This is carried out by passing the gelled emulsion through a sieve, shredding it into noodles. Washing now commences swelling the noodles and removing the harmful by-products.

Emulsion noodles

Digestive ripening

Once again the emulsion is heated, melting down the noodles to a liquid state. Because of the absence of the silver halide solvents this second application of heat does not alter the size of silver halide grains, but forms sensitivity centres

Light sensitive emulsions

on each individual grain during a period of ripening at a constant temperature. This reaction is aided by the addition of gelatine rich in sulphur content and sensitizing agents, such as gold compounds. The amount of sensitizing compound added must be rigidly controlled as it also increases chemical fog. The sensitizing, e.g. sulphur compounds, become absorbed on the surfaces of the silver halide grains.

Digestion is concluded once sufficient silver ions have reacted with the active sulphur compounds to form silver sulphide at certain points on the surfaces of the silver halide grains. These points of silver sulphide become the sensitivity centres, where during exposure to light the formation of the latent image begins.

Silver halide grain: after digestive ripening
1. Sensitivity centres

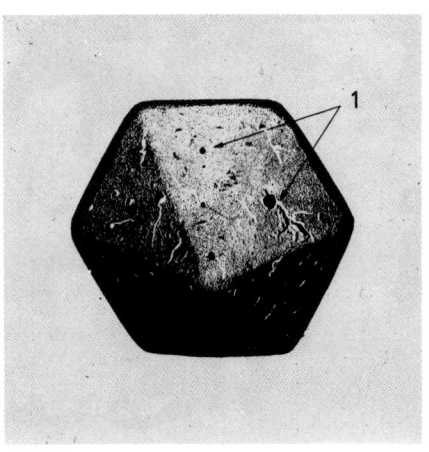

This is the final mixing of active ingredients culminating in an emulsion possessing properties which will do the work it has been formulated for. The finished emulsion is usually a mixture of a number of emulsions, each having a desirable quality, but handicapped by an accompanying disadvantage. By correct mixing the different emulsion have a cancelling-out effect on some of the disadvantages, e.g. an emulsion of extremely high sensitivity (fast speed) will have a low gamma factor (number of tonal gradations recorded and maximum density). This is overcome by mixing this emulsion with an emulsion of low sensitivity which is capable of a high gamma factor. In short, you will get the best of both worlds.

At this point the emulsion has the correct sensitometric properties but lacks other qualities. These are achieved by including chemicals which 'doctor' the emulsion.

Manufacturing process

Up to this stage in the process the emulsion is only sensitive to ultra-violet and blue light. By adding a small quantity of dye the emulsion's sensitivity may be extended into the green, yellow, red and infra-red regions, e.g. dyes which extend the sensitivity of the emulsion into the green region are red, because a red pigment absorbs green light. The dye *adsorbs* itself on to the silver halide grain and passes on the light energy it has absorbed to the silver halide grain.

To ensure that the emulsion retains its characteristic properties throughout its stated life, stabilizers, preservatives and hardening agents are added. Finally, to help produce a thin, even coating on the base material, spreading agents, which reduce the surface tension of the emulsion, complete the mixture.

Sensitivity dyes:
1. Certain light is absorbed by the dye
2. Without the dye this light is transmitted

Coating

The emulsion is now coated on to its base material which will be paper, film or glass, depending on whether the emulsion is for bromide printing papers, process films or dimensionally stable glass plates. In the production of films the base material will be either a tri-acetate or polyester material. The emulsion is helped to adhere and function on the base by the inclusion of substratum, anti-stress, and anti-halation layers (see diagram below).

Monochrome or non-register work may be carried out quite satisfactorily on the tri-acetate material. This base has a tendency to absorb water and moisture; its dimensional stability is therefore affected by relative humidity and alterations in room temperature.

Light sensitive emulsions

Colour separation and precise register work demands a stable base material. For many years glass plates satisfied these demands, but in recent years polyester film has proved to be as good for work of this type, without the obvious handling difficulties encountered with glass plates. In addition to the choice of base material, the thickness of the base may be altered to suit the stability required.

Emulsion structure

The structure of a completed light-sensitive emulsion will, when viewed as a cross-section, be similar to the following diagram:

Emulsion structure: a cross-section
1. Anti-stress layer
2. Emulsion layer
3. Substratum
4. Base material
5. Substratum
6. Anti-halation backing

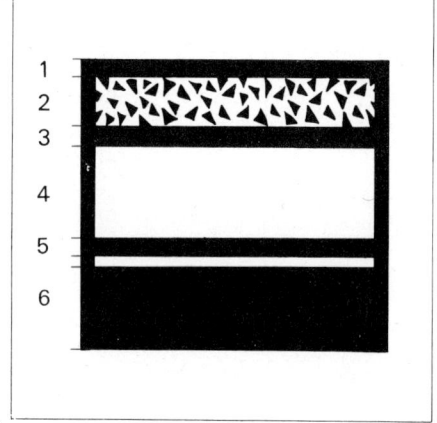

1. Anti-stress layer

This thin top coating or supercoat reinforces the emulsion layer, helping to minimize scratches, abrasions and Newton's rings, while allowing more robust handling during the photographic process.

2. Emulsion layer

This is the gelatine layer holding the light-sensitive silver halide grains. In some cases—colour films, tone correcting films and X-ray films—the base material is multi-coated with two, three or more, differing emulsions. Common to the many diverse emulsions available there is a basic relationship between the grain-size and the sensitometric properties of the particular emulsion.

(a) An emulsion containing large grains will react faster and more easily with light rays because of the larger surface areas (more sensitivity centres) of the light-sensitive grains.

Emulsion structure

This emulsion will have a high sensitivity, but accompanied by a coarse grain effect throughout the recorded image.

(b) An emulsion capable of recording an original which has a long range of tonal gradations must possess grains of different sizes, resulting in an image of relatively lower contrast. This is because of the fact that some grains, both large and small, will become completely black and represent tones of high, medium and low density. Conversely, an emulsion consisting of grains of virtually equal size will not record many tonal gradations. The grains will become either completely blackened or remain unexposed. This type of emulsion is used when an image of extreme contrast is required.

4. Base material

There are many materials used in each category, but generally speaking papers are the result of the appropriate emulsion being coated on to baryta paper—good quality paper coated with barium sulphate in gelatine—films on to tri-acetate or polyester materials and plates on to good quality glass sheets.

3 & 5. Substratum layer

'Subbing' is generally a weak mixture of gelatine and base material solvent. This substance is employed wherever adhesion is required.

6. Anti-halation backing

Selected dyes are dissolved in gelatine and coated on to the back of the base material. These dyes absorb any light

Irradiation and halation:
A. Without halation backing
   1. Halation
   2. Irradiation
B. With an anti-halation backing
   2. Irradiation

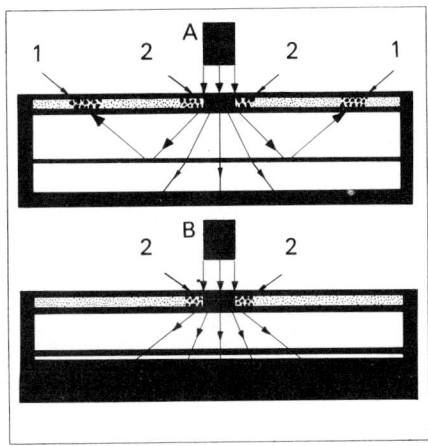

Light sensitive emulsions

which has passed through the emulsion, arresting the reflection of these rays which would result in a second exposure being recorded on the emulsion. This backing layer also stops the film curling and in doing so, ensures that the emulsion lies flat during exposure. In some cases the absorption of the unwanted light is increased by tinting the base material a grey colour. The definition of an image is degraded by halation effects. This is usually seen as a ring or halo encircling a strong highlight area such as street lamps photographed at night. Irradiation is another degrading effect which will be encountered when relatively thick emulsions are coupled with long exposure times, causing a final exposed area to be wider than the incident light beam.

Emulsion data

When light-sensitive emulsions are used it is essential to be aware of their ideal handling conditions. A detailed account of these conditions for all the emulsions available at the present time would be out of place in this book. It is always a wise procedure to obtain a list of detailed handling conditions from the manufacturer for the particular emulsions in question.

In outline the handling conditions refer to:

1. Ideal relative humidity and permissible deviations.
2. Room temperature and permissible deviations.
3. Acclimatization periods in deviated conditions.
4. Storage conditions, batch numbers and organized replenishment.
5. Speed, this may be denoted by a numerical value in ASA ratings— American Standards Association—where a doubling of the number indicates twice the speed and a consequent halving of the previous exposure time; or alternatively in DIN values—Deutsche Industrie Normen—in which the numbers are logarithmic, so that an increase of 3 units indicates twice the speed. In graphic reproduction the emulsions used are so slow that ASA and DIN ratings are not practical and give way to relative exposure times. The relative exposure factor is produced under standard conditions by the manufacturers for all their emulsions, e.g. a camera operator has found that 50 light counts is correct for an emulsion with a relative exposure factor of 5. If this film is changed to one having a factor of 2, exposed and processed under the same conditions, the new exposure would be

$$\frac{50 \text{ light counts} \times 2}{5} = 20 \text{ light counts}$$

Emulsion data

6. Sensitometric properties, the emulsion's characteristic curve, which is a graph showing the emulsion's behaviour during the photographic process with particular regard to gamma (possible contrast ranges).

Sensitometric properties:
1. High contrast emulsion
2. Low contrast emulsion

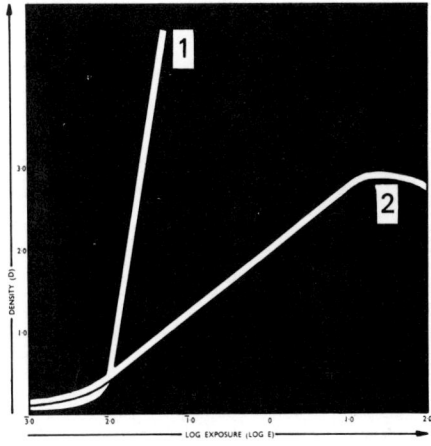

7. Colour sensitivity—a wedge spectrogram. This is a curve indicating the regions of the spectrum to which the emulsion is sensitive.
8. Filter factors and ratios—the amount by which a normal white light exposure is increased when a filter is introduced.
9. Processing and drying conditions.

Colour sensitivity:
1. Ordinary emulsion
2. Orthochromatic emulsion
3. Panchromatic emulsion

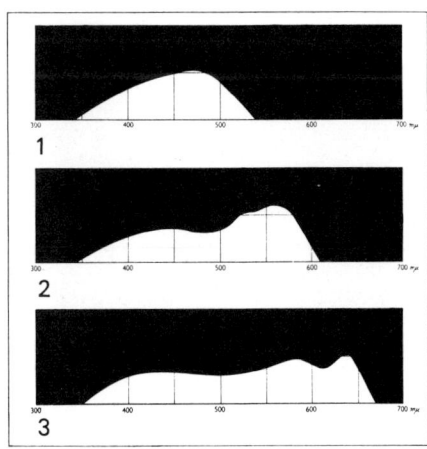

Light sensitive emulsions

Latent image theory

When light-sensitive silver halides are exposed to light a reaction occurs within the light-struck grains. This reaction forms a latent (hidden or invisible) image, which although invisible is capable of development. The development process carries on the work of the light rays and produces a visible image in the exposed areas.

The interaction of the light energy or quanta and the silver halide grain resulting in the formation of a latent image has been a bone of contention for many years, surrounded by numerous theories. The amalgamation of the common themes, prevalent in all these various ideas, and the perfection of the electron microscope in 1930, enabled Professor N. F. Mott and Doctor R. W. Gurney, after a great deal of dedicated work, to propound the Gurney–Mott Electro-chemical Theory in 1938.

This theory postulates that the silver halide grain undergoes photolytic decomposition when exposed to light quanta. This decomposition produces a coagulation or concentration of minute particles of silver at the sensitivity centres created in the manufacturing process. Once these particles of silver have reached a stable state at these centres the latent image is formed. Development begins at these centres converting the whole grain into visible, black metallic silver. During this period of photolytic decomposition the bromide ions gradually diffuse into the surrounding gelatine.

Gurney and Mott propounded that concentration of silver ions at the sensitivity centre suggests that the action of light energy releases a negatively-charged electron from a halide ion. This mobile electron unites with the sensitivity centre—silver sulphide—and then because it is negatively-charged attracts a silver ion (interstitial ions, which increase in number as the temperature rises) from inside the grain. With a prolonged exposure or an increase in intensity this reaction or electron shift is repeated, building up a concentration of silver atoms at the sensitivity centre to await development. The interstitial silver ion remains in a stable state at the sensitivity centre because of the neutralization of its electrical charge once it reaches the centre.

This theory has been widely accepted because it provides feasible answers not only for the usual photographic reactions, but for the more uncommon phenomena which become apparent under certain exposing conditions, e.g. solarization, reciprocity failure and desensitization. Investigations dealing with such complex and minute chemical reactions will, of course, undergo modification as more experiments facilitated

Latent image formation

by new equipment provide more enlightening facts. Continuous exploration of this type is ideally summed up by Professor Mott's own words, 'One could not ask of theory that it should be perfect; one could only ask that it should be useful.'

Latent image formation

In the light of what has just been written it would seem appropriate to review the fundamental stages in the formation of a latent image as they are accepted at the present time. These stages may be considered in the following order:

1. Crystal lattice and sensitivity centres

Let us take as an example a relatively fast emulsion consisting of silver bromide grains plus a small proportion of silver iodide. The silver bromide grains take on a triangular shape which may be interpreted diagrammatically as a cube-shaped, crystal lattice containing a regular pattern of small silver ions, positively-charged because they are silver atoms minus one electron, and larger bromide ions, negatively-charged because they are bromine atoms plus one electron.

Due to the surplus of alkali halide in the emulsion manufacture, a small amount of unwanted bromide ions adsorb themselves to the grain surface, providing it with a negative charge, an essential feature when development takes place.

The silver bromide grain is not a correctly-formed crystal. The precise lattice pattern is interrupted and divided by minute cracks. These cracks, containing sensitizing compounds, are what we have been calling 'sensitivity centres'. The cracks reach the surface where an assembly of sensitizing compounds have formed other atoms and ions in areas where

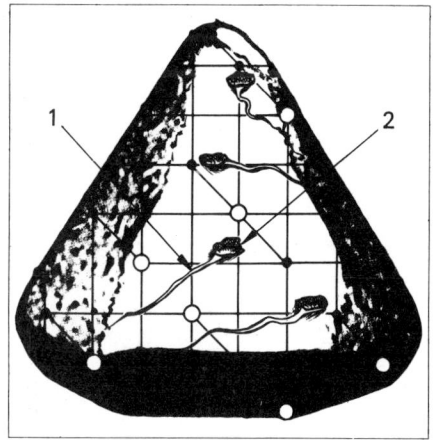

Silver halide viewed diagrammatically
1. Sensitivity centres
2. Irregular pits

Light sensitive emulsions

the lattice pattern is uneven (the crystal's ions have their mathematical structure disrupted and their cohesion forces become dissatisfied) producing irregular pits on the crystal's surface.

It may be seen from this that sensitivity centres leading to irregular pits are formed by the presence of other ions, such as iodide ions and sulphur compounds which are not complementary to the lattice pattern, and therefore alter and subdivide its structure.

2. Exposure and absorption of quanta

The silver halide crystal contains positively-charged interstitial silver ions which are free and capable of wandering throughout the crystal structure. When light energy is absorbed by a silver bromide grain it liberates a negatively-charged electron from a bromide ion. This electron shifts about inside the crystal until it is captured by one of the sensitivity centres. This is rather like a loose billiard ball finally falling into a corner pocket. As the exposure is increased more liberated electrons are held at the sensitivity centres, increasing their negative charge. This negative charge attracts the positively-charged interstitial silver ions which are in a free state. These silver ions gradually make their way to the sensitivity centres where they are neutralized and remain as stable silver atoms.

Formation of a latent image:
1. Silver halide grain
2. Absorption of light quanta—the silver ions move to the sensitivity centres
3. The silver ions become silver atoms and the sensitivity centres change to development centres

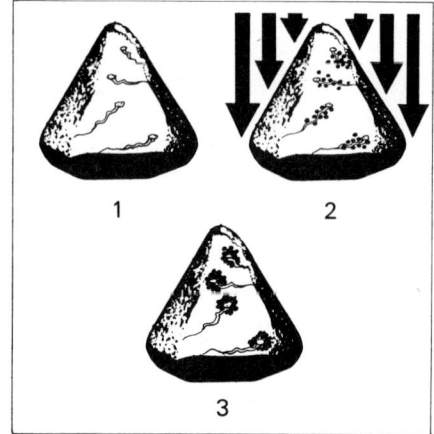

Returning to our billiard ball analogy, because the pockets are full the table corners become wedged with other balls which have been directed there. This build-up of silver atoms changes the sensitivity centres into development centres.

Latent image formation

3. Reciprocity failure and other effects

Correct exposure is when the desired degree of blackening has taken place within the emulsion after light of an acceptable intensity has affected the silver halide grains during an appropriate time period. This is consistent with Hurter and Driffield's and Bunsen and Roscoe's definition of exposure— Intensity × Time—as long as the emulsion's response to light remains a constant factor, e.g. if the intensity is halved and the time doubled the same degree of blackening should occur.

Anyone with experience of electronic flash illuminants will know that this reciprocal process will break down when high intensities are used in conjunction with short exposure times. Conversely, prolonged exposure times with low intensities will also disturb the balance.

In the case of high intensity the light energy is so great that the amount of silver actually activated will partially disperse, reducing the amount of silver at the sensitivity centres. In doing so the number of latent image specks formed are less than those created by an exposure of lower intensity.

In its initial state the latent image is rather unstable and if a light source of relatively low intensity is used the first silver atoms forming the latent sub-image have time to break away, resulting in only a few latent image specks being formed after a considerable exposure time.

Reciprocity failure:
A. Short exposure to an intense light = decrease in predicted densities
B. Long exposures to a weak light = predicted densities correct

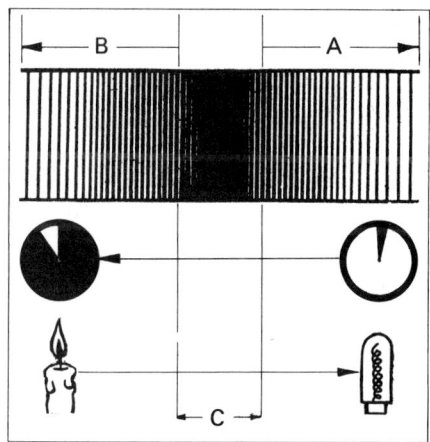

When a photographer is faced with no alternative to the two extreme light sources previously mentioned, he must try to overcome his dilemma by intensifying the image by *latensification*. If the source is too intense then an after-

Light sensitive emulsions

exposure to a weak light will increase the latent image specks, while pre-exposure to a suitable illumination will help the formation of latent image specks when the subject is lit by a weak source.

The Bunsen and Roscoe reciprocity law has now been superseded by laws formulated from work carried out in 1899 by Karl Schwarzschild, a German astronomer. Schwarzschild stated that, effective exposure = $I \times t^p$. The figure p is a constant which is obtained from standards being maintained with regard to emulsion type, light intensity, wavelength, temperature and development. The value p only remains a constant over a very limited range of intensities.

While working on effective exposures Schwarzschild found that an intermittency effect occurred when an effective exposure time was given in intermittent exposures. This resulted in a lower density being formed. With electronic flash illumination at high intensities this intermittency effect reverses and is called the reverse Schwarzschild effect.

Intermittency effect

This effect is related to reciprocity failure and occurs when a relatively long exposure is divided into a series of short exposures, say 100 light counts broken up into a series of 50 individual exposures of 2 light counts. This intermittent exposure will produce less silver density than the single, prolonged exposure of 100 light counts.

Callier effect

This can be seen in graphic reproduction photography when a positive is made from a negative illuminated by

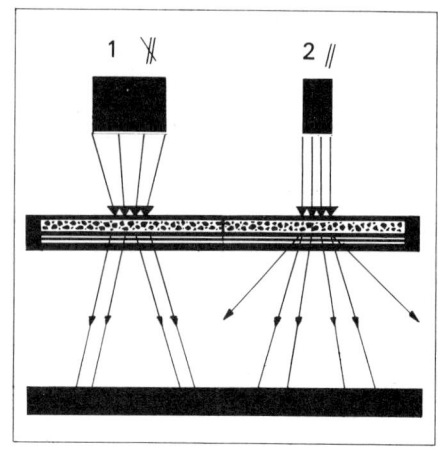

Callier effect:
1. Diffuse illumination
2. Specular illumination

Latent image formation

Herschel effect: Positive to positive or negative to negative (but note reversal)

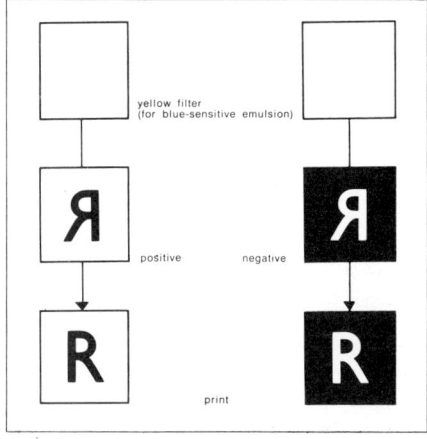

specular reflections from white paper and then repeated with diffuse light through a sheet of opal glass. It is also apparent when comparison positives are exposed on an enlarger-type camera using in the first instance a point-light source projecting specular rays through a condenser and then diffuse fluorescent illumination via an opal glass screen. It will be quite apparent in both cases that the positive made with specular light rays contains greater image contrast and is slightly sharper. The difference is caused by the behaviour of the light rays as they pass through the silver grain structure of the negative. In both cases a small portion of the incident light will be absorbed by the silver grains, while other rays are reflected from the grain surfaces scattering the light and reducing its intensity. This light scatter is not so noticeable with diffuse lighting, because the incident light is scattered to begin with, but with specular illumination the light rays will only scatter as they leave the negative. The outer rays diverge so much that they do not reach the light-sensitive emulsion. This loss of light will occur more in the dense areas of the negative than in the lighter tones producing an increase in the contrast of the resulting positive.

Herschel effect

During the year 1840, Sir John Herschel, an English astronomer and pioneer photographic chemist, recorded that when an exposed, undeveloped emulsion is exposed again, this time to wavelengths to which it is not sensitive the previously formed latent image will break down and eventually disappear. This effect is employed to give the positive to positive or negative to negative results achieved in auto-

Light sensitive emulsions

reversal or duplicating films. These emulsions are received with a latent density already formed, so if development was carried out without exposure an overall black density would appear. Exposure must be to a wavelength to which the emulsion is not sensitized. The emulsion's latent image will fade out in the areas struck by the light. This method allows direct positives and direct negatives to be made from positive and negative images.

7. PROCESSING

Historical introduction

Early attempts to capture pictures framed in the camera's focusing screen relied on the action of light alone. These extremely prolonged exposures produced weak images, the result of light energy decomposing the silver halide, producing dark metallic silver in the exposed areas. Photographers persevering with this method rarely achieved what we would now consider to be a photograph. Even their proudly held successes had to be viewed in a dim light because of their inability permanently to fix the image.

The first permanent photograph by Joseph Nicéphore Niépce in 1826 was an ingenious combination of photography in the exposing stages and photomechanical treatment in producing the image. A pewter plate was coated with light-sensitive bitumen of Judea and after an exposure of at least eight hours the latent image was developed by washing away the soft unexposed areas of the bitumen-based coating with oil of lavender, leaving highlights in bitumen and shadow tones revealed as bare pewter.

Niépce's discovery divided photography into two operations: *exposure*—a time period during which a light-sensitive emulsion is undergoing a change because of the action of light energy; and *processing*—a latent image is developed by the application of certain chemicals, even at a later date. Although Niépce sowed the seed it was left to Daguerre and Fox Talbot to harvest the crop.

Daguerre's introduction to latent image development in 1835 is now a classic story of an accident producing the mysterious hidden link. After a series of unsuccessful attempts to delineate an image by prolonged exposures to light, Daguerre, in a rather disheartened mood, threw the blank, apparently unexposed silver iodized copper plates into a storage cupboard. When Daguerre returned after a period of time to clean off these copper plates for re-use, to his utter amazement and delight images of his previous experiments were present on their surfaces. By repeating the experiment and systematically eliminating the many chemicals housed in a lower part of the cupboard he discovered that mercury

Processing

vapour from a broken thermometer had condensed on to the exposed areas of the silver-iodide emulsion.

During the following years of Daguerreotype mania Fox Talbot worked on the rather different Calotype process and along with the Rev. J. Bancroft Reade successfully developed a latent image formed in an emulsion of silver chloride. Development was achieved by applying more silver nitrate in a gallic acid solution, thereby carrying out physical development—reinforcing the surface of the latent image with extra silver. By this time the previous day-long exposure times were being reduced to thirty minutes. With the advent of Scott Archer's Wet Collodion process physical development continued, but pyrogallic acid was replaced by an iron sulphate developer, introduced in 1844 by Robert Hunt, an English mineralogist and photographic research worker.

When Maddox presented his gelatine dry plates in 1871 the reaction of the developer was changed from a physical nature to a chemical reduction. Previously, in 1862, Major Charles Russell, an English amateur photographer, produced an alkaline developer for chemical reduction in the form of pyro-ammonia and also stated the restraining properties of potassium bromide. This developer's working life was improved in 1882 by the addition of the preservative sodium sulphite, almost at the same time as Sir William Abney, an eminent English photo-chemist, introduced the vigorous reducing agent, hydroquinone. These processing innovations were finalized by the efficient fixing agent, hyposulphite of soda (now known as sodium thiosulphate) proposed earlier by Sir John Herschel, the distinguished scientist.

The turn of the century saw the discovery of a number of new developing agents which had been further isolated from benzene, mainly as a result of the work of the German scientist, Momme Andresen. These included amino compounds, such as para-aminophenol. These discoveries were complemented by the work of another German scientist, Bogisch, and the list of aminophenol reducing agents was concluded with metol.

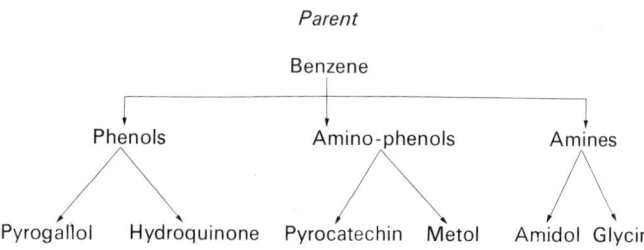

Chemical introduction

Gradually these chemicals, along with preservatives, accelerators and restrainers, were formulated into numerous developing solutions, some containing more than one reducing agent. They enable the photographer to produce a diverse range of images with regard to gamma (contrast), grain and density characteristics. At the present time development, together with the rest of the photographic process, is under constant investigation and research, which produces such advancements as paraformaldehyde, the basic ingredient of the extreme contrast developer for 'lith' type emulsions and dye-coupled development used to produce coloured images.

Chemical introduction

Photographic processing entails the application of solutions containing specific substances which will in turn reduce the exposed silver grains to a black, metallic silver state, arrest the developer's action, convert the unexposed silver halide grains to water-soluble salts and finally, remove these unwanted salts, leaving the image delineated in black silver ready for after-treatment or drying. Before reviewing the processing stages a basic look at the background of these solutions and their chemical reactions will help.

The action of the developer

After exposure and the formation of the latent image specks, development takes place revealing these minute areas of metallic silver. These latent image specks, because of the action of the developer, are reinforced with more silver atoms, so that gradually, if development is prolonged, the whole grain will be converted to black metallic silver, producing a visible image. Therefore it can be taken that the latent image is revealed by the addition of more silver. This quantitative build-up of silver may be achieved in two diverse ways.

1. **Physical development**

In this case the extra silver depositing on to the latent image specks comes from the developing solution, which contains a developing agent plus silver atoms. Their reaction results in the precipitation of the silver atoms. When this solution comes into contact with the exposed silver halide grains, the metallic silver at the latent image specks has a catalytic effect upon the developing solution, accelerating the precipitation and deposition of the silver on to the exposed areas. Therefore during physical development the latent

Processing

Elements and pH scale

ELEMENTS

Elements react to produce compounds and mixtures which may be dissolved in water to produce chemical solutions. Every element has been given a descriptive symbol, e.g. Silver-Agentum = Ag, Potassium-Kalium = K, Copper-Cuprum = Cu, and Iron-Ferrum = Fe, and so on.

Simple chemical compounds with molecules of identical reactive powers (valency) are written as, Ag(silver) + Br(bromide) = AgBr(silver bromide). More complicated combinations can be written to indicate the valency characteristics of the compounds, e.g.

$K_2S_2O_5$ (potassium metabisulphite) = 2 atoms of potassium have combined with 2 atoms of sulphur and 5 atoms of oxygen to produce potassium metabisulphite; or to show the ionic property of the compound, e.g. potassium metabisulphite is a salt and would appear as—$K^+\ S_2O_5^{--}$
$$K^+\ S_2O_5^{--}$$

HYDROGEN-ION CONCENTRATION

Water, H_2O, is fractionally ionized—one molecule in every ten million is divided into a hydrogen ion H^+ and a hydroxyl ion OH^-. This minute ionized fraction of water may be termed a constant when measured at a specific temperature in gram equivalents per litre; this would be equal to 0·0000001 or 10^{-7} (H^+) × 10^{-7} (OH^-) = 10^{-14}. Therefore the figure 10^{-14} becomes our constant for pure water. If a solution is mixed and the ingredients lower the hydrogen ion (H^+) content then the hydroxyl ion (OH^-) proportion will increase; if the solution's substances have the opposite effect the situation will be reversed. To provide lower numbers the logarithmic 10 is omitted and as the relationship—is reciprocal, the ionic state of a solution is measured in pH—the power (logarithmic negative number) of hydrogen ions concentrated in the solution. The hydrogen ion concentration determines the acidity or alkalinity of a water-based solution and may be shown diagrammatically as the pH scale

```
  0   1   2   3   4   5   6   7   8   9   10   11   12   13   14
  |_____ACIDITY_____|NEUTRAL|_____ALKALINITY_____|
```

pH Scale

The action of the developer

```
Acid          Acid    acid     Wetting  Borax       Metol      Lith-      Hydro.
hardening     fixing  stop     agent    devel-      hydro-     type       quinone
fixing bath   bath    bath     washing  lopers      quinone    develop-   caustic
                               water                develop-   ers        developers
                                                    ers
```

ACIDS

are substances which in solution-form have a pH number lower than 7, because this number is a negative logarithm; the low figure denotes a high concentration of hydrogen ions.

BASES

are substances which will increase the amount of hydroxyl ions in a solution. An ALKALI is a soluble substance which will react as a base in solution form.

SALTS

are compounds produced by ionic-reaction, e.g. chloride, iodid, are formed by the negative ion of an acid. Ammonium, sodium, etc. are produced by the positive ion of a base.

PHYSICAL FORM

Basically chemicals come in three physical forms, crystal, granular and powder. Powdered or granulated chemicals dissolve readily and are more convenient to handle and store. A number of chemicals are naturally formed into a crystalline structure. Some contain a quantity of water, termed water of crystallization. This may be removed by a drying process which converts the crystalline chemical into a free-flowing, granulated form. Grinding may now take place, breaking down the granules into a fine powder. This anhydrous (water of crystallization removed) powdered chemical is more active weight for weight than the same chemical in crystal form, therefore substitution tables have to be consulted to make allowances whenever there is a need to use one chemical form in place of another

OXIDATION AND REDUCTION

Reduction is the removal of oxygen or equivalent, such as the addition of hydrogen or more applicable in photography the gaining of electrons. *Oxidation* is the reverse process, the loss of electrons.

CATALYSTS

Slow reactions can be accelerated by the use of a catalyst, another additional substance which produces chemical conditions conducive to the reaction required. The reverse of this operation is the inclusion of inhibitors which have the opposite effect of a catalyst.

BUFFERS

Buffers are chemicals which help to stabilise pH.

COMPLEX IONS

Stages of the photographic process rely on the formation of complex ions. This occurs when a number of ions react to form larger complex ions. These complex ions have different chemical properties.

Processing

image undergoes a process of silver intensification. This type of development was employed in Fox Talbot's Calotype process, and in a slightly modified form during Wet Plate development. Physical development has the advantage of producing images which retain fine definition and detail, because of the fact that the image is really a deposit of minute silver atoms on the emulsion surface, resulting in an extremely fine grain effect.

2. Chemical development

Chemical development relies on the silver halide grains in the emulsion to produce the image-forming silver. The developing solution's major ingredient is the *reducing agent*, which donates the electrons that convert the exposed grains to a metallic silver state. This process begins when an exposed silver halide grain is attacked by the developing solution, which results in the lattice structure of silver ions and bromide ions being converted to silver atoms as a result of the addition of an electron and hydrobromic acid by the uniting of a hydrogen ion with a bromide ion.

A developer's efficiency revolves round its reducing agent. The reducing agent's potential is reliant on its ability to donate electrons. The greater the donation, the higher the potential factor of the agent. This efficiency factor is also related to the activity of the agent's electrons. Some substances have an efficiency factor which is too high and would develop the unexposed as well as the exposed grains.

When manufacturers are formulating a new developer the reducing agent's efficiency is of the utmost importance and may be measured in terms of reduction and oxidation capabilities, producing a *redox* value for that particular reducing agent. Even with a correctly-formulated developer, if the development time is prolonged the unexposed grains will be reduced as well as the exposed grains. Correct development relies on the catalytic properties of the latent image specks— attracting and accelerating the reduction of its parent grain. Development now becomes selective, attacking the exposed grains first. In short, the exposed grains have a head start on the unexposed.

Effective development

Effective development is when the amount of metallic silver formed is proportional to the amount of light quanta received. This state of affairs can only be achieved with a correctly-formulated developing solution. Let us once again, in the interests of explanation, turn to a simple analogy.

The developer's constituents

The photographic process may be modified and altered to produce many variations in the contrast, maximum density, graininess, etc., of the resultant image. It is rather like dealing with motor-cars. Light sensitive emulsions are akin to fuel, the action of light provides the ignition, while the developing solution furnishes the machinery to make the reaction go. You may select a photographic reaction to suit your requirements. This is determined by the length and intensity of exposure, sensitivity of the emulsion and the reducing power of the developer. Again this is rather similar to our analogy, e.g. the choice between a high-speed, racy sports car or a sedate limousine, both achieving the same end, but satisfying different requirements on the way. A developing solution may be separated into its constituent parts as listed below, each with its own particular function.

Developing solution

Major constituents and *Motor car analogy*

1. REDUCING AGENT. The reducing agent is the engine, providing machinery to convert the exposed grains to metallic silver.

2. PRESERVATIVE. A preservative protects the reducing agent from oxidation, as an oil protects and enhances the working of an engine.

3. ACCELERATOR. The accelerator provides the reducing agent with an increased range of performance, increasing the reaction between the engine and the fuel.

4. RESTRAINER. Similar to a motor car, the developer can get out of hand and a brake is needed to control its speed and direction.

5. SOLVENT. This is the bodywork, incorporating all the ingredients into a workable structure.

The developer's constituents

Now, after listing their fundamental functions, we will take a closer look at the developer's constituents.

Reducing agent

As the name implies it reduces (gives up electrons to) the exposed silver halide crystals converting them to a black, metallic silver state. To provide effective development with the correct degree of contrast and latitude, two reducing agents are usually mixed together, e.g. metol + hydroquinone or phenidone + hydroquinone. If an image of high contrast is required a relatively greater amount of hydroquinone is

Processing

Reducing agent: A positive print produced by the reducing alone e.g., metol

Negative curve: Indicating the action of the reducing agent

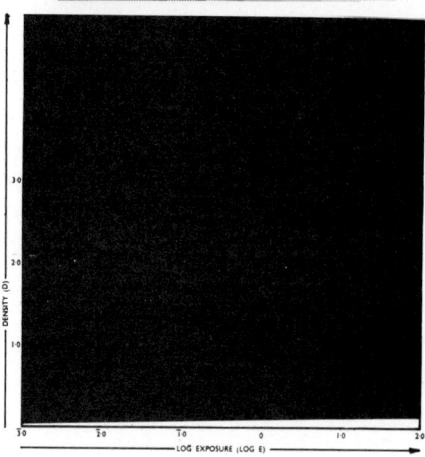

used in conjunction with a small amount of metol. For the development of images containing low or soft contrast the reverse obtains—large amounts of metol or phenidone with a small amount of hydroquinone. The majority of photographic reducing agents are compounded from benzene and may be tabulated in the following manner.

METOL. Used with an alkaline accelerator to produce images of very low contrast. Normally mixed with hydroquinone in a developer formula for general continuous tone work.

HYDROQUINONE. This agent is capable of developing images to an extremely high density (therefore suitable for line and halftone reproduction). Its high-contrast producing properties may be controlled by the addition of another, less violent reducing agent.

The developer's constituents

PHENIDONE. Is an Ilford product which will produce medium contrast. It is more effective when mixed with hydroquinone to give a very practical continuous tone developer.

PYROCATECHIN. A reducing agent which is very similar to hydroquinone, but can be more easily controlled to give a range of contrast.

Preservatives

When the reducing agent is mixed with the alkaline accelerator it is very prone to oxidation. Oxidation of any type retards the developing action. Therefore a substance which has a greater affinity for oxygen than the other ingredients is included to protect the reducing agent and its developing

Preservative: e.g., Sodium Sulphite. This increases the developer's working life. A positive print produced by the metol, sodium sulphite and a second reducing agent, e.g. Hydroquinone

Negative curve: Showing the combined action of the two reducing agents and a preservative

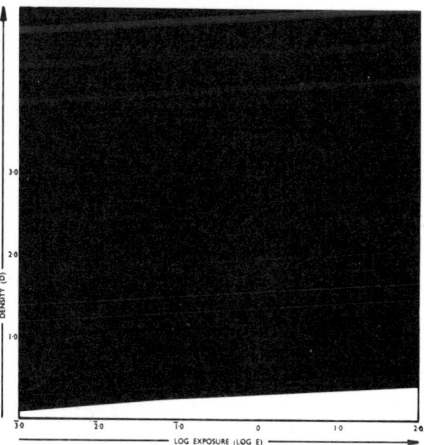

Processing

action, which is amplified by the alkaline accelerator. It has been found that sulphite and its derivatives impeded the harmful of oxidation during the development period.

SODIUM SULPHITE. This is generally used in all continuous tone developers bought in powder form.

POTASSIUM METABISULPHITE. Because of its high solubility this preservative is included in high contrast developers and the more concentrated liquid developers.

Accelerators

Reducing agents have their developing powers amplified by the addition of an alkali. The rate of development and the final density achieved are influenced by the alkalinity of the

Accelerator:
A positive print illustrating the amplifying properties of Sodium carbonate

Negative curve:
Indicating the density increase caused by the accelerator

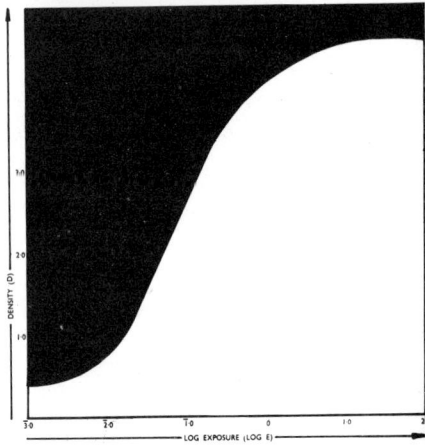

The developer's constituents

developing solution. Both properties increase with a rise in the pH value of the solution.

BORAX. This has a very mild accelerating action and is widely used in fine-grain, low contrast developers.

SODIUM CARBONATE. An anhydrous salt selected for many continuous tone developers. It provides a range of images containing medium and lower contrast.

SODIUM HYDROXIDE. An extremely violent accelerator, used exclusively for high contrast developers employed in line and halftone work.

FORMALDEHYDE. Used with hydroquinone to produce the 'Lith' type developers. This substance is classified as an accelerator but is a reducing agent responsible for the extreme contrast characteristics of the 'Lith' type emulsions and will be viewed as a separate topic later.

Restrainer (anti-foggant)

Effective development occurs when only the exposed grains are reduced. Any tendency to reduce the unexposed grains which will produce chemical fog in the unexposed areas must be restricted within the specified development time. Potassium bromide has proved to be the most effective restrainer and is used in almost every developing solution.

Solvent

Tap water in most areas is generally quite adequate, but if impurities causing harmful reactions are encountered then filtering, or in the last resort distilled water, is the answer. If the tap water is particularly hard or atmospheric conditions extreme, then the following additions may be included.

SODIUM HEXAMETAPHOSPHATE (Water softener). Usually obtained under the name 'Calgon'; prevents calcium and magnesium salts from precipitating on to the surface of the emulsion.

SODIUM SULPHATE (Hardening agent). Used in tropical developers; stops the gelatine from swelling to an extent where 'frilling', or the separation of the gelatine from the base, would occur.

TEEPOL (Wetting agent). Reduces the surface tension of liquids; may be used while washing the emulsion, but not in the developer as emulsions contain their own wetting agents.

Processing

As with the choice of sensitive emulsions, when dealing with a particular type or range of work it is wiser to consult the manufacturers on the best developer for the job. Therefore an exhaustive list of processing formulae and working conditions is not included here. Tabulated below are the basic formulae of the four grades of developer normally used.

Table 2. Developer constituents

Grade	Reducing agent(s)	Preservative	Accelerator	Restrainer	pH Value
Fine-Grain Low contrast	Metol and Hydro-quinone	Sodium Sulphite	Borax	—	8·5
Medium contrast	Metol and Hydro-quinone or Hydro-quinone and Pheni-done	Sodium Sulphite	Sodium Carbonate	Potassium Bromide (Pheni + Hydro-organic Restrainer)	10·5
High contrast	Hydro-quinone	Potassium metabi-Sulphite	Sodium Hydroxide	Potassium Bromide	11·5
Lith extreme contrast	Hydro-quinone	Sodium Sulphite and Potassium Metabi-Sulphite	Para-formalde-hyde	Phenosa-franine (desensi-tiser)	10·0

Restrainer:
This positive print is the result of the reducing and accelerating properties of the previous chemicals being controlled by the restrainer e.g., Potassium Bromide

Development theory

Negative curve:
This characteristic curve is the product of the correct developer formulation

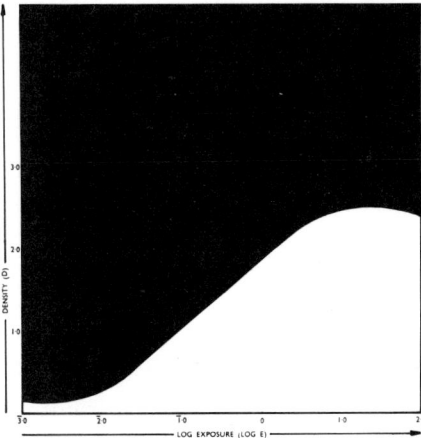

Development theory

When the exposed emulsion is immersed in a developing solution molecules of this solution permeate through the porous gelatine, finally adsorping themselves on to the surfaces of the silver halide grains. These molecules of developing solution contain a negative charge and their penchant for donating electrons is rebuffed by the negative charge present on the grain's surface.

The developer's pressure to transfer electrons is finally satisfied when a latent image speck is discovered by the mobile molecules. These minute specks of metallic silver possess a positive charge and become a chink in the armour, inviting, in a catalytic manner, the developing molecules into the crystal lattice. Once the developing molecules enter the silver bromide lattice they give up their electrons to the silver ions, reducing them to silver atoms. This is because the reaction is basically an electrical transference and the negatively-charged electrons will go the positive conductor.

The latent image speck now becomes the slowly expanding neck of a rapidly growing, mushroom-shaped body of black, metallic silver atoms moving within the grain. The bromide ions are released from their lattice positions and are free to move out into the surrounding solution, and many of them combine with hydrogen ions to form hydrobromic acid.

This process continues at an ever-increasing rate as the reducing agent in the developing solution perpetuates the supply of negatively-charged electrons to the silver ions within the grain through the catalytic latent image speck, gradually reducing the whole grain to stationary, insoluble silver atoms.

Processing

Photographic development:
1. An exposed silver halide grain containing sensitivity centres converted to development centres
2. The developer enters the grain through the development centres
3. The developer denotes electrons and reduces the whole grain to metallic silver atoms

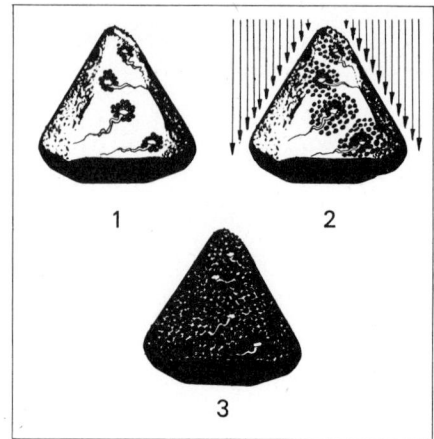

Although fast and slow development is produced by the constituents of the developing solution, it can be seen that development relies fundamentally on atomic motion, i.e. the movement of atoms, electrons and ions. Their energy and resultant reactions depend a great deal upon temperature. Their energy increases or decreases with a rise or fall in temperature; e.g. hydroquinone, a most vigorous reducing agent, becomes almost inactive at temperatures lower than 60°F.

The next topic to review is the working conditions which produce or hinder effective development.

Variants in development

Throughout the photographic process there are numerous operations, each one having its own built-in variant—in fact the process is fraught with too many fluctuating factors. What can we do about this? The 'Job's comforters' of the photographic world would say that you have just got to accept them and struggle on producing one photograph of high standard for at least two of inferior quality. This is the wrong approach. First of all, acknowledge the difficulties and the variants, and then endeavour, with undiminishing perseverance, to overcome and standardize them with logical thought and systematic deed.

As far as development is concerned, for the moment we will assume standardized exposure (a large assumption, but

Variants in development

this will be tackled later). Development procedure is complicated by six major variants, which are:

1. development time;
2. temperature of the developer;
3. dilution of the developer;
4. agitation of the developer;
5. exhaustion of the developer; and
6. method of development.

Let us now take up the fight and see what standardization can be brought to bear on these six factors.

Development time

Development begins with a slow induction period and then quickly increases, until after a period of time maximum density is reached. This relationship between time and density may be plotted as a curve showing the gradual increase in density and contrast (gamma) until all the exposed silver grains are fully reduced and gamma-infinity ($\gamma \infty$) is reached.

This curve, indicating the relationship between development time and the gamma achieved at any one point during this time, may be plotted for all the light-sensitive emulsions in your working range. It almost goes without saying that development time must be controlled by a precise, luminous alarm clock.

Time gamma curve

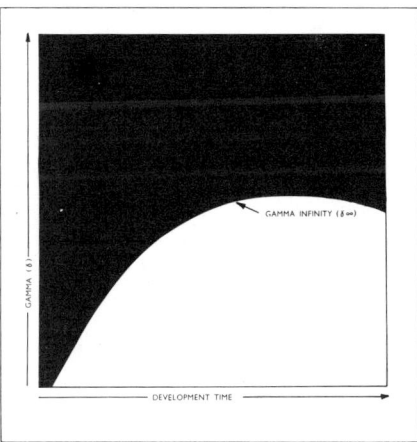

Temperature

It is essential, because of the considerable effect temperature has on the behaviour of a developer, to maintain a rigid

Processing

working temperature of 20°C (68°F). This is, of course, easier said than done without the aid of temperature controlled sinks, water jackets and ultimately, automatic processing, but even a lowly dish can be placed inside a larger dish containing water at a suitable temperature to maintain the developing solution in the smaller dish at 20°C ± 1°C for $2\frac{1}{2}$ min, and a deviation chart may be produced. Carry out a series of tests, producing the same end-densities on a step-wedge, altering the temperature and its related time period to achieve this, e.g. 17°C for 3 min, 19°C for $2\frac{3}{4}$ min, 20°C for $2\frac{1}{2}$ min, 21°C for $2\frac{1}{4}$ min, 23°C for 2 min; and so on. Now plot these temperatures against the time periods to produce an oblique line, as in the following diagram.

These tests and resultant deviation chart may be produced for the different developments needed for your particular range or work.

Temperature deviation chart

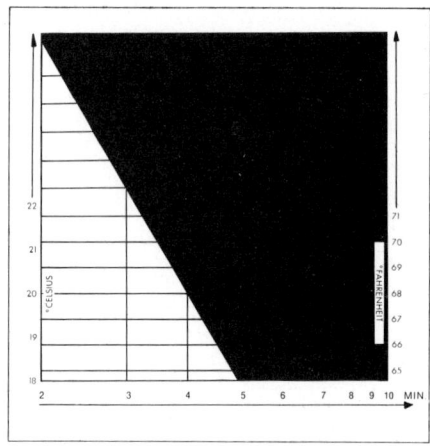

Dilution

The stock developing solution may be diluted to a weaker working strength by the addition of more parts of water. The gamma of an image may be lowered this way as long as the development time remains constant and the degree of dilution is not taken to an extreme. When using a medium range continuous tone developer a limit is usually set at 1:6 parts of water, while for high contrast developers not more than 1:1 part of water. Once again the diluting effect may, after a series of tests, be represented by a characteristic curve. By plotting density against development time a family of curves can be shown, each one being a different dilution.

As with all chemical mixing, standardization in the

Variants in development

Dilution curves:
1. Neat developer
2. One part of developer to two parts of water
3. One part to four parts of water
4. One part to six parts of water
(N.B. in each case the development time is constant)

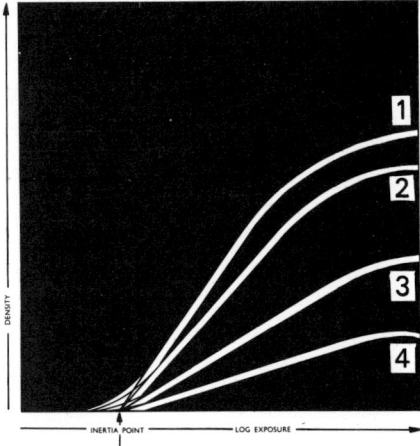

developer's working strength is more likely to be maintained if the solutions are prepared in bulk.

Agitation

A standardized agitation rate is the key to constant repeatability of good results and therefore the most difficult to achieve without spending a good deal of money. The developing solution must be agitated to ensure that the layer of exhausted molecules and oxidation products are being constantly removed from the emulsion and replaced with fresh molecules.

Continuous, vigorous agitation of the developing solution results in rapid development, while 'still bath' techniques, in which the exhausted developing molecules and oxidation

Variations in agitation rate:
1. Continuous and vigorous agitation
2. Intermittent agitation
3. Still bath technique

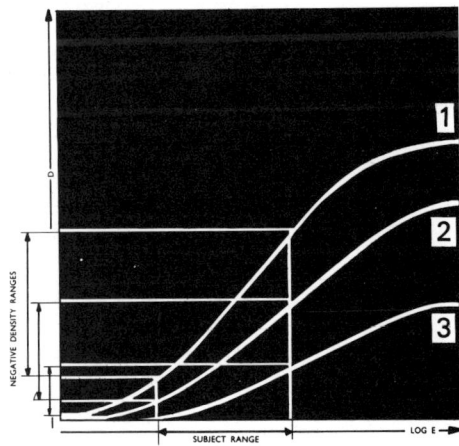

177

Processing

products are allowed to remain undisturbed on the emulsion, severely retard development. Uncontrolled intermittent agitation produces wild fluctuations in the density ranges of a number of identically exposed emulsions.

To a photographer working with unsophisticated equipment the obstacles seem almost insuperable, when the ideal agitation is a uniform movement of the developing solution with an irregular flow pattern. With this in mind the sixth factor—methods of development—has been included, indicating the extent of standardization possible with each method.

Exhaustion

A developing solution loses some reducing power with every emulsion that is passed through it. Exhaustion is caused by many contributory factors:

1. The quantity of reducing agent is diminished as it converts the exposed grains to silver.
2. Liberated bromide ions and hydrogen ions retard the developing action.
3. As the developed emulsion is removed, developer is lost by 'drag-out', although this can be minimized by adhering to a constant draining-time before removing the emulsion completely.
4. Additional to these losses, there is the continuous oxidation process taking place between the developer and the surrounding air. This is more serious with dish development than tank development because of the greater contact area.

If dish development is employed then the only answer is to replace the old solution with fresh developer after what is considered to be the maximum number of emulsions for that particular quantity. With nitrogen burst tanks or an automatic processor exhaustion can be overcome by systematic replenishment. This regeneration of the developer is usually carried out after a predetermined number of films have been developed, or after a period of time in which the developing solution has not been used and oxidation has taken place (this can be reduced by excluding the air with tank lids). Generally the regenerating amount is 1 litre (l) for every 1 square metre (m^2) of developed emulsion or 125 millilitres (ml) for a 380 mm × 305 mm tank (1 quart of replenisher for every square yard or 4 oz for a 15 in × 12 in tank) for each standing day. A more systematic approach is the pre-exposed step-wedge test. Boxes of identically exposed step-wedges are obtainable from the emulsion manufacturers. Each morning one of these control-wedges is developed, and by plotting its developed densities a curve is drawn and com-

Methods of development

Pre-exposed stepwedge test:
1. First test wedge (densities too low)
2. Standard test wedge
3. Second test wedge (after replenishment)

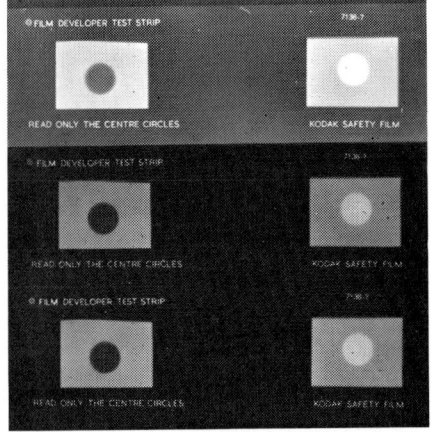

pared against the standard curve. From the degree of deviation the correct amount of replenisher may be added, this in turn being checked by passing a second control-wedge through the developing solution.

Methods of development
Dish

A developing bench with a glass top housing a suitable range of safelights underneath, capable of shining through a transparent developing dish is desirable, so that inspection of high contrast emulsions, such as 'lith' materials, may be carried out during the development period. An immediate, smooth entry of the emulsion into the solution is essential to obviate streak marks. A different technique is required to suit the base material—when using films a more uniform development takes place if flat-bottomed dishes are employed.

Agitation is provided by the following motions:

1. ROCKING. The dish is lifted sideways and lengthways in an alternating manner. A constant rhythm must be maintained and this is helped by the whistling or singing of a suitable tune, even if it is disconcerting to the other photographers.

2. BRUSHING. The dish and emulsion are left untouched and a nylon brush of suitable width lowered into the developing solution just below the surface. Development commences with rhythmic brushing motions lengthways and sideways, taking care that the bristles of the brush do not come into contact with the emulsion.

Processing

3. DISH ROCKER. There are a number of good dish rocking machines on the market, which for a small outlay provide a good step forward in the march of standardization. Some models incorporate a variable speed control so that a suitable agitation rate can be selected for any particular emulsion.

Nitrogen burst tanks

The developing solution is contained in a narrow tank, at the base of which lies a pipe or a collection of pipes with small holes drilled at intervals along the underside of their length. Pressurized nitrogen gas is released through these holes, and on its entry into the developing solution it forms bubbles which rise up through the solution causing turbulence. Nitrogen is chosen because of its inert reaction with the developing solution. Air, of course, would create oxidation problems.

The pressure, duration and interval of burst are usually controllable, so that continuous turbulence, or bursts at regular intervals may be applied to the emulsion hanging inside the tank. If a constant development time is maintained, images of different gamma (contrast) values are produced by altering the gas pressure, burst duration and burst interval. An increase in the gas pressure will produce images of high gamma, but a pressure limit will be found, which if exceeded produces uneven development. A similar gamma increase can be achieved by lengthening the burst duration, but once again a limit will be encountered. Conversely, if the following adjustments are made individually: pressure reduced, burst duration shortened and the burst interval increased; the gamma value will be lowered.

Nitrogen burst installation:
1. Developer
2. Stop tank
3. Fixing tank
4. Washing tank
5. Nitrogen burst pipes
6. Nitrogen gas cylinder valve
7. Reducing valve
8. Pressure gauge
9. Timers for duration and interval of bursts
10. Temperature control for the water jacket

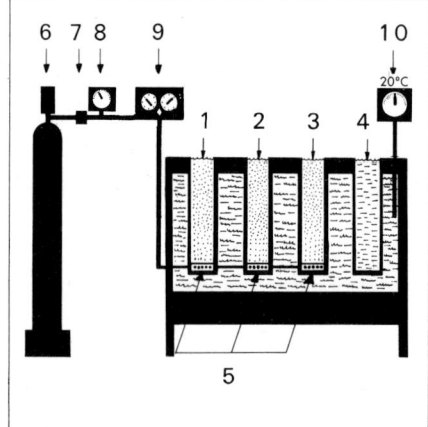

Methods of development

From this exhaustive list of variants the previous statement, that there are too many fluctuating factors, can be understood. Nitrogen burst development will prove to be less of a permutation riddle, if the gas pressure, burst duration and burst interval are kept at standard readings which produce the most vigorous agitation possible, yet provide uniform development. Then the desired gamma value may be achieved by altering the development time.

This method of development with its systematic replenishment is particularly applicable to continuous tone development, but has certain disadvantages when used with lith materials, as 'adjacency' effects (development is greater on one side of the image than the other), as a result of the one way direction of the developer, may appear. Nitrogen burst units are available in a range of models, varying in sizes and price, with controllable bubble turbulence in the developing, stop bath, fixing and washing tanks.

There are numerous camera studios which cannot afford this type of equipment. Nevertheless a simple one-tank installation for developing only is not beyond the capabilities of any resourceful photographer. This would prove adequate for continuous tone emulsions, while perhaps a dish rocker could be acquired for lith materials.

Automatic processing

For the more opulent, there are a number of rapid developing machines carrying out the entire process of development, stop bathing, fixation, washing and finally, transferring the dried film into a delivery tray housed in a normally lit room. It is possible to buy a relatively cheap processor which gives

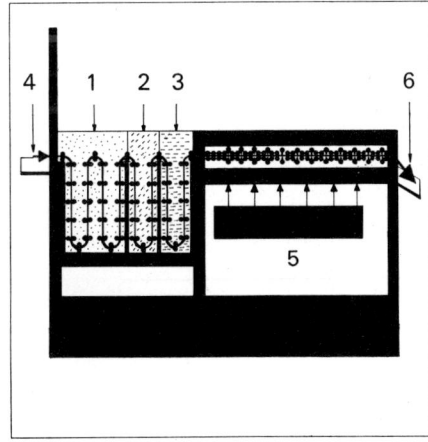

An automatic processor:
1. Developer
2. Fixing tank
3. Washing tank
4. Feed tray
5. Heaters
6. Delivery

Processing

rapid and standardized results, but the film is just lowered into a fixing tank; washing and drying have to be carried out in the normal manner.

When contemplating a complete transition to automatic processing, it will be seen that not one, but two processors are really necessary. One is for continuous tone development —processors designed for this work usually have two developing tanks. When high gamma is required the film passes through both of them; for lower gamma only one tank is employed. On some machines there is an additional gamma control; the speed by which the film travels through the developing solution is variable. This, of course, is an alteration of the development time. The second processor would be employed for line and halftone work on lith materials.

Both processors need to be rigidly controlled by the step-wedge system, although while in operation replenishment is automatic, related to the film size as it enters the machine. The internal tanks are temperature-controlled, and uniform development is produced by the squeegeeing and oscillating movements provided by numerous rollers as they move the film through the developing solution. Treatments in the stop bath and fixing departments are carried out in a similar manner. The developed film is dried as it passes over heaters on its final journey to the delivery tray.

The acquisition of two automatic processors was envisaged in the model camera studio layout. They would be expected to serve four camera operators, eight retouchers and two planners working cameras, enlargers, electronic scanners and contact frames. A more conservative step would be a nitrogen burst installation for the development of continuous tone emulsions and automatic processing for the lith material.

The aim of a progressive photographer, no matter what his company's resources are, should be to standardize processing and exposure completely, so that the only variant is the original copy which is difficult to standardize as it normally comes from outside sources.

Final stages

After development the emulsion is stop-bathed, fixed, washed and dried, and if necessary after-treatments can be carried out. These operations render the unexposed, undeveloped silver halide grains insensitive to photochemical reaction and produce an emulsion ready for retouching and platemaking.

Stop bath

This is a weak acid solution having an approximate pH

Final stages

value of 4. Its function is to stop the developing action on the surface and inside the swollen emulsion layer by neutralizing the alkaline developer. An effective stop bath is produced by preparing a 5% solution of glacial acetic acid in water. A stop bath of this type will prevent residual developing solution clinging to the emulsion from being carried into the fixing bath and reducing its fixing powers.

Fixing bath

Permanent fixation is carried out in a 40% solution of sodium thiosulphate (hyposulphite of soda) or the more rapid ammonium thiosulphate. Other chemicals can be added to provide fixing baths with special properties:

Acid-fixing baths. The inclusion of a weak acid salt, e.g. potassium metabisulphite, arrests dichromic fog.

Acid-hardening fixing baths. The addition of potassium alum with a buffer of acetic acid retards the swelling of gelatine.

When an undeveloped silver halide grain comes into contact with sodium thiosulphate a two-part reaction takes place. Initially the grain becomes a complex compound of partially-soluble silver sodium thiosulphate. In the presence of an excess of sodium thiosulphate the reaction continues until this grain is converted to silver sodium dithiosulphate, a water soluble salt. If this second state is not reached, then there is a danger of the partially-soluble silver sodium thiosulphate salts decomposing, causing unwanted stains.

In a normally working bath the earliest an emulsion can be removed is after the same amount of time has elapsed as it took for the creamy appearance to disappear from the unexposed areas. The working life of a fixing bath is definitely prolonged with the use of a stop bath and careful drainage of the emulsion between baths. Exhaustion of a fixing bath is virtually determined by the number (subject to emulsion thickness and area of unexposed grains) of developed emulsions passed through it. Generally speaking 1 l of fixing solution will cope with a maximum of 1 m^2 of emulsion. The state of a bath can be ascertained by pouring a sample through a funnel containing a filter paper; if the paper becomes ringed with a line of black silver, then the bath is exhausted.

Washing

This is really the third stage of fixation—the removal of the water-soluble salts now present in the unexposed areas of the emulsion. A copious supply of clean, running water is needed, flowing into a dish or tank housing the processed emulsion, the most important action being a complete change

Processing

of water every 5 min, ensuring that unwanted salts are being continuously removed. Water changing can be carried out by a water-syphon or simply by means of an old dish with small holes drilled in its bottom corners. Washing should continue for a minimum period of 30 min ending with the addition of a few drops of wetting agent to facilitate even drainage. The washed emulsion can be squeegeed or wiped carefully with a chamois leather.

Drying

To retain dimensional stability in the base material drying should take place in a circulation of cool air. Rapid drying may be achieved in a drying cabinet with a maximum internal temperature and relative humidity of 40°C (104°F) and 55%.

Activation stabilization system

This is a rapid method of processing photographic material. Usually known as stabilization or rapid access papers they have a normal silver halide emulsion to which a developing agent has been added, usually hydroquinone. Rapid processing in a roller processor with an alkaline activator, e.g. 5–10 per cent sodium hydroxide with sodium sulphite, will induce development by activating the developing agent incorporated in the paper. This is followed by stabilization in place of fixing and washing. A chemical treatment which converts unexposed silver halide into more or less light-stable, colourless compounds. Thiourea, sodium thiosulphonate or thiocyanate all act as stabilizers. The processed print emerges slightly damp from the processor within an access time of typically, 15 sec., a minimum time achieved when absolute permanence is unimportant.

Diffusion or chemical transfer

A negative is exposed in the usual manner and processed in contact with a positive receptor layer through a roller processor. The positive is produced on paper or film coated with plain gelatine by the silver halide not used in forming the negative image. The paper or film which receives the positive image contains a small quantity of hypo to dissolve the silver salt and colloidal silver sulphide to act as nuclei for the formation of black metallic silver reduced by the developer. The two sheets are separated after a few moments, the transfer paper or film bears a positive image which need not be rinsed and is ready for immediate use. Positive papers coated on both sides are also available which when processed in contact with two negatives allow the production of double-sided copies.

After-treatment

Although the dictum, 'the negative should be made in the camera—not in the sink!) is an excellent one to adhere to,

After treatment

Reduction

occasionally the variable qualities of the original reach such extremes that to produce an acceptable result reduction or intensification is necessary. Reduction in this sense must not be confused with the reduction process responsible for development, and in camera studios this operation is referred to as 'etching' or 'cutting' the image.

The image consisting of insoluble silver is immersed in a solution which gradually converts this silver into a soluble salt. Basically three types of reduction are used, unproportional, proportional and super-proportional.

Unproportional reduction

This is carried out in Farmer's reducer, a two-solution reducer made from mixing one part of the potassium ferricyanide solution with two parts of the sodium thiosulphate solution in three parts of water. The lighter tones, highlights or the thinnest lines are reduced first. The effect of this reducer will be increased or decreased by changing the amounts of potassium ferricyanide and sodium thiosulphate in relationship to one another, an excess of ferricyanide having the more violent reducing action.

Proportional reduction

This takes place when a solution's reaction is in proportion to the amount of silver. Shadow areas, densely populated areas of black silver are reduced more than the highlight areas. A most useful proportional reducer is old Tri-mask or Multi-mask bleach-fixing solution, which may be diluted to achieve the desired degree of proportional reduction.

Super-proportional reducers.

These rely on a catalytic reaction, which is increased in proportion to the amount of silver present. A very effective reducer of this type can be prepared in a two-stock solution form by mixing ammonium persulphate in water. This is the reducing solution, while the second solution of sodium sulphite is used as a stop bath, arresting the reducing action.

Intensification

There are numerous intensification methods using lead, copper, silver and mercury compounds to reinforce the initial silver image which appears too weak for subsequent printing-down operations. The most popular compound seems to be mercuric chloride. The emulsion may be 'cut' before intensification (in some cases the need for intensification is due to excessive reduction), then thoroughly washed in warm water. The emulsion is now immersed in the mercuric chloride, potassium bromide solution where it will be seen to bleach-out, producing a white appearance throughout the depth of the image areas. Once this has been achieved the

Processing

Photographic requirements:

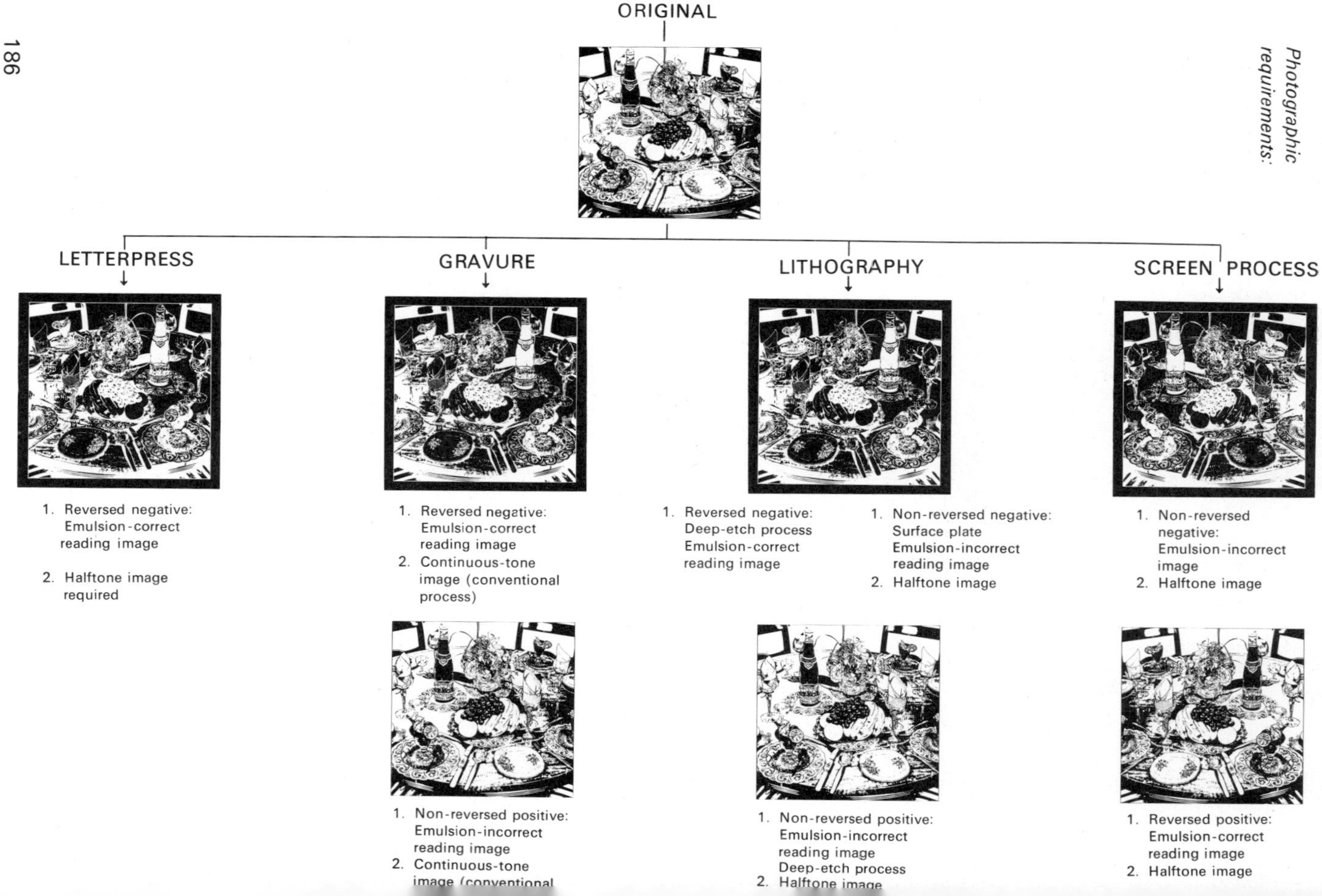

ORIGINAL

LETTERPRESS →
1. Reversed negative: Emulsion-correct reading image
2. Halftone image required

GRAVURE →
1. Reversed negative: Emulsion-correct reading image
2. Continuous-tone image (conventional process)

1. Non-reversed positive: Emulsion-incorrect reading image
2. Continuous-tone image (conventional

LITHOGRAPHY →
1. Reversed negative: Deep-etch process Emulsion-correct reading image

1. Non-reversed negative: Surface plate Emulsion-incorrect reading image
2. Halftone image

1. Non-reversed positive: Emulsion-incorrect reading image Deep-etch process
2. Halftone image

SCREEN PROCESS →
1. Non-reversed negative: Emulsion-incorrect image
2. Halftone image

1. Reversed positive: Emulsion-correct reading image
2. Halftone image

186

emulsion is washed and finally reblackened in a 10% solution of ammonia.

Suitable formulae for all these solutions can be found in the manufacturer's handbook related to the emulsions used in the studio.

Photographic control

We have now studied the five basic factors which go to make up the photographic process used to service the graphic reproduction industry. When linked together we can readily see their position in the complete process.

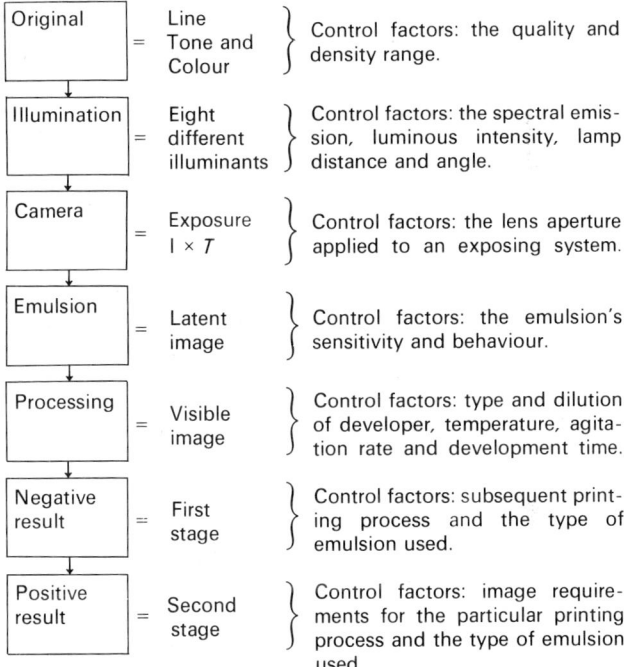

The skill of the photographer is to control each of these stages by employing systematic working, so that the best possible photograph is produced. Each printing process has its own photographic requirements which suit the printing method. The first requirement is that the final photographic film should be in the correct form (negative or positive) and its image must appear in the following manner on the emulsion side. The other requirements such as tonal range, halftone dot size and contrast will be discussed in the chapters on line, tone and colour reproduction.

8. SYSTEMATIC WORKING

Historical introduction

Nearly all the early photographers used purely empirical methods to obtain their results, although they realized that light was responsible for the blackening effect and that an increase or decrease in exposure time would result in a darker or lighter photographic image. It appears that this was the extent of their systematic approach. The reciprocal relationship between exposure and image blackness, its predictability and precise application escaped them because of the inconsistency of available emulsions with regard to sensitivity, grain-size and image characteristics. This lack of standardization was exaggerated by the absence of an instrument (photometer) capable of measuring the amount of light exposing the emulsion. These photographic deficiencies forced these pioneers to work to a principle of 'suck it and see'. This was not only a wasteful system to apply to a process using expensive materials, but also one which caused extreme frustration as their results fluctuated to say the least!

This state of affairs did not suit the analytical minds of Ferdinand Hurter, 1844–1898, a Swiss chemist working in

H and D curve

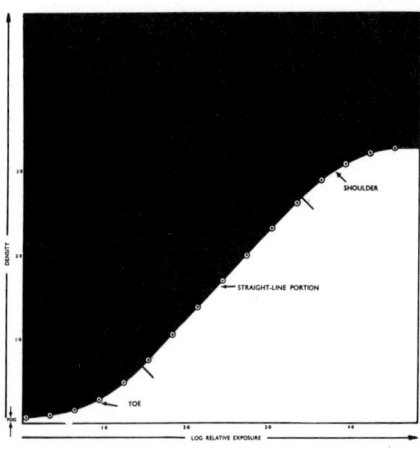

Sensitometry

England and his English colleague, Vero Charles Driffield, 1848–1915, a fellow chemist.

The H and D curve was the foundation of their study, and on this they based methods of calculating emulsion speed or sensitivity, image characteristics and the reciprocal link between exposure and resultant density. Hurter and Driffield gave all their contemporary photographers a base to work from; light intensity was now being measured and related to the emulsion's sensitivity so that an accurate exposure time could be calculated. These two men of science extended their investigations into the theory of the latent image, its development and the control of gamma (image contrast). They could see that exposure and development were the two dominant factors in the photographic process and needed to be controlled and standardized if consistent results were to be achieved. Hurter and Driffield defined these two factors as:

Exposure = Intensity × Time and

Development = Dilution × Agitation × Temperature × Time

They stated that exposure mainly controlled the bottom end of their H and D curve, while development time mainly affected the top end, producing variations in the steepness of the curve—in other words altering the gamma value. Their complete works were published in 1890 and won them the Royal Photographic Society's Progress Medal in 1898. Although a great deal of Hurter and Driffield's work is not applicable to today's emulsions these two men set a fine example in determination and illustrated vividly what logical thought and systematic deed can achieve in a process relying on a series of successive operations each one containing a variable.

Sensitometry

The aim of graphic reproduction is to produce, by using photographic methods, a printed copy which is as near a facsimile reproduction of the original as is possible with modern materials. A perfect reproduction may be defined as one containing tones and colours which have the same proportional relationship to one another as those appearing in the original picture.

Placing aside for the moment the limitations of the photographic process and the major printing processes, the initial success of any reproduction process relies entirely upon the photographic images being exact renditions of the original's tonal and colour values. This ideal situation can only be

Systematic working

achieved if strict control is maintained when exposing and developing the blackened image. The specific conditions determining this relationship between exposure and image blackness are gathered together under the term *sensitometry*. The importance of understanding sensitometry is that the action of light can be *measured* and therefore its effect can be predicted accurately.

The sensitometric curve

The best way to appreciate the reciprocal relationship between exposure and image blackness is to take a strip of continuous tone film (normal negative emulsion) and give it a series of progressive time exposures using a light source of constant intensity. Each exposed step should bear a constant ratio to its neighbour. This can be achieved by doubling the exposure time in each case, e.g. 0, 1, 2, 4, 8, 16,

Reciprocal relationship

0 1 2 4 8 16 32 64 128 256 512 sec

32 sec, etc. After suitable development the strip of film will appear as a photographic stepwedge.

This relationship becomes even more meaningful if it is presented graphically as a sensitometric curve.

This typical sensitometric curve follows a particular shape and can be divided into three main parts. The first is produced by the very short exposures and curves upwards—this is called the *toe*. (The extremely short exposures have no effect on the emulsion. The light used here is not great enough to overcome the inertia of the emulsion, so the blackness produced is no higher than the *fog* produced by developing this small unexposed part of the curve.) The second part is the result of medium exposures and takes on

Sensitometric terms

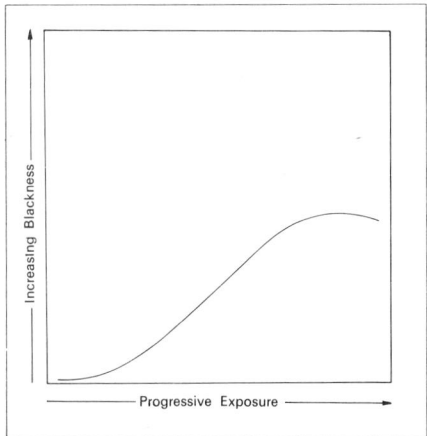

Sensitrometric curve:
Divided into its constituent parts

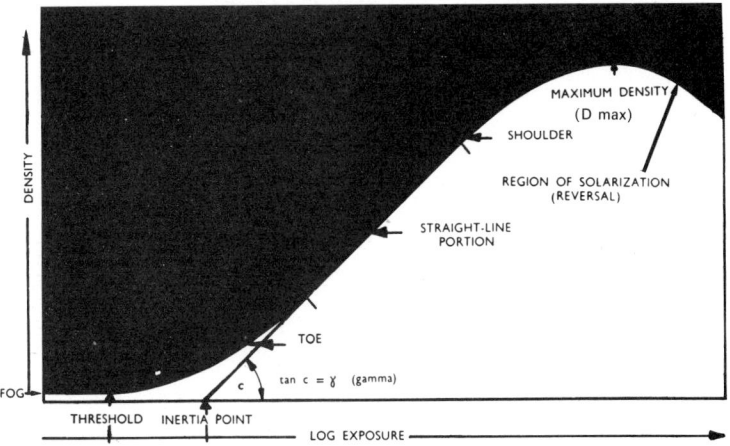

the shape of a *straight line*. The third part corresponds to the long exposure times and begins to curve downwards—this is termed the *shoulder*.

Sensitometric terms

Before proceeding any further we must consider some terms and definitions used in sensitometry:

Exposure

A light-sensitive photographic emulsion responds to the total amount of light it receives, i.e. the product of intensity I × time T, which is IT (discounting reciprocity failure), and termed exposure E. Total exposure can be used in two ways. The first is as a *Time* scale exposure, where the intensity of

Systematic working

the light source is kept at a constant, while the time period is varied to suit the conditions. The second method is the complete reverse of the first. In *Intensity* scale exposures time becomes the constant while the intensity of the light source is varied to obtain the required result.

Once a photographic emulsion has been exposed and developed a degree of blackening is obtained in the emulsion layer. The next step is to establish the relationship between the exposure given and the degree of blackening obtained. This is carried out by transmitting a light beam through, or reflecting it off the blackened image areas. By recording the loss of light due to absorption in the blackened areas the following information is obtained.

Transmission

This is the ratio of the amount of light that the developed silver grains allow to pass through (transmitted light) to the light which strikes the image (incident light).

$$\text{Transmission} = \frac{\text{Transmitted light}}{\text{Incident light}} \quad \begin{matrix} \text{always less than 1} \\ \text{when expressed as} \\ \text{a fraction.} \end{matrix}$$

e.g. $\quad \text{Transmission} = \dfrac{10 \text{ cd.}}{20 \text{ cd.}}$

∴ $\quad \text{Transmission} = \dfrac{1}{2}$ or 50%

Transmission: 50%

$$T = \frac{T}{I}$$

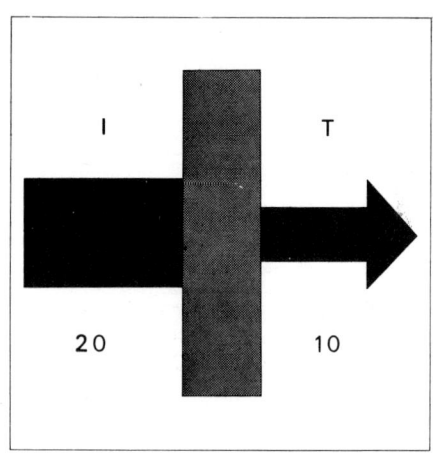

Sensitometric terms

Reflectance

In the case of images on white paper or any other opaque surface, this is the ratio of the amount of light that the image allows to pass on (reflected light) to the light which strikes the image (incident light).

$$\text{Reflectance} = \frac{\text{Reflected light}}{\text{Incident light}} \Big\} \text{ always less than 1 when expressed as a fraction.}$$

e.g. $\text{Reflectance} = \dfrac{10 \text{ cd.}}{20 \text{ cd.}}$

∴ $\text{Reflectance} = \dfrac{1}{2}$ or 50%

Reflectance: 50%

$R = \dfrac{R}{I}$

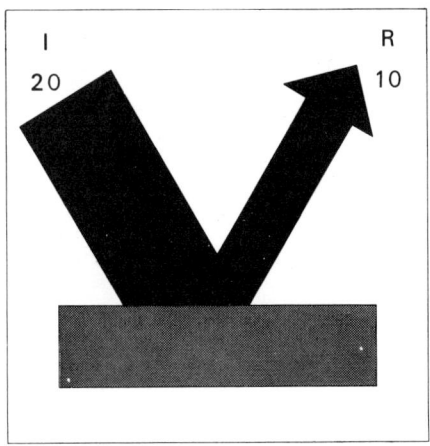

Opacity

This term indicates the light-stopping power of the image and is by definition the reciprocal of the transmission. If the transmission is $\frac{1}{2}$ then the opacity is 2. Photographic images have opacity values up to 10,000. If opacity is plotted against exposure the curve becomes very compressed and unrealistic to work with. Opacity is found by dividing the transmitted or reflected light into the incident light.

$$\text{Opacity} = \frac{\text{Incident light}}{\text{Transmitted light}} \Big\} \text{ always greater than 1.}$$

e.g. $\text{Opacity} = \dfrac{20 \text{ cd.}}{10 \text{ cd.}}$

∴ Opacity = 2

193

Systematic working

Density

This is represented by a density reading—a numerical value which is the logarithm of the opacity. When opacity values vary from 1 to 10,000 the density numbers only range from 0 to 4.

Table 3. Density, opacity and transparency

Density (log)	Opacity (No.)	Transparency	Density (log)	Opacity (No.)	Transparency
0·00	1·00	1·00	1·55	35·5	0·028
0·05	1·12	0·89	1·60	39·8	0·025
0·10	1·26	0·80	1·65	44·7	0·022
0·15	1·41	0·71	1·70	50·1	0·020
0·20	1·59	0·63	1·75	56·2	0·018
0·25	1·78	0·56	1·80	63·1	0·016
0·30	2·00	0·50	1·85	70·8	0·014
0·35	2·24	0·45	1·90	79·4	0·0126
0·40	2·51	0·40	1·95	89·1	0·0112
0·45	2·82	0·36	2·00	100	0·0100
0·50	3·16	0·32	2·05	112	0·0089
0·55	3·55	0·28	2·10	126	0·0080
0·60	3·98	0·25	2·15	141	0·0071
0·65	4·47	0·22	2·20	159	0·0063
0·70	5·01	0·20	2·25	178	0·0056
0·75	5·62	0·18	2·30	200	0·0050
0·80	6·31	0·16	2·35	224	0·0045
0·85	7·08	0·14	2·40	251	0·0040
0·90	7·94	0·126	2·45	282	0·0036
0·95	8·91	0·112	2·50	316	0·0032
1·00	10·0	0·100	2·55	355	0·0028
1·05	11·2	0·089	2·60	398	0·0025
1·10	12·6	0·080	2·65	447	0·0022
1·15	14·1	0·071	2·70	501	0·0020
1·20	15·9	0·063	2·75	562	0·0018
1·25	17·8	0·056	2·80	631	0·0016
1·30	20·0	0·050	2·85	708	0·0014
1·35	22·4	0·045	2·90	794	0·0013
1·40	25·1	0·040	2·95	891	0·0011
1·45	28·2	0·036	3·00	1,000	0·0010
1·50	31·6	0·032	4·00	10,000	0·0001

Using the previous example of a strip of film being exposed to a series of exposures, thus producing a photographic stepwedge, it becomes obvious that if we transmit a light beam through the step that received no exposure and therefore has not really blackened, the transmission will be 100%, the opacity 1 and the density reading 0·00. As the image blackness increases the opacity and density value rise, while the transmission figures gradually diminish.

Exposure in sec	0	1	2	4	8	16	32	64	128	256	512
Density	0.00	0.05	0.10	0.20	0.30	0.45	0.70	1.00	1.30	1.60	1.90
Opacity	1.00	1.12	1.26	1.59	2.00	2.82	5.01	10.0	20.0	39.8	79.4
Transmission	1.00	0.89	0.80	0.63	0.50	0.36	0.20	0.100	0.050	0.025	0.0126

Total effective opacity

If three identical pieces of developed film, each having an opacity of 2, are interposed so that each piece transmits ½ of the incident light falling upon it, then starting with an incident beam of 20 cd. each film will only allow ½ of the incident light to pass through, so the final transmitted beam will only be ⅛ (2.5 cd.) of the original incident light.

Total effective opacity

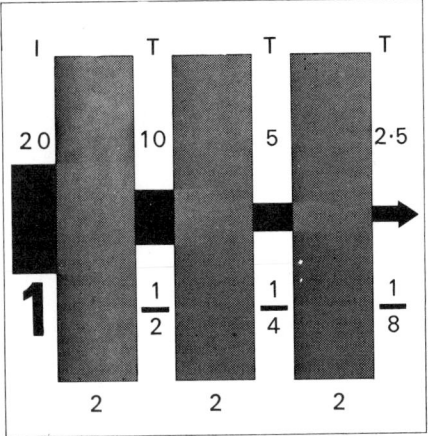

The total effect of adding the three opacities is an eightfold reduction in the intensity of the transmitted light. If we just add the three separate opacity values together the sum is only 6. Therefore the total effective opacity is found by *multiplying* the opacities 2 × 2 × 2 = 8. This can be simplified in practice by replacing opacity values by their logarithms, i.e. the density readings, (2 = 0.3010). To arrive at the total effective opacity we just *add* the three logarithms together (0.9030) and convert the total back to natural units by finding the antilog, which in this case equals 8.

Logarithms

By now it will have become clear that logarithms are used a great deal in sensitometry and densitometry. With most students this appears to be the moment of departure! This need not be—logarithms are merely a form of mathematical shorthand. Their purpose is to quicken and simplify multipli-

Systematic working

cation and division problems. By using logarithms we can represent any number in terms of another number in decimals —which is the *Power* to which 10 has to be raised to obtain our original number. If we consider the following numbers, series B becomes the easiest to remember and apply.

A:　$1=10^0$, $10=10^1$, $100=10^2$, $1000=10^3$, $10,000=10^4$
B:　　0,　　　1,　　　2,　　　3,　　　4

The simple numbers in series B are the logarithms of the corresponding numbers in series A. The logarithm of 10,000 is 4 because 10 must be raised to the fourth power to obtain 10,000.

$10 \times 1 = 10 \times 10 = 100 \times 10 = 1000 \times 10 = 10,000$
　　1　　　　2　　　　3　　　　　4

The number 10 is called the *Base* and the figure 4 the *Index* or *Power* number, 10^4. Any other number, say x becomes $x = 10^y$; y is our logarithm.

Using logarithms

The logarithm of any number is in two parts,
The　　　　　　Whole number · and a *Series* of numbers
e.g. log of 32 =　　1　　·　　50

The whole number preceding the decimal point of the logarithm is known as the 'characteristic' (1 in the example) and the decimal fraction as the 'mantissa' (·50 in the example). The characteristic whole number can be found easily as it is the number of figures before the decimal point of the original number *Less* 1 (example 32: therefore $2 - 1 = 1$ · *mantissa*). The mantissa must be looked up in a table of logarithms (example 32 = log 1·50). This can also be learnt by going through a set procedure.

1. *Finding the characteristic*

Look at the number and decide whether it falls,
　(a) between　　0 and 10
or (b) between　　10 and 100
or (c) between　　100 and 1000
or (d) between 1000 and 100,000

Having decided this find the characteristic whole number part of the logarithm. This characteristic number becomes,

0 if the number is　　0–10　　　　= 10^0
1 if the number is　　10–100　　　= 10^1
2 if the number is　　100–1000　　= 10^2
3 if the number if　　1000–10,000　= 10^3
4 if the number is 10,000–100,000 = 10^4

Logarithms

2. Completing the logarithm

Let us do this by finding the logarithms of the following numbers:

	4	44	504	1240
First step, the characteristic whole number	0·mantissa	1·mantissa	2·mantissa	3·mantissa
Second step, find the mantissa via logarithmic tables	·6021	·6435	·7024	·0934
Complete the logarithm in each case	0·6021	1·6435	2·7024	3·0934

3. Conversion back by antilogarithms

After solving problems in logarithmic terms we must convert the answers back into our original units by consulting the table of antilogarithms.

Example. log 3·0934

The first step is to take the mantissa and look this up in the antilog tables:

$$·0934 = 1240$$

We now have to decide where to place the decimal point. The characteristic whole number in this example is 3, or in other words,

$$10^3 = 1000$$

This indicates to us that the final figure must be between 1000 and 10,000 and is therefore, 1240.

Applying logarithms

The log of a product (a number resulting from *multiplication*) is the sum (added value) of the logarithms of each factor, e.g.:

$$\log (x \times y) = \log x + \log y$$

The log of a quotient (a number resulting from *dividing* one number by another) is the difference between the log of the integer (whole number) and the log of the denominator (divisor in vulgar fraction), e.g.:

e.g. in general $x \div y = x/y$
∴ $\log (x/y) = \log x - \log y$

To sum up, when we wish to multiply we *add* the logarithmic values and when we want to divide we *subtract* the logarithms.

Systematic working

Multiplication

Once again we will turn to our example numbers, using them as multiplication problems.

Examples.   4 × 44   1240 × 44

Find the logarithm in each case,

$$0.6021 + 1.6435$$

$$3.0934 + 1.6435$$

$$\overline{2.2456}$$ $$\overline{4.7369}$$

Consult the antilog tables,

2456 = 1760   7369 = 54560

Place the decimal point,

Characteristic is 2,   Characteristic is 4,
therefore between   therefore between
100 and 1000   10,000 and 100,000

Answer = 176·0   *Answer* = 54,560

Division

Let us now consider the use of logarithms through two example numbers. In the following examples we are going to encounter *minus* characteristic whole numbers. Again there are system numbers to use.

(a) if 0·1   −1·0   then characteristic whole number is $\bar{1}$
(b) if 0·01   −0·1   then characteristic whole number is $\bar{2}$
(c) if 0·001   −0·01   then characteristic whole number is $\bar{3}$
(d) if 0·0001   −0·001   then characteristic whole number is $\bar{4}$

Examples. $4 \div 44 = \dfrac{4}{44}$ $1240 \div 44 = \dfrac{1240}{44}$

Find the logarithm in each case,

$$0.6021 - 1.6435$$ $$3.0934 - 1.6435$$

$$\overline{2}.9586$$ $$1.4499$$

Consult the antilog tables,

9586 = 9090   4499 = 2818

Place the decimal point,

Characteristic is $\bar{2}$,   Characteristic is 1,
therefore between   therefore between
0·01 and 0·1   10 and 100

Answer = 0·09090   *Answer* = 28·18

Logarithms

Logarithms of decimal numbers

Finally, we must find the logarithms of decimal numbers. For decimal numbers the first step in finding the characteristic whole number is to remember that the whole number is 1 *more* than the noughts showing *after* the decimal point.

Consider the following examples:

	0·4	0·044	0·0562	0·004127	0·0008
First step, the characteristic whole number	$\bar{1}$·mantissa	$\bar{2}$·mantissa	$\bar{2}$·mantissa	$\bar{3}$·mantissa	$\bar{4}$·mantissa
Second step, find the mantissa via logarithmic tables	·6021	·6435	·7497	·6156	·9031
Complete the logarithm in each case	$\bar{1}$·6021	$\bar{2}$·6435	$\bar{2}$·7497	$\bar{3}$·6156	$\bar{4}$·9031

We can now easily multiply or divide complicated decimals.

Examples.   0·4 × 0·004127   0·0562 × 0·004127

Find the logarithm in each case,

$$\bar{1}\cdot 6021 + \bar{3}\cdot 6156 \qquad \bar{2}\cdot 7497 + \bar{3}\cdot 6156$$

$$\overline{\bar{3}\cdot 2177} \qquad\qquad \overline{\bar{4}\cdot 3653}$$

Consult the antilog tables,

2177 = 1650   3653 = 2319

Place the decimal point,

Characteristic is $\bar{3}$,   Characteristic is $\bar{4}$
remember this is 1
more than the noughts
showing *after* the
decimal point

Answer = 0·001650   Answer = 0·0002319

Example. $\dfrac{4 \times 0\cdot 0562}{504 \times 0\cdot 0008}$

First step, calculate   4 × 0·0562
the integer by using   0·6021 + multiply by
logarithms $\bar{2}$·7491   adding the logs

$$\overline{\bar{1}\cdot 3518}$$

Systematic working

Second step, calculate the denominator in a similar fashion	504×0.0008 2.7024 $\overline{4}.9031$ ――― $\overline{1}.6055$	+
Third step, divide the denominator into the integer	$\overline{1}.3518$ $\overline{1}.6055$ ――― $\overline{1}.7463$	divide by subtracting the logs
Finally, convert back to original units	Consult the antilog tables, $7463 = 5576$ Place the decimal point, Characteristic whole number is 1 *Answer* = 0·5576	

Log of exposure

Now that we have a better understanding of logarithms it will become easier to see why so many photographic results are best represented by means of a curve from which, by means of its shape, a photographer can obtain vital information.

From the previous curves which were obtained by comparing the basic concepts—progressive exposure and increasing blackness—it is obvious that exposure is plotted against density. Density values are obtained by using a densitometer, an optical instrument which uses transmitted or reflected light to produce direct density readings. (See Section on Densitometry.)

In practice log exposure is used instead of exposure. Referring to our first experiment of producing a photographic stepwedge, it will be seen that exposure increases from 1 to 32,768 sec, while the resultant densities increase from 0·05 to 3·00. In order to plot these two lots of information an exceedingly short exposure scale would be needed and then it becomes difficult to follow the curve, especially the section corresponding to very short exposure times (keen students will try to plot such a curve).

However, if we take the logarithms of the exposure times, the exposures will range from 0 to 4·5 and the same scaling can be used for both exposure times and density readings. Because we progressively increased the exposure time by $\times 2$, the straight-line portion of the curve will clearly show us its reciprocal properties.

By using logarithms, $\log(x \times y) = \log x + \log y$, so instead of recording successive exposures as,

The characteristic curve

1, 2, 4, 8, 16, 32 sec, etc.,

they can be written as,

1, 1 × 2, 2 × 2, 4 × 2, 8 × 2, 16 × 2, etc.

clearly indicating that each exposure is double the previous one. Each successive exposure increase will be seen as the log of $2 = 0.3$.

Log E = **0** 0·3 0·6 0·9 **1** 1·2 1·5 1·8 **2** 2·1 2·4 2·7

Corresponding
exposures 1 2   4   8   16 32 64   128 256 512
in seconds

Log E = **3**   3·3 3·6   3·9   **4** 4·2   4·5

Corresponding
exposures 1024 2048 4096 8192   16,384 32,768
in seconds

The characteristic curve

From the previous work it is quite obvious that the density produced in a photographic emulsion does not always change in proportion to the exposure time. The relationship between between exposures and corresponding densities can be shown clearly in the shape of a curve. This curve is called the 'characteristic curve'.

A curve of this type may be plotted from the following readings obtained from our photographic stepwedge. The

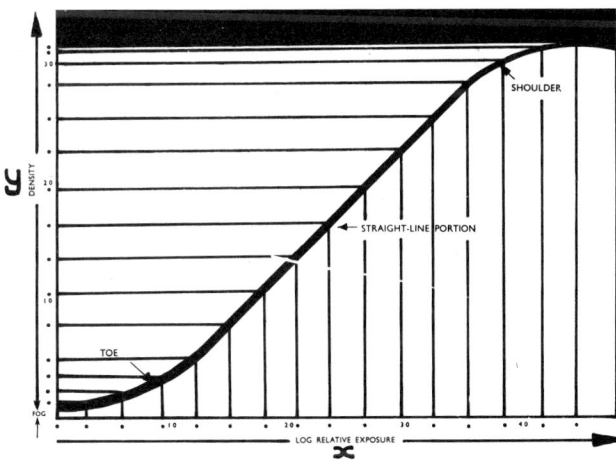

Characteristic curve

Systematic working

log E values are plotted along the 'abscissa' (or x axis). The corresponding density readings are recorded on the 'ordinate' (or y axis). When the points of intersection, formed by horizontal and vertical lines drawn from the values plotted on the abscissa and ordinate, are carefully joined up, a curve in the shape of an elongated and oblique 'S' is produced. This is the 'characteristic curve'.

Table 4. Log E and density readings (obtained from photographic step-wedge)

Exposure in sec	1	2	4	8	16	32
Log E	0	0·3	0·6	0·9	1·2	1·5
Density	0·05	0·1	0·2	0·3	0·45	0·7
Density difference		0·05	0·1	0·1	0·15	0·20
	A	Under-exposure				B

Exposure in sec	64	128	256	512	1024	2048
Log E	1·8	2·1	2·4	2·7	3·0	3·3
Density	1·0	1·3	1·6	1·9	2·2	2·5
Density difference	0·3	0·3	0·3	0·3	0·3	0·3
	B	Correct-exposure				C

Exposure in sec	4096	8192		16,384	32,768
Log E	3·6	3·9		4·2	4·5
Density	2·7	2·85		2·95	3·0
Density difference	0·2	0·15		0·10	0·05
	C	Over-exposure			D

The characteristic curve consists of three main sections—the area of under-exposure AB (the *toe* in our early curve), the region of correct exposure BC (the *straight-line portion*) and the area of over-exposure CD (the *shoulder*). The region of correct exposure is the all important part. Along this line an increase in density is always directly proportional to an increase in exposure. Taking our curve as an example, the increase in density is 0·3 which is directly proportional to an exposure increase of ×2.

Correct tone reproduction only occurs on the straight-line portion. This is when the relationship between the original's tonal values is transferred as an exact rendition on to the photographic emulsion.

Proportional relationship

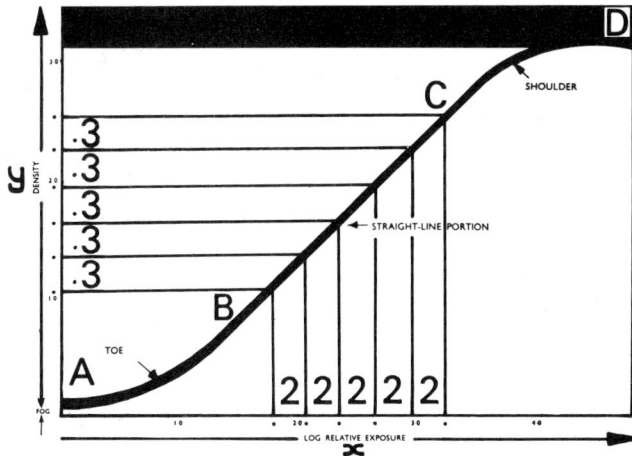

The manufacturer's characteristic curve

The manufacturer's characteristic curve

On opening a new box of film the first thing which comes to hand is the information data sheet. This contains a diagram of the manufacturer's characteristic curve. The most important step here is to remember that this curve is *not* characteristic of the emulsion, but characteristic of the complete photographic operation carried out—emulsion's sensitivity, exposure conditions and processing procedure. The photographic operation to be carried out may not be exactly the same, so although the information data sheet is an extremely helpful guide to the behaviour of the emulsion, it is better to produce 'house' characteristic curves under the prevailing conditions.

A characteristic curve for testing new batches of film can be produced in two ways.

1. Exposure = Intensity *constant* × Time *varying*

    This first system uses time scale modulators in the form of a 'falling plate' or a 'rotating sector wheel'. These modulators have cut-out sections that produce exposed steps which have a constant ratio to one another.

    A strip of the film to be tested is placed behind the modulator. The constant intensity light source is switched on and the modulator is either lowered or rotated. The resultant film contains a series of exposures from which an accurate and reproducible characteristic curve can be plotted

2. Exposure = Intensity *varying* × Time *constant*

    This system is more practical than the first one, because the normal studio conditions of exposure are used combined with the light sources that will be used for the subsequent

Systematic working

Time scale modulators

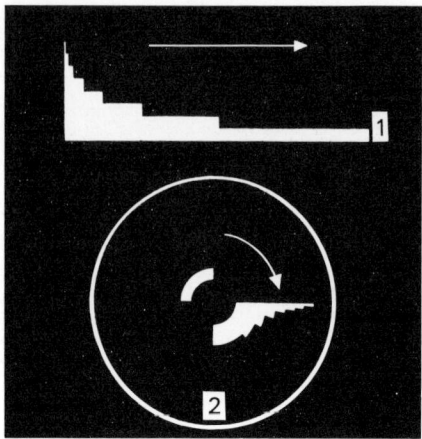

photographic work. The variation in intensity is set up by exposing via a stepwedge—a transmission stepwedge in the case of contact work and a reflection stepwedge for camera work.

The film to be tested is positioned behind or in front of the stepwedge (dependent on the work in hand). Once again the exposed and developed film strip will contain the required density steps.

Standard stepwedges:
1. Transmission
2. Reflection

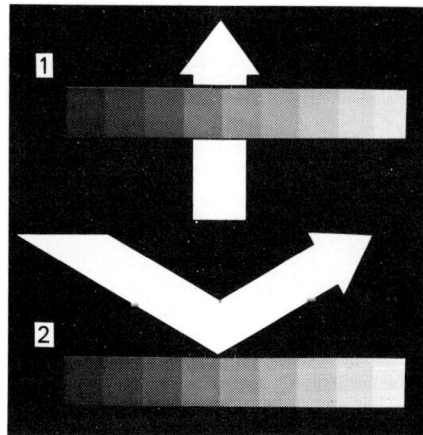

Original's density range

If we take a black and white tone original and measure the light intensities reflecting from the whitest highlight and blackest shadow (using a reflection densitometer), we will, of course, record a considerable difference. The maximum

Original's density range

difference usually encountered with reflection originals is a reflection value of 89·2% from the highlight area and 1·00% from the shadow areas, but once again it is advantageous to convert these values into reflection *density* readings. The highlight area would now become 0·05 and the shadow 2·00. In the case of transmission originals, e.g. colour transparencies, the original transmission densities may be anything from 0·02 to 3·0.

The next step is to transfer this reflection or transmission ratio on to the photographic emulsion. The camera image obtained will invariably have a brightness ratio which is lower than the original's. This is mainly the result of *lens flare*. Some of the illumination coming from the highlight and lighter tones scatters over the emulsion and in doing so, degrades the shadow detail. This flattening effect is accentuated if the scattered highlight illumination is reinforced by stray external light and interior reflections from the camera bellows.

Table 5. The degrading effect of a normal lens flare factor of 1%

Original density	Reproduction density
3·00	1·96
2·50	1·88
2·00	1·70
1·50	1·38
1·00	0·96
0·50	0·49
0·05	0·05
0·00	0·00

After recording the original's density range we must decide what density range we require on the photographic copy—a negative image in most cases. Although many factors determine this required image density range (we will tackle these a little later on) the usual maximum is a 1·4 density range, with a highlight density of 1·8 and a shadow density reading of 0·4, just on the start of the straight-line portion of the characteristic curve.

To obtain this result, especially with colour transparencies, the original's density range may have to be lowered by a contrast reducing mask—a light photographic image which is placed on top of the original and thereby subdues its contrast. Once we have a suitable range of original densities, we have to position them on the straight-line portion of the characteristic curve. Our success in doing this is governed by the original's density range, lens flare factor, lens aperture employed and time of exposure.

Systematic working

Exposures on the characteristic curve

The time of exposure can be varied to correspond with the reflection or transmission density range of the original. It is this ratio between the highlight and shadow areas of the original which needs to be conserved in the negative if the subsequent printed reproduction is to give a facsimile impression to the viewer. This 'faithful translation' will normally be via a negative from which a positive must be made also giving a 'true' rendition. In other words, the negative becomes the original with luminosities corresponding to its densities and the ratio of these densities must be reproduced in the photographic positive, printing plate and printed reproduction.

Now that we are plotting negatives made from either reflection or transmission originals, the resultant curves

Characteristic curve:
Produced from the original's reflection densities

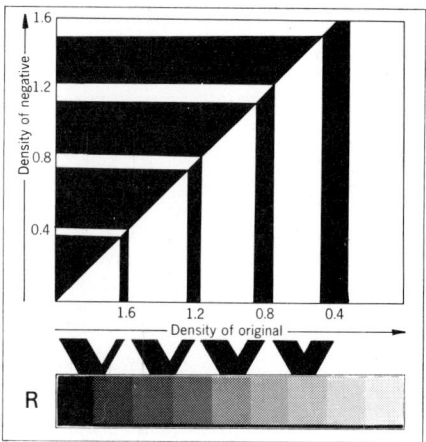

Characteristic curve:
Produced from the original's transmission densities

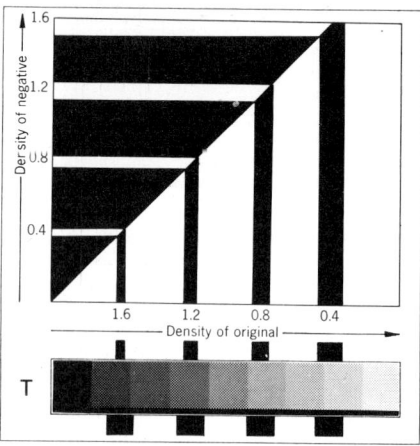

Exposure on the characteristic curve

1. Negative curve:
   Very short exposure—TOE. Positive print

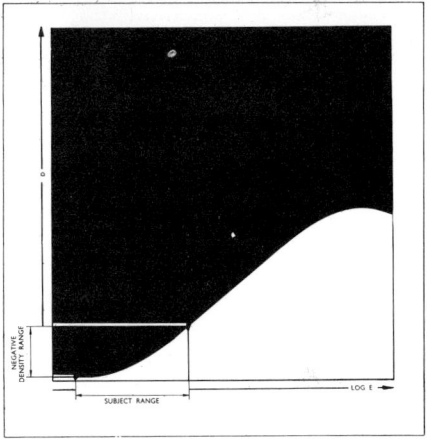

become more accurate if we substitute log E for the original's reflection density readings or transmission density readings. When a stepwedge is photographed and the original stepwedge densities plotted against the photographic image densities a characteristic curve of the photographic process being employed is obtained. If two original stepwedges are used, one transparent, exposed in contact with the emulsion and a reflection stepwedge photographed in the camera. The first curve will be a fairly accurate characteristic curve of the emulsion working with the particular exposing and developing method, while the second curve will illustrate the effect of lens flare and be characteristic of the camera conditions. Using original densities in place of log exposure is acceptable because density is a logarithm. A stepwedge becomes a

Systematic working

2. Negative curve:
Short exposure—Lower part of slope. Positive print

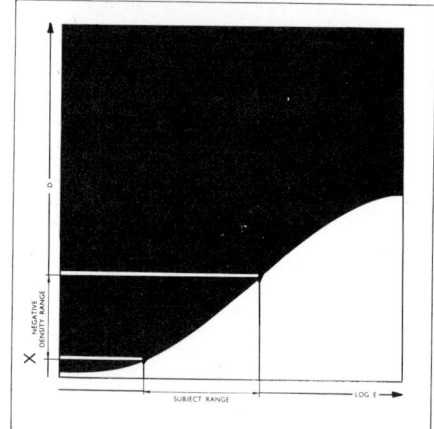

usually ·3

relative log exposure scale. The axis containing these densities is captioned, *Relative Log E*.

If the exposure time is so short that the highlight areas receive less exposure than the inertia point on the log *E* axis, the exposure will lie to the left of this point and the resultant negative would appear without a trace of an image. By increasing the exposure time the image will gradually move up the characteristic curve.

Correct exposure time

From the above diagrams it will be seen that in order to obtain a 'perfect' negative image, one which corresponds to our early definition of a 'perfect reproduction', it must lie exactly on the straight-line portion between B and C (diagram 3). This accounts for the descriptive term—region of correct

Correct exposure time

3. Negative curve:
Medium exposure—Straight–line–portion
Positive print

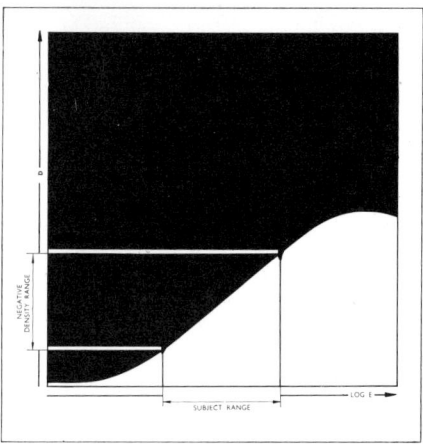

exposure. This is absolutely true if our aim is only to produce 'perfect' negative images, but a negative in graphic reproduction is only one of many intermediate image duplications between the original and the final printed reproduction. Each successive stage in the reproduction process introduces its own tonal distortion, so the production of a 'perfect' negative does not result in a 'perfect' reproduction. In fact it has been found that negative images recorded as having the deepest shadow on the toe of the curve (diagram 2) tend to counteract the tonal distortions introduced by the subsequent stages of the process. So it can be said that the correct exposure time is the minimum time period which will produce the shadow detail required and normally on negative continuous tone emulsions records at position X in diagram 2.

Systematic working

4. Negative curve:
Long exposure—shoulder. Positive print

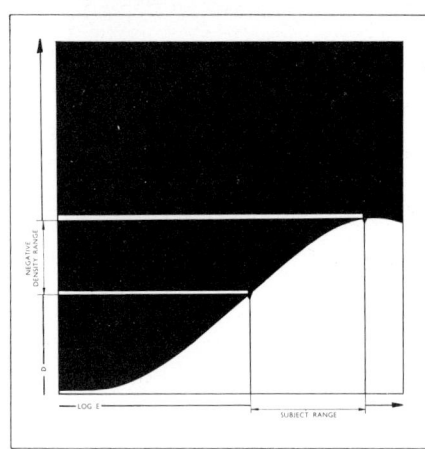

Image characteristics

Once an accurate characteristic curve has been produced under studio conditions a picture of the practical behaviour of the emulsion in question will be available. From this picture such photographic properties as fog level, sensitivity, maximum density, exposure range and contrast can be deduced.

Fog level

This is the density produced by development only in unexposed areas of the emulsion. This density includes the density value of the glass or film base. By scratching away a small section of the developed density it will be possible to read the density of the base material. This will allow subtraction of this base density from the developed density. The

Image characteristics

result will be the fog level of the emulsion. Generally speaking, this level should not exceed 0·15.

Sensitivity or speed

As stated before in emulsion manufacture there are two major systems of calculating the emulsion speed—ASA ratings and DIN values. With emulsions employed in graphic reproduction it is preferable to use one manufacturer's range of photographic materials. Then ASA and DIN ratings can give way to the more practical relative exposure factors. Each emulsion has its own exposure factor number which is used as an exposure time multiplier when changing from one emulsion to another.

Maximum density

This represents the highest density reached by the straight-line portion of the curve, just before it shoulders. The density reading obtained will, of course, vary considerably with the contrast range of the particular emulsion.

Exposure range

If we take an original reflection density range of, say, 1·4 (0·05 highlight to a 1·45 shadow) and a normal continuous tone negative emulsion with a straight-line portion of 0·5–3·0, the original's density range could be either exposed just on to the straight-line portion at 0·5–1·9 or taken to the limit of 1·6–3·0. If we link the lowest density in each case with its log E value, 0·5 to log E 1·25 and 1·6 to log E 2·4, then the exposure range is their displacement, i.e. 2·4–1·25 = 1·15. (This is not taking into account minimum exposure times on the *toe* section to counteract tonal distortion.)

Contrast

The contrast of a photographic image is measured by the degree of steepness produced in the slope of the straight-line portion (SLP) of the characteristic curve. This can be readily seen if we produce four photographic images of the same original on four different emulsion types.

The density difference between the highlight and shadow areas will, in the examples above, become gradually greater. The lith-type emulsion (Illustration 4) has the greatest density difference and therefore produces images of extreme contrast. Image contrast may be determined by dividing the original density range into the image range achieved in the emulsion. To do this we extend vertical lines up the abscissa, departing from the highlight and shadow densities of the original until

Systematic working

1. Negative curve:
Pan masking film. Angle of S.L.P. = 27°
γ0.50. Positive print

2. Negative curve:
Fine grain ordinary film. Angle of S.L.P. = 45° γ1.00. Positive print

Image characteristics

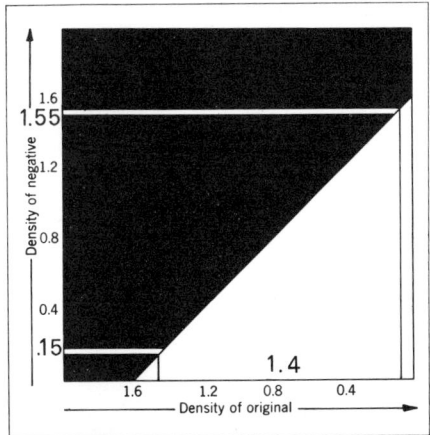

3. Negative curve:
Process (thin film halftone) film. Angle of S.L.P. = 68° γ 2.50

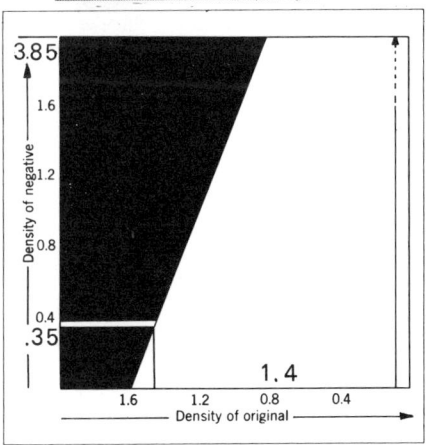

Systematic working

4. Negative curve:
Lith film. Angle of S.L.P. = 73°. γ 3.20.
Positive print

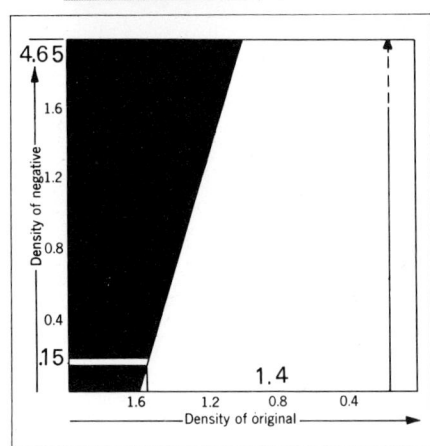

they intersect the characteristic curve. Horizontal lines are now drawn at right angles to the verticals and are extended until they cross the ordinate, thereby indicating the density range of the image (see Illustrations 1–4). The image contrast is found in each case by the equation,

$$\text{image contrast} = \frac{\text{image density range}}{\text{original density range}}$$

The image contrast values of the four examples are:

Illustration 1   Image contrast $= \dfrac{0\cdot 7}{1\cdot 4} = 0\cdot 5$

Illustration 2   Image contrast $= \dfrac{1\cdot 4}{1\cdot 4} = 1\cdot 0$

Illustration 3   Image contrast = $\dfrac{3\cdot 5}{1\cdot 4} = 2\cdot 5$

Illustration 4   Image contrast = $\dfrac{4\cdot 5}{1\cdot 4} = 3\cdot 2$

After careful observation of the above pictures and diagrams it will become clear that the differences in image density are indicated by the steepness of the straight-line portion of the characteristic curve.

Gamma

The term gamma, denoted by the symbol γ the Greek letter for C, is used to describe the contrast of the photographic image compared to the contrast of the original. Gamma may be defined as a numerical system of illustrating the contrast of a photographic material by measuring the slope of the straight-line portion of the characteristic curve. The gamma value is found by protracting the straight-line portion of the characteristic curve until it cuts the abscissa. The tangent of the angle Θ formed by the intersection becomes the gamma value. The tangent can be found in two ways, firstly by extending a vertical line (A) from the abscissa so that it cuts the straight-line portion. Now we have a right-angled triangle and the tangent of the angle $\Theta = A/B$. The second method is to find the degree of angle with a protractor and look up the natural tangent of this angle in a table of logarithms. Of course it must not be forgotten that the correct exposure for the shadow areas of the original often lies on the toe of the characteristic curve, so that the image contrast would be lower than the emulsion's gamma value. The true contrast

Gamma:
Extend the S.L.P. until it cuts the base-line A, angle A γ = tangent of $\angle$ A.
Average gradient:
Join up the lowest and highest *image* densities with a line which also cuts the base-line X, angle B. $\bar{G}$ = tangent of $\angle$ B.

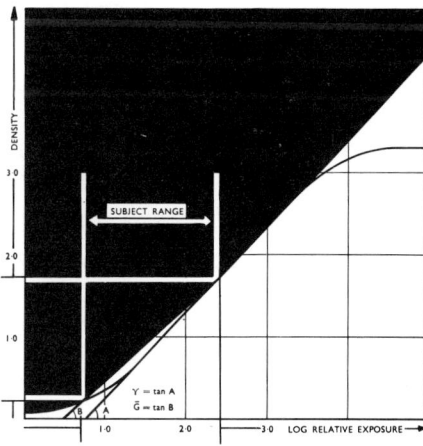

Systematic working

value, that is the image's density range compared to the original's density range, in this case would be found from an average gradient—a line (X) drawn between the lowest and highest densities on the characteristic curve, e.g. 0·30 and 1·70.

There are other considerations to bear in mind when discussing image contrast—lens flare, emulsion sensitivity and the condition of the original do have a considerable influence on the photographic image produced, so when the term gamma is being used image contrast produced by development alone is being referred to.

Time/gamma/curve

With continuous tone emulsions, if the development time is increased the gamma value will also increase until it finally reaches a limit, the point of maximum contrast, termed gamma infinity $\gamma \infty$. This relationship can be seen and used by producing a family of characteristic curves related to precise increases in development time. The gamma value is found for each curve and related to the development time.

We can now plot these two pieces of information against one another and produce a Time/Gamma curve.

This puts us in the position, as long as the developing conditions are kept constant, of being able to predict the development time for a particular gamma value when the continuous tone emulsion in question is being used.

Example. If we are confronted with a tone original which has a reflection density range of 2·0 (lens flare factor accounted for) from which a continuous tone negative possessing a 1·4

Family of curves:
Constant exposure, varying dev time
1 min = < A 25° γ 0·47
2 min = < B 41° γ 0·86
4 min = < C 51° γ 1·24
6 min = < D 53° γ 1·33
8 min = < E 53° γ 1·33

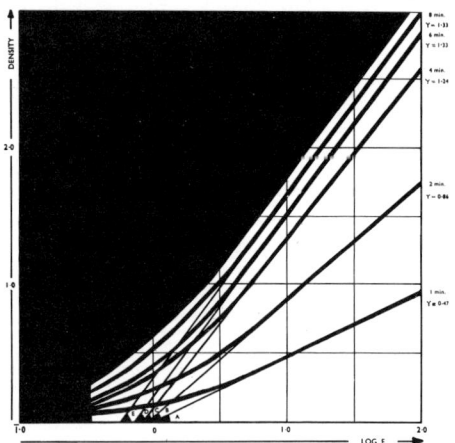

Densitometry

Time gamma curve:
1 min dev = γ 0·47
2 min dev = γ 0·86
4 min dev = γ 1·24
6 min dev = γ 1·33
8 min dev = γ 1·33
(N.B. In graphic reproduction it is more practical to use $\overline{G}$ values because the toe of the characteristic curve is used to counteract tonal distortions)

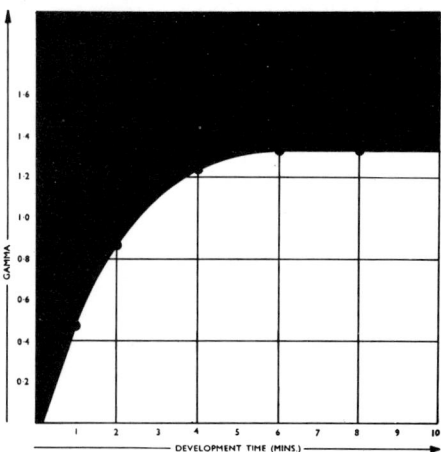

density range is required, then the gamma value would be determined by the equation,

$$\text{Negative gamma} = \frac{\text{density range of the required negative}}{\text{density range of the original}}$$
(lens flare factor accounted for)

e.g.

$$\gamma = \frac{1\cdot 4}{2\cdot 0} = 0\cdot 7$$

Consult the time/gamma curve,

Development time for a required gamma of $0\cdot 7 = 1\frac{1}{2}$ min

Densitometry

A photographic image and any subsequent modifications to its tonal range can only be intelligently appraised if we know the density values in each case. This knowledge allows us to plot tonal curves, work out total effective opacities and predict exposure and development times.

After exposure and development the amount of silver deposited in the emulsion layer is measured by a *densitometer*, an optical instrument especially designed to calculate the opacities of photographic images and give the answers as direct density readings—the logarithms of the opacities in question. Before we proceed any further and run the risk of becoming engulfed by the many diverse types of photoelectric densitometers available at the present time, it is a very good thing to look at the basic functions of a transmission and

Systematic working

Lummer–Brodhum cube:
1. Glass prism with a silvered circle
2. Two glass prisms forming the cube which brings two images together

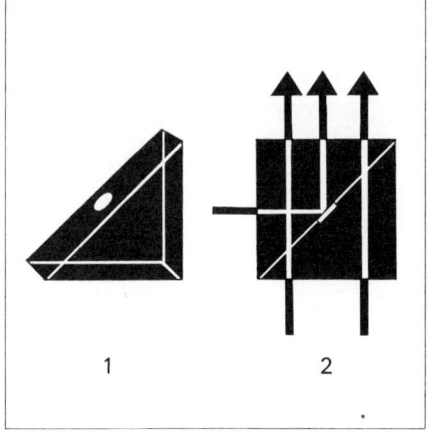

Known and unknown densities:
1. Incident light (X) of 10 cp projected through the known density A of 0·3, therefore a reflected ray of 5 cp is received
2. By using the Lummer–Brodhum cube known densities A can be compared to unknown densities B

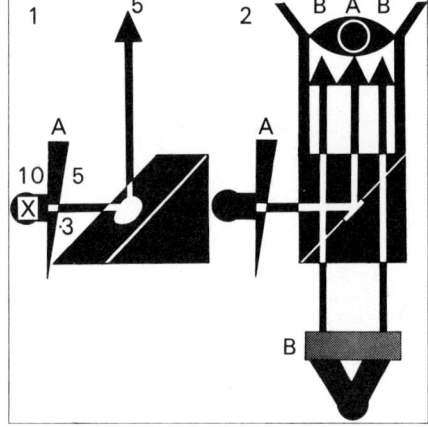

reflection densitometer (a keen student will build a simple working model).

We are faced with the problem of finding the density of a particular photographic image area. The easiest way to do this is by a visual comparison against an image of known density. To do this we must bring the two image areas together for comparison. This is aptly carried out by the Lummer–Brodhum cube made from two glass prisms; one of the prisms has a small silvered circle in the centre of its longest side. This silvered circle reflects the image of the known density.

If we place a graduated wedge containing known densities in position A we will be capable of matching, by moving the wedge up and down, the density of a previously unknown image area. A simple visual densitometer built on this principle would look like this:

Specular and diffuse densities

Lummer–Brodhum visual densitometer:
1. Reflection density readings
2. Transmission density readings

A Known densities
B Unknown densities
C Mirror
D Lummer–Brodhum cube build into an eye-piece
X Lamp
I Incident light=10 cp
T Transmitted light=5 cp
R Reflected light=5 cp

A Known densities
B Unknown densities
C Mirrors
D Lummer–Brodhum cube build into an eye-piece
X Lamp
I Incident light=10 cp
T Transmitted light=5 cp
T Transmitted light=5 cp

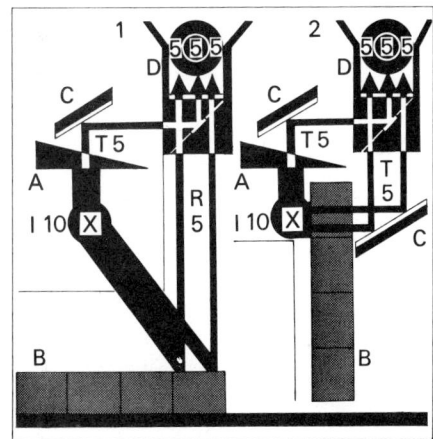

Photo-electric densitometer:
1. R Reflection density readings
2. T Transmission density readings

I Incident light=10 cp
B Unknown density
R Reflected light=5 cp
D Photoelectric cell
E Electrical current in proportion to the reflected light
F Ammeter

I Incident light=10 cp
B Unknown density
T Transmitted light=5 cp
D Photoelectric cell
E Electrical current in proportion to the transmitted light
F Ammeter

Reflectance $=\dfrac{R}{I}=\dfrac{5}{10}=\dfrac{1}{2}$

Transmission $=\dfrac{T}{I}=\dfrac{5}{10}=\dfrac{1}{2}$

Opacity $=\dfrac{I}{R}=\dfrac{10}{5}=2$

Opacity $=\dfrac{I}{T}=\dfrac{10}{5}=2$

Density: log of opacity = 0·3

Density: log of Opacity = 0·3

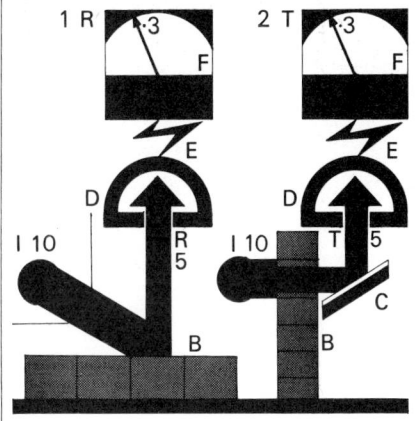

While building a visual densitometer of this type gives students academic pleasure, it is more realistic if the working model is taken a stage further and the density-equalizing combination of the Lummer–Brodhum cube plus a human eye is replaced by a photo-electric cell which will generate an electrical current in proportion to the light intensity received. The electrical current (usually after amplification) can be used to work a simple ammeter marked off in density readings. The basic densitometer would now appear as in the following diagram.

Specular and diffuse densities

Accurate transmission density readings of a photographic image made up of minute silver grains are made difficult by

Systematic working

the fact that when light falls on the image it divides roughly into three parts. The first part is transmitted through undeviated, the second absorbed by the black silver grains and the third part is scattered sideways. Now if we measure the transmitted light with a photo-cell placed a little distance away from the image, position SP in the diagram, we only measure the parallel light which has been transmitted in a specular fashion, but in doing so we will be measuring less transmitted light than was actually transmitted through the image. Therefore the opacity value and the corresponding density reading will be *higher* than they should be. This density reading is termed the *'Specular density* $(D\backslash\backslash)$*'*, and would be greater than the density of the same image area if it was read again with the photo-cell placed against the image surface at position *DF*,

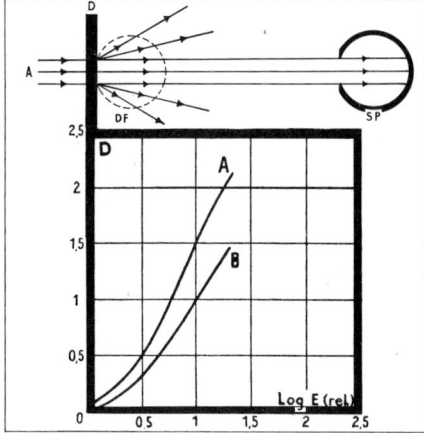

Specular and diffuse densities:
A   Incident light
D   Density
DF Photocell measuring diffuse density
SP Photocell measuring specular density
A   Specular density readings higher than,
B   Diffuse density readings

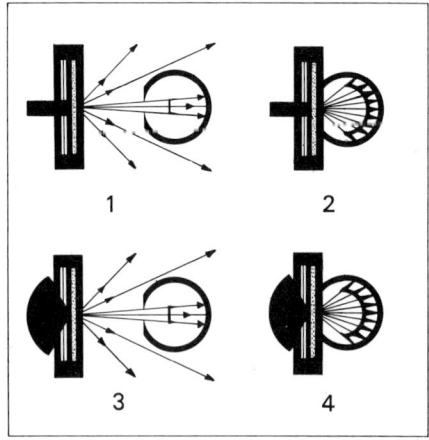

Types of density:
1. Specular density
2. Diffuse density
3. Diffuse density
4. Doubly diffuse density

220

Photo-voltaic or barrier-layer photo-cell

so that the whole of the transmitted light is collected. The reading obtained here is called the *'Diffuse density (D⊬)'*. The ratio $D \backslash\backslash / D ⊬$ is known as the *'Callier Coefficient'* and is generally in the region of 1·5.

In densitometry it is usual to measure both reflection and transmission densities in diffuse density. This is produced in two ways; lowering the photo-cell head as we stated earlier, or illuminating the image area to be measured with completely diffuse light and measuring the parallel part of the transmitted light. We can in fact illuminate and measure an image area with four different types of light movement.

Types of density

The resultant density reading is defined by the manner in which the opacity (incident light)/(transmitted light) is measured. The density is then the logarithm of the quotient.

1. SPECULAR DENSITY. Measure parallel incident light and divide by parallel transmitted light.

2. DIFFUSE DENSITY. Measure parallel incident light and divide by all the transmitted light.

3. DIFFUSE DENSITY. Measure diffuse incident light and divide by the parallel transmitted light.

4. DOUBLY DIFFUSE DENSITY. Measure all diffused incident light and divide by the diffuse transmitted light.

Both diffuse and specular density values are used in graphic reproduction photography, depending upon the optical system in which the density is to be used. Diffuse density readings are the more accurate when the negative is being exposed in *contact* with another light-sensitive emulsion, but specular readings are needed when the negative is exposed in an enlarger-type camera with a condenser light source. These specular readings must be defined in conjunction with the optical system used for the measurement and the particular optical design of the enlarger being used, complicated by the fact that only one lens aperture is used in practice, resulting in the need for a new exposure time for each alteration in magnification. This arises because the density values and contrast of the negative increase as the lens is stopped down.

Photo-voltaic or barrier-layer photo-cell

This is the simplest of all photo-electric cells. The photo-cell consists of a thin light-transmitting metallic coating, a layer of selenium and a thick metal base plate. These photo-

Systematic working

voltaic cells produce an electric current because the thin light-transmitting metallic coating attracts electrons which are released by the selenium when it is struck by light. The electric current produced by the photo-cell without the need for any external supply of electricity can be connected to a meter by taking leads from the thick metal base plate and a conducting ring which is placed in contact with the thin light-transmitting metallic coating. The meter will now give direct density readings proportional to the amount of light received by the photo-cells' surface.

Vacuum emission photo-cell

The photo-cell is made up from two metal parts in vacuum, inside a glass bulb. The first metal part is the photo-cathode which is photo-sensitive, i.e. it releases electrons when struck by light. The second metal part is the anode. When light strikes the cathode it releases electrons into the vacuum surrounding it. These electrons are negatively-charged and when a positive voltage is applied to the anode, it collects the electrons released as a result of the action of light and an electric current flows through the circuit to the meter giving the density readings. When the light intensity is reduced or increased, the number of electrons released is altered, therefore varying the voltage entering the meter. From this it can be seen that the density readings are correlated to the light intensity received by the photo-cathode.

The photomultiplier cell

A photomultiplier cell is used when the intensity of light to be measured is extremely low and if a simple photo-cell was employed the electrical current produced would be so weak as to be useless. The photomultiplier is a photo-cell, usually the vacuum emission type, plus an electron multiplier. If a weak pencil of light falls on to the photo-cathode and releases one electron, it is possible to bounce this electron off a series of electron multiplying metal plates (electrodes), so that after the first deflection up to six electrons are released. When the second electrode is hit by six electrons up to thirty-six electrons are produced and so on. A multiplication of a million times is not uncommon in a photomultiplier containing ten electrodes. On emerging from the multiplier the electrons are collected by an anode and an electric current is produced. Once more this current will activate a meter which is specially calibrated for these high density readings.

Densitometers

Barrier-layer cell:
1. Light
2. Selenium
3. Metal base
4. Ammeter
5. Transparent metallic coating

Photomultiplier cell:
1. Light
2. Photocathode
3. Anode

Vacuum emission cell:
1. Light
2. Photocathode
3. Anode
4. Glass bulb
4. Multiplying electrodes
5. Collector

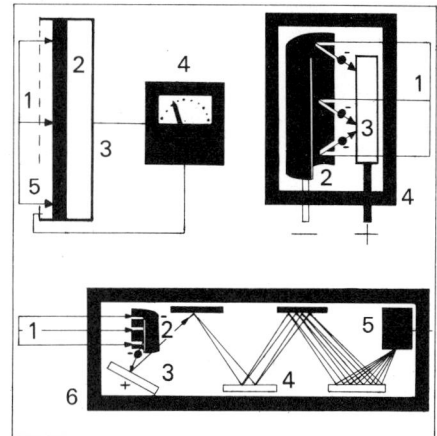

Densitometers

The camera operator has many designs to choose from, ranging from visual models costing under twenty pounds to photo-electric instruments selling at hundreds of pounds. The selected densitometer must be capable of producing accurate reflection and transmission density readings from the class of work in question, in a reasonably short period of time. The densitometer should cover a range of density readings from 0·00 to 3·00. The density scale needs to be calibrated in increasing densities of 0·02. The accuracy of the instrument should be checked at regular intervals by using a standard transmission and reflection stepwedge. Densitometers can be divided into two main groups, the robust visual types and the more sensitive photo-electric models.

Visual densitometers

By using a comparatively simple optical system the unknown tone area is brought into a viewing position adjacent to a control tone. By turning the disc holding the control stepwedge the operator decreases or increases the density of the control tone until he sees, using his eye as an equalizing agent, that the unknown tone area matches a particular control tone. The density of the unknown tone is then the same as the control tone and may be read off from a calibrated scale. Visual densitometers are certainly inexpensive and robust, but are bought at the cost of eye-strain after twenty or more readings have been taken and inaccurate, unreproducible density readings when high densities are encountered.

Systematic working

Visual densitometer:
1. Lamp
2. Lamp
3. Control stepwedge-known densities
4. Unknown density
5. Mirror with a central aperture
6. Lens
7. Known densities calibrated
8. Eye piece—e.g. central circles correct

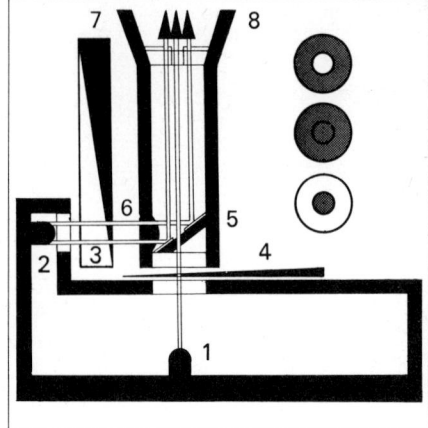

Photoelectric densitometer:
1. Lamp
2. Condensing lens
3. Filters: neutral density of:
   2·00 for readings 0·00 to 1·00
   1·00 for readings 1·00 to 2·00
   0·00 for readings 2·00 to 3·00
4. Mirror
5. Lens
6. Unknown density
7. Photo-electric cell
8. Ammeter

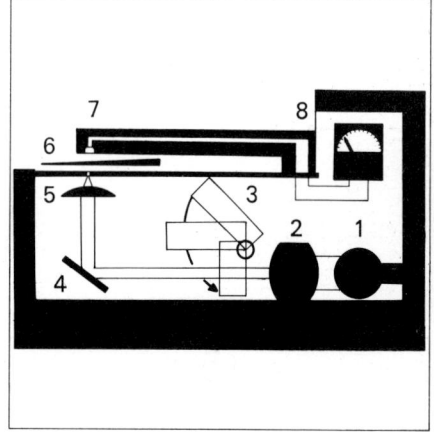

Photo-electric densitometers
　　　　The accuracy of this type of densitometer is governed by the type of photo-cell used in the instrument. The simplest and cheapest models, using a barrier-layer cell connected to a galvanometer, are extremely practical in their design, but accuracy becomes questionable for densities over 2·00. For density readings up to 3·00 the vacuum photo-cell is widely used, but for readings higher than this, especially in the region of 4·00, the transmitted or reflected light becomes so weak that a photomultiplier is needed to give accurate density readings.

Reproduction quality
　　　　One of the most difficult tasks in graphic reproduction is to

maintain a constant standard of result. Although reproductions of the highest quality should be aimed for, most companies are far happier with a constant output of results of acceptable quality, than with superlative reproductions one moment and poor, unacceptable copies the next. The quality of printed reproductions is closely related to the condition of the original, the equipment used and the most important factor—the technician's ability and approach. These ingredients need to be neatly wrapped up into a system of working.

The need for a system

In modern industry work produced must be uniform, economic and of acceptable quality. These are obtained as a direct result of the technician's approach to his work.

A printing technician must be able to apply his knowledge in a systematic manner keeping the following aim points in mind:

1. Conservation of energy and economic use of materials.
2. Standardization of operations and variable factors.
3. An organized pattern of work to produce uniformity in the end products.
4. Comprehensive records of work, so that results can be easily duplicated after a period of time.
5. A conscientious approach complemented by a determination to keep abreast of modern techniques and methods.

Standardization

A workable system normally uses one variant in each operation as its versatility factor. This variable is manipulated so that a seemingly uncontrollable input ingredient may be arrested and altered into an output result of constant quality. For example, the reflection or transmission density is the uncontrollable input ingredient. At the exposure stage the illumination is constant, the intensity of the light passing through the lens is constant, and the emulsion's sensitivity is constant. The versatility factor is the exposure time, which is altered so that a constant density is produced on the emulsion from differing original densities. Also, at the development stage, the developer's dilution, agitation rate and temperature are kept constant, while the time of development becomes the versatility factor. These systems only work if they are built on sound foundations. These foundations are the constant factors, e.g. uniformity of illumination, emulsion sensitivity, temperature of the developer, etc. All these factors need to be rigidly standardized. This can be done in the following manner.

Systematic working

Emulsions

Use one manufacturer's products and become familiar with their current information and data. Order in bulk, asking for emulsions with the same batch number. When a new batch number is encountered test its sensitivity by carrying out a test exposure on a standard stepwedge. Wherever possible, plot 'house' characteristic curves and time/gamma curves.

Illumination

Whenever possible use the same lamp distance and angle. Check the uniformity of the illumination over the original's area by using a luxmeter. Smooth out lamp fluctuations by exposing through a light-integrating meter, turning exposure time into light counts.

Chemicals and solutions

The best edict here is—buy in liquid form, buy in bulk, mix in bulk. Development needs to be carried out in such a manner that a replenisher becomes the only additional solution

Quality

This somewhat elusive creature can be captured with the aid of instruments.

Table 6. Quality control

Control	Instrument
Image densities	Reflection and transmission photoelectric densitometer with a density probe
Illumination	Luxmeter, exposure meter and a light integrating meter
Dot sizes and formation	Measuring microscope and a comparator microscope
Colour measurement	Spectrophotometer

Photographic procedure

When a camera technician is carrying out photographic work for all the graphic reproduction processes his work-pattern can be broken down into:

1. *Line reproductions on lith-type emulsions*

This involves the production of negatives and positives on an emulsion of extreme contrast. The photographs only contain two tones, the black image having a maximum silver

Photographic procedure

density of 4·00 and the non-image areas consisting of transparent film having a minimum density of 0·02.

2. *Line reproduction on continuous-tone emulsions*

The negatives and positives are produced on an emulsion capable of recording tonal variations; the maximum silver density can be varied from 0·2 to 3·00. The photographs only contain two tones, the image area which contains a high silver density that suits the printing process selected for the reproduction and the non-image areas of low density.

3. *Tone reproductions on lith-type emulsions using a halftone screen*

This type of result, in which the original's tones are translated into dots of varying size, is needed for printing methods which can only impart one weight of ink. The continuous-tone appearance of the original is simulated by the use of an optical illusion—small black dots surrounded by large expanses of white paper appear to the eye as areas of light grey, while large black dots surrounding minute areas of white paper appear as expanses of dark grey. This translation of the original's tones into dots is performed by a halftone screen positioned in the camera back (see chapter 10). Negatives and positives obtained in this manner contain dots of maximum silver density surrounded by transparent film having a minimum density of 0·02. This is really a line situation, just two tones.

4. *Tone reproductions on continuous-tone emulsions*

This photographic procedure can be interpreted as a method of matching a picture tone for tone. The emulsion can record an extensive range of tonal gradations. The aim here is to produce negatives and positives which possess density ranges and tonal gradations that represent the original, but also suit the printing process.

5. *Colour separations on panchromatic emulsions*

This work entails the photographic separation of a coloured original into its basic colours by exposing through coloured filters. Once the original has been classified as a line or tone reproduction the photographic work follows the previous line, halftone or continuous tone procedures. Of course, a negative and positive is required for each component colour complicated by the fact that most colour reproductions involve a combination of all these photographic procedures.

Once this segmented work-pattern is understood, it is a wise step to learn the underlying conceptions upon which each

Systematic working

operation is built. Therefore let us take a simple, mathematical look at these conceptions, before we actually tackle line, tone and colour reproduction.

Fundamental relationships

Exposure technique must be constant and repeatable. The two important factors here are:

Luminous intensity—which can be standardized in all cases by the use of a luxmeter and a light integrating meter.

Colour temperature—the spectrum emitted; this attribute fluctuates with intensity, so it is advisable to keep an illuminant at full intensity and if an intensity reduction is needed employment of neutral density filters will provide the reduction without altering the colour temperature. For camera and enlarger work pulsed xenon and single flash xenon lamps have proved to be the most constant type of light source, while for contact work single flash xenon and quartz-iodine lamps fit the requirements.

The camera technician mainly exposes for a calculated time to produce the required density and develops for a predetermined time to achieve a certain gamma value. Exposure time and development time are closely related; both exposure time and the required gamma value are calculated from a numerical comparison of the original's densities and the photographic densities required on the emulsion. A calculated exposure for a gamma value of 1·0 needs to be increased for gamma values lower than 1·0 and decreased for gamma values greater than 1·0. A camera technician must become aware of the relationship between exposure time and development time by conducting a series of experiments on the range of emulsions he needs to cover all the types of reproduction encountered. In order to keep this information data concise the number of different emulsion-types must be kept as low as possible.

The value of these experiments is immeasurable because the results have been produced under the prevailing conditions. The calculated exposure times must be carried out with constant and repeatable light intensities. From the photographic results produced under these controlled conditions the following information can be collected—the emulsion's sensitivity, tonal gradations, gamma values, minimum and maximum densities. From this recorded data maximum density

Controlled exposure

and time gamma curves may be constructed, condensing into a concise picture the image characteristics of the particular emulsion in question.

The following methods and equations for calculating exposure and development times only serve as fundamental systems founded on basic concepts. They provide an elementary stepping-stone to the more sophisticated exposure systems provided by most emulsion manufacturers. Once the selection of a single manufacturer to supply all materials and emulsions has been made, a complete, comprehensive system becomes feasible.

Controlled exposure

Once an exposure has been given and the desired photographic image achieved, to repeat this we must recreate the same situation all over again. The total effective exposure, that is $I \times T$, must be an exact recreation of the first exposure. The amount of light falling on the emulsion should be identical; link this with the same processing conditions and the second photographic image will closely resemble the first. To help us achieve this in graphic reproduction photography we have two exposing systems to choose from:

1. The v/ratio system

This system uses a lens aperture which is always a proportion of the camera extension (v). This is an attempt to keep the exposure time constant for a given original density photographing on to a particular emulsion.

$$\text{Exposure} = \underset{\text{(varying)}}{\overset{\text{(Lens aperture)}}{\text{Intensity}}} \times \underset{\text{(constant)}}{\text{Time}}$$

This v/ratio system is controlled by the equation,

$$\frac{1}{\text{Ratio}} = \frac{\text{Lens aperture}}{\text{Camera extension}}$$

The ratio numbers used in practice have a basic factor number of 2 and like f/numbers when compared to the preceding number allow half the amount of light through the lens. Because the diameter of the lens aperture has to be precisely set, the lens barrel is marked off in millimetres.

Ratio Numbers = $v/32$, $v/48$, $v/64$, $v/96$, $v/128$.

Systematic working

Example. When using a camera with a lens of 65 cm focal length at a magnification factor of 0·25 the camera extension would be 81·28 cm. By employing the above equation the millimetre lens apertures can be found for the following v/ratio numbers.

v/32 ratio

$$\frac{1}{\text{ratio}} = \frac{\text{Lens aperture}}{\text{Camera extension}}$$

$$\frac{1}{32} = \frac{LA}{81 \cdot 28 \text{ cm}}$$

$$\frac{81 \cdot 28 \text{ cm}}{32} = \frac{LA}{1}$$

Convert to millimetres by moving the decimal point back one place

$$25 \cdot 4 \text{ mm} = \frac{v/32 \text{ Lens Aperture}}{}$$

v/48 ratio

$$\frac{1}{\text{ratio}} = \frac{\text{Lens aperture}}{\text{Camera extension}}$$

$$\frac{1}{48} = \frac{LA}{81 \cdot 28 \text{ cm}}$$

$$\frac{81 \cdot 28 \text{ cm}}{48} = \frac{LA}{1}$$

$$16 \cdot 76 \text{ mm} = \frac{v/48 \text{ Lens Aperture}}{}$$

It is extremely useful to construct a lens aperture table for each v/ratio.

v/64 ratio

$$\frac{1}{\text{ratio}} = \frac{\text{Lens aperture}}{\text{Camera extension}}$$

$$\frac{1}{64} = \frac{LA}{81 \cdot 28 \text{ cm}}$$

$$\frac{81 \cdot 28 \text{ cm}}{64} = \frac{LA}{1}$$

$$1 \cdot 270 \text{ cm} = \frac{LA}{1}$$

$$12 \cdot 70 \text{ mm} = \frac{v/64 \text{ Lens Aperture}}{}$$

v/96 ratio

$$\frac{1}{\text{ratio}} = \frac{\text{Lens aperture}}{\text{Camera extension}}$$

$$\frac{1}{96} = \frac{LA}{81 \cdot 28 \text{ cm}}$$

$$\frac{81 \cdot 28 \text{ cm}}{96} = \frac{LA}{1}$$

$$0 \cdot 8380 \text{ cm} = \frac{LA}{1}$$

$$8 \cdot 38 \text{ mm} = \frac{v/96 \text{ Lens Aperture}}{}$$

v/128 ratio

$$\frac{1}{\text{ratio}} = \frac{\text{Lens aperture}}{\text{Camera extension}}$$

$$\frac{1}{128} = \frac{LA}{81 \cdot 28 \text{ cm}}$$

$$\frac{81 \cdot 28 \text{ cm}}{128} = \frac{LA}{1}$$

$$0 \cdot 6350 = \frac{LA}{1}$$

$$6 \cdot 35 \text{ mm} = \frac{v/128 \text{ Lens Aperture}}{}$$

Example. When using a camera with a lens of 25 in. focal length at a magnification factor of 0·28 the camera extension would be 32 in. By employing the above equation the millimetre lens apertures can be found for the following v/ratio numbers

v/32 ratio

$$\frac{1}{\text{ratio}} = \frac{\text{Lens aperture}}{\text{camera extension}}$$

$$\frac{1}{32} = \frac{LA}{32 \text{ in.}}$$

$$\frac{32 \text{ in.}}{32} = \frac{LA}{1}$$

$$1 \text{ in.} = \frac{LA}{1}$$

convert to millimetres × 25·4

v/48 ratio

$$\frac{1}{\text{ratio}} = \frac{\text{Lens aperture}}{\text{camera extension}}$$

$$\frac{1}{48} = \frac{LA}{32 \text{ in.}}$$

$$\frac{32 \text{ in.}}{48} = \frac{LA}{1}$$

$$0 \cdot 66 \text{ in.} = \frac{LA}{1}$$

Controlled exposure

It is extremely useful construct a lens aperture ble for each v/ratio.

$v/32$
$25\cdot4$ mm = Lens aperture

$v/48$
$16\cdot76$ mm = Lens aperture

64 ratio

$$\frac{1}{\text{ratio}} = \frac{\text{Lens aperture}}{\text{camera extension}}$$

$$\frac{1}{64} = \frac{LA}{32 \text{ in.}}$$

$$\frac{2 \text{ in.}}{64} = \frac{LA}{1}$$

$$5 \text{ in.} = \frac{LA}{1}$$

$v/64$
$12\cdot70$ mm = Lens aperture

$v/96$ ratio

$$\frac{1}{\text{ratio}} = \frac{\text{Lens aperture}}{\text{camera extension}}$$

$$\frac{1}{96} = \frac{LA}{32 \text{ in.}}$$

$$\frac{32 \text{ in.}}{96} = \frac{LA}{1}$$

$$0\cdot33 \text{ in.} = \frac{LA}{1}$$

$v/96$
$8\cdot38$ mm = Lens aperture

$v/128$ ratio

$$\frac{1}{\text{ratio}} = \frac{\text{Lens aperture}}{\text{camera extension}}$$

$$\frac{1}{128} = \frac{LA}{32 \text{ in.}}$$

$$\frac{32 \text{ in.}}{128} = \frac{LA}{1}$$

$$0\cdot25 \text{ in.} = \frac{LA}{1}$$

$v/128$
$6\cdot35$ mm = Lens aperture

Table 7. Magnification f/numbers

Magnification	f/Stop sizes			
	$v/32$	$v/48$	$v/64$	$v/96$
0·0	f/32	f/48	f/64	f/96
0·1	29	44	58	87
0·2	27	40	53	80
0·3	25	37	49	74
0·4	23	34	46	69
0·5	21	32	43	64
0·6	20	30	40	60
0·7	19	28	38	56½
0·8	18	26½	36	53
0·9	17	25	34	50½
1·0	16	24	32	48
1·1	15	23	30	46
1·2	14½	22	29	44
1·3	14	21	28	42
1·4	13½	20	27	40
1·5	13	19	26	38
1·6	12½	18½	25	37
1·7	12	18	24	36
1·8	11½	17	23	34
1·9	11	16½	22	33
2·0	10½	16	21	32

Systematic working

*Table 8. Stops proportional to camera extension**

Camera Extension = v		v/48	v/64	v/96
in.	mm	⌀ mm	⌀ mm	⌀ mm
20	508	10·5	8·9	5·0
22	558	11·0	8·5	5·5
24	609	12·5	9·5	6·0
26	660	14·0	10·5	7·0
28	711	14·5	11·0	7·0
30	762	16·0	12·0	8·0
32	815	16·5	12·5	8·0
34	863	18·0	13·5	9·0
36	914	18·5	14·0	9·0
38	965	20·0	15·0	10·0
40	1016	20·5	15·5	10·0
42	1067	22·0	16·5	11·0
44	1118	23·0	17·5	11·5
46	1169	24·0	18·0	12·0
48	1219	25·0	19·0	12·5
50	1270	26·0	19·5	13·0
52	1320	27·0	20·5	13·5
54	1372	28·5	21·5	14·0
56	1422	29·0	22·0	14·5
58	1437	30·5	23·0	15·0
60	1524	31·0	23·5	15·5
62	1575	32·5	24·5	16·0
64	1626	34·0	25·5	17·0
66	1676	34·5	26·0	17·0
68	1728	36·0	27·0	18·0
70	1778	36·5	27·5	18·0
72	1828	38·0	28·5	19·0
74	1879	39·0	29·5	19·5
76	1931	40·0	30·0	20·0
78	1981	41·0	31·0	20·5
80	2032	42·0	31·5	21·0
82	2083	43·0	32·5	21·5
84	2133	44·5	33·5	22·0
86	2184	45·0	34·0	22·5
88	2235	46·5	35·0	23·0
90	2286	47·0	35·5	23·5
92	2336	48·5	36·5	24·0
94	2387	49·0	37·0	24·5
96	2439	50·5	38·0	25·0
98	2489	51·0	38·5	25·5
100	2540	52·5	39·5	26·0

* ⌀ mm = Diameter of stop in millimetres.

Controlled exposure

2. The inverse system

The treatment here is the reverse of the v/ratio system. The lens aperture becomes constant, set to an f/number which has been found to give the best field of coverage. This f/number is selected after a series of comparison negatives have been made using a lens testing card enlarged to the maximum working area of the camera. The exposure time is related to the camera extension (v) This is an attempt to keep the intensity of light constant, so that the time factor can be calculated in accordance with the original's density range and the length of the camera extension.

$$\text{Exposure} = \underset{\text{(constant)}}{\underset{\text{Intensity}}{\text{(Lens aperture)}}} \times \underset{\text{(varying)}}{\text{Time}}$$

This inverse system is controlled by the equation,

$$\text{New exposure time} = \text{Old exposure time} \left[\frac{(M_N + 1)^2}{(M_O + 1)^2} \right]$$

M_N = New magnification factor.
M_O = Magnification factor corresponding to the old exposure time.

The inverse system is applicable to equipment using light, e.g. a condenser-type enlarger.

Example. It has been found that at a magnification factor of 0·5 the correct exposure time is 10 light counts. To find the new exposure time when the camera is changed to 0·75 the equation is used in the following manner:

$$\text{New } E = \text{Old } E \left[\frac{(M_N + 1)^2}{(M_O + 1)^2} \right]$$

$$\text{New } E = 10 \left[\frac{(\tfrac{3}{4} + 1)^2}{(\tfrac{1}{2} + 1)^2} \right]$$

$$\text{New } E = 10 \left(\frac{49/16}{9/4} \right)$$

$$\text{New } E = 10 \left(\frac{196}{144} \right)$$

$$\text{New } E = 10 \times 1\cdot 3$$

$$\text{New } E = 13 \text{ light counts.}$$

Systematic working

A table of exposure multiplication factors linked with magnification figures is usually the fulfilment of this system.

Table 9. Exposure multiplication factors

Magnification (M)	Exposure factor (E)
20	110
19	105
18	90
17	81
16	72
15	64
14	56
13	49
12	42
11	36
10	30
9	25
8	20
7	16
6	12
$5\frac{1}{2}$	10
5	9
$4\frac{1}{2}$	7·6
4	6·3
$3\frac{1}{2}$	5·3
3	4
$2\frac{1}{2}$	3·1
2	2·3
$1\frac{3}{4}$	1·9
$1\frac{1}{2}$	1·6
$1\frac{1}{4}$	1·2
s/s	1·0
$\frac{3}{4}$	0·76
$\frac{1}{2}$	0·56
$\frac{1}{3}$	0·44
$\frac{1}{4}$	0·39
$\frac{1}{5}$	0·36
$\frac{1}{6}$	0·34
$\frac{1}{8}$	0·32
$\frac{1}{10}$	0·30
$\frac{1}{12}$	0·29

First find the exposure time at s/s and then multiply by the factor (E) corresponding to the magnification (M) required.
Equation for calculating values other than given below:

$$E = \frac{(M_N + 1)^2}{4}$$

Controlled exposure

Exposing continuous-tone emulsions

The problem is that we are confronted with an original which has a certain density range. This density range must be either compressed or expanded to a negative density range which suits the printing process. The best way to start is to take an average original density range, say 1·4 taking into account the flare factor, and carry out a series of test exposures, developing to a gamma of 1·0 until a negative is produced which has the required shadow density. Once this has been achieved 1·4 becomes the old density range and the correct exposure time found by the tests is termed the old exposure time. Now we have a base to work from. When a new original is encountered possessing a different density range than 1·4, the new exposure time can be calculated after the required gamma value has been found by the equation,

$$\text{Gamma value} \quad \gamma = \frac{\text{Required negative density range}}{\text{Original's density range}}$$

$$\therefore \gamma = \frac{1 \cdot 4}{2 \cdot 0} \rightarrow \text{N.B. Remember the degrading effect of an average flare factor of 1\%. The original density of 2·0 will become 1·70 in the camera.}$$

$$\text{Corrected value } \gamma = \frac{1 \cdot 4}{1 \cdot 7}$$

$$\gamma = 0 \cdot 82$$

The effect of a 1% lens flare factor can be appreciated by turning to the lens flare table 25 (Appendix). If the lens flare factor is greater than 1%, then the degradation of the original densities may be found in the following way:

Example. The reflection original has a shadow density of 2·0 the light reflected by this shadow tone is 1·0%. Add the flare factor number, in this case 4% thus producing a total reflection of 5%. The camera density becomes log 100/5·0 = 1·3. This can be applied to the highlight density. By subtracting the highlight density from the shadow density, the camera density range is found.

The correct development time for a gamma value of 0·82 can be found by consulting the appropriate Time/Gamma or Time/$\overline{G}$ curve.

When the development gamma is altered it has a slight effect upon the film speed. A general film speed correction value is, with a decrease of 0·1 γ, increase the log E units by

Systematic working

0·1. Similarly, with an increase of 0·1 γ, decrease the log E units by 0·1. Once a family of curves representing a constant exposure time related to different development times and their gamma values has been constructed. The film speed correction factor may be checked by drawing a horizontal line across the curves from a point on the ordinate which represents the required shadow density. At each point of intersection with the curves, perpendicular lines are drawn down to the relative log E values. The antilogs of the displacements between the relative log E values become exposure factors indicating the decrease or increase of the exposure time related to each gamma value.

| Log of the New exposure time | = | Log of the Old exposure time | + | New shadow density | − | Old shadow density |

N.B. Film speed correction value.

It will be found that if the new shadow density is less than the old, the log of the old exposure time is subtracted. Conversely, if the new shadow density is greater than the old, then the log of the old exposure time is added. This equation is also a practical proposition for calculating the exposure times when continuous-tone positives are being made.

Example. With a shadow density of 1·4 (including the degrading effect of a 1% flare factor) it has been found that an exposure time of 20 light counts gives the required negative shadow density of 0·3, while developing to a gamma of 1·0 produces a highlight density of 1·7, and a negative density range of 1·4. A new original is received which has a density range of 2·0, a camera density range of 1·7. The same printing process is being used so a negative density of 1·4 is still required. Using the previously mentioned equation the required gamma value would be 0·82 linked with the appropriate developing time. The next step is to calculate the correct exposure time, taking into account lens flare, the original's shadow density of 2·0 becoming 1·70.

Table 10. Exposure factors

									Base
Shadow densities of original	1·00	1·05	1·10	1·15	1·20	1·25	1·30	1·35	1·40
Opacity	10·0	11·2	12·6	14·1	15·9	17·8	20·0	22·4	25·1
Exposure factor	0·40	0·45	0·50	0·56	0·63	0·70	0·80	0·89	1·00
Log E difference	−0·40	−0·35	−0·30	−0·25	−0·20	−0·15	−0·10	−0·05	0·00

Controlled exposure

$$\frac{\text{Log of the New exposure time}}{\text{Log of the new } E} = \frac{\text{Log of the Old exposure time}}{\text{Log of the old } E} + \frac{\text{New shadow} + \text{density}}{+ Ds} - \frac{\text{Old shadow} - \text{density}}{- Ds}$$

$$\text{Log of the new } E = \frac{\text{Log of 20 light counts}}{1 \cdot 3} + 1 \cdot 7 - 1 \cdot 4$$

$$\text{Log of the new } E = 1 \cdot 3 + 0 \cdot 3$$

$$= 1 \cdot 6 \text{ Consult antilog tables}$$

Log of the
new E = 39·81
        1·51 +
       41·32

$$\frac{\text{New exposure}}{\text{time}} = \frac{41}{\text{light counts}}$$

N.B. Film speed correction value. $\gamma = 0.82$
Correction factor for the influence of development time on the film speed is, $\gamma\, 1\cdot00 - 0\cdot82 = 0\cdot18$ log E units consult antilog tables.
Correction factor = 1·51 light counts.

At the first view this may seem an enormous increase, but this is necessary if the required negative densities are going to be achieved in the short development time predicted for a gamma value of 0·82.

The above equation provides a very good guide to exposure times related to a new original density range. Taking our exposure of 20 light counts for a shadow density of 1·4 as a base of 1·00. It would be necessary to alter the exposure for every new original possessing a different shadow density. A table of general exposure factors may be constructed relying on the fact that exposure changes are for the most part proportional to opacity (the antilog of density). Density differences may also be recorded because logarithmic exposure counters are becoming quite common in camera design.

This table does not take into account such important factors as the emulsion's characteristics, reciprocity failure

1·45	1·50	1·55	1·60	1·65	1·70	1·75	1·80	1·85	1·90	1·95	2·00
3·2	31·6	35·5	39·8	44·7	50·1	56·2	63·1	70·8	79·4	89·1	100·0
1·10	1·20	1·40	1·60	1·80	2·00	2·20	2·50	2·80	3·00	3·50	4·00
0·05	+0·10	+0·15	+0·20	+0·25	+0·30	+0·35	+0·40	+0·45	+0·50	+0·55	+0·60

Systematic working

or the camera's flare factor. These deficiencies are considered in greater detail on page 248 under continuous-tone development.

Exposing lith-type emulsions

In the production of line and half-tone negatives the image is recorded on a 'lith' emulsion of extreme contrast. The lines and dots in the image must obtain maximum density. Therefore development can only be altered over a comparatively short time period, say $1\frac{3}{4}$ min. Alterations in line width are brought about mainly by variations in exposure time. Similar line negatives can be produced from originals of different density ranges by the employment of the following equation. This equation can also be applied to line contact work.

Log of the New exposure time = Log of the Old exposure time + New density range − Old density range

Example. An exposure time of 15 light counts has been found to give the required line result from an original having a density range of 1·60 (flare factor accounted for). A new original has a camera density range of 1·80, but the resultant line negative must be similar to the first. Calculation of the required exposure time is carried out in the following manner.

$$\frac{\text{Log of}}{\text{new } E} = \frac{\text{Log of}}{\text{old } E} + Dr_N - Dr_O$$

$$\frac{\text{Log of}}{\text{new } E} = \frac{\text{Log of 15 is}}{1\cdot 18} + 1\cdot 80 - 1\cdot 60$$

Log of new E = 1·18 + 0·20

Log of new E = 1·38 Consult antilog tables

$$\frac{\text{New exposure}}{\text{time}} = 24 \text{ light counts}$$

It must be stated here that the above formula only works over a short range of original densities; a minimum and maximum exposure time will be found related to a particular development time or technique. For example, if 15 light counts are found to be the correct exposure for a density range of 1·60 this would take up a central position when compared to a wide range of line originals.

These rather restrictive conditions are brought about by the peculiarities of lith emulsions coupled with the unique lith development. With this in mind please turn to page 275 on line reproduction.

Controlled exposure

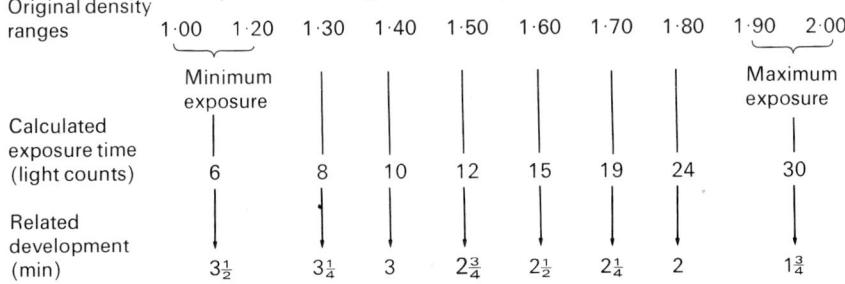

Example. Taken from a base of 1·60 = 15 light counts exposure with 2½ min development.

Exposing lith-type emulsions via a halftone screen

The problem of compressing or expanding the original's density range into a required negative density range is still with us in halftone work. The average maximum ink density ranges capable of being printed by the major printing processes can be tabulated.

Table 11. *Printing processes (density ranges)*

Process	Average ink reflection density range							Maximum ink density—single ink film
Lithography Average number of tonal steps	0·05, 1	0·1, 2	0·2, 3	0·4, 4	0·6, 5	0·8, 6	1·0, 7	1·4
	1·25, 8	1·3, 9	1·45 10					
Letterpress Average number of tonal steps	0·05, 1	0·15, 2	0·3, 3	0·45, 4	0·65, 5	0·85, 6		1·6
	1·05, 7	1·2, 8	1·35, 9	1·45, 10	1·55, 11	1·65 12		
Gravure Average number of tonal steps	0·05, 1	0·1, 2	0·2, 3	0·4, 4	0·6, 5	0·8, 6	1·0, 7	1·8
	1·2, 8	1·3, 9	1·4, 10	1·5, 11	1·6, 12	1·7, 13	1·8 14	
Screen process Average number of tonal steps	0·00, 1	0·2, 2	0·4, 3	0·6, 4	1·0, 5	1·2, 6	1·4, 7	2·0
	1·6, 8	1·8, 9	2·0 10					

Gravure printing in its conventional form uses continuous-tone emulsions, but in recent years halftone positives have been incorporated into the gravure *'invert halftone'* methods. Screen process can print the maximum ink density, but intermediate tonal graduations are limited.

Systematic working

The table indicates that if a faithful reproduction is to be achieved the original's maximum density range must compare favourably with the maximum density capability of the printing process selected for the reproduction. If the original's maximum density is greater than the maximum density range of the printing process the camera technician must lower the original's density range and produce negatives and positives possessing density ranges which suit the particular printing process. Of course, in doing so the reproduction will contain two deficiencies. Firstly the maximum shadow tone will appear as a visually lower density when compared to the original density, and secondly the intermediate tonal gradation will be lower in number and visually less distinct. These alterations are brought about by changing the density range and characteristic screening curve of a contact halftone screen, altering the lens aperture or screen distance if a glass halftone screen is being used (see page 315), varying the main exposure and increasing or decreasing the supplementary flash exposure. This short exposure to white light can expand or drastically compress the negative density range.

To summarize, any original tone lower than the maximum density of the printing process must be represented by the correct sized halftone dot, while any tonal value higher than the maximum density of the printing process should appear as the maximum halftone dot size, or in cases of extreme deviation as a solid area. Ideally, the camera operator should have a range of screens at his disposal, each one possessing a different reproducible density range, so that one halftone screen could be selected to match the maximum density range printed by the particular printing process on the quality of paper chosen for the reproduction.

The following equations give the camera technician a good idea of how much the main exposure and flash exposure should be altered to accommodate originals possessing different tonal ranges.

Main exposure

$$\frac{\text{Log of the}}{\text{New exposure time}} = \frac{\text{Log of the}}{\text{Old exposure time}} + \text{New highlight density} - \text{Old highlight density}$$

Flash exposure

$$\frac{\text{New}}{\text{flash exposure}} = \frac{\text{Basic}}{\text{flash exposure}} \left[1 - \text{antilog} \left(\frac{\text{screen's basic density range}} - \frac{\text{new original density range}} \right) \right]$$

Controlled exposure

Example. Once again a test has been conducted with a tone original possessing reflection densities of 0·05–1·45, a density range of 1·4. With the halftone screen arranged to produce dot sizes covering a normal tonal range, it was found that the correct main exposure was 100 light counts complemented by a supplementary flash exposure of 12 sec. On encountering a new original with reflection densities of 0·08–1·98, a density range of 1·9, the problem of finding the new main exposure and flash will be satisfied by the following equations.

Main exposure

$$\frac{\text{Log of new } E}{} = \frac{\text{Log of Old } E}{} + \text{Highlight } D_N - \text{Highlight } D_O$$

$$\frac{\text{Log of new } E}{} = \frac{\text{Log of 100 is}}{2·0} + 0·08 - 0·05$$

$$\frac{\text{Log of new } E}{} = 2·0 + 0·03$$

$$\frac{\text{Log of new } E}{} = 2·03 \quad \text{Consult antilog tables}$$

New exposure time = 107 light counts.

Flash exposure

$F = F_b [1 - \text{antilog} (D_S - D_O)]$
$F = 60 \text{ sec} [1 - \text{antilog} (1·3 - 1·9)]$
$F = 60 \text{ sec} [1 - \text{antilog} (-0·6)] = \overline{1}·4$. Consult antilog table $= 0·2512$
$F = 60 \text{ sec} [1 - 0·2512]$
$F = 60 \text{ sec} [0·7488]$
Flash Exposure $= 45$ sec

Screen distances for glass halftone screens

To produce dots of optimum quality with a glass screen there has to be a precise distance between the screen lines and the emulsion surface. When the light passes through the openings in the screen it is forced into cone shapes. The

Systematic working

apexes of these cones must be focused exactly on the emulsion surface of a contrast range similar to the original's is to be obtained on the negative and subsequent printed copies. This distance is determined by four factors:

1. The screen ruling—the number of lines per inch or centimetre.
2. The v/ratio number—v/32, v/48, v/64, v/96 or v/128.
3. The thickness of one face plate.
4. The contrast range required in the negative.

Once the screen ruling has been selected (see page 324) the screen distance required for a particular v-ratio number can be determined by the equation,

$$\frac{1}{\text{Ratio}} = \frac{\text{Screen Aperture}^* (SA)}{\text{Screen Distance} (SD)}$$

$$^*\text{Screen Aperture is } \frac{1}{\text{Twice the screen ruling}}$$

The screen gear like the lens barrel is marked off in millimetres to facilitate the setting of precise distances, so the answer needs to be converted into millimetres.

Example. A good quality paper is being used so a fine screen of 150 lines per in. can be employed. Exposure times have been determined on the v/64 ratio. so all that remains is the calculation of the screen distance.

$$\frac{1}{\text{Ratio}} = \frac{SA}{SD}$$

$$\frac{1}{64} = \frac{\frac{1}{300} \text{ in.}}{SD}$$

$$\frac{SD}{1} = \frac{1}{300} \times \frac{64}{1}$$

$$\frac{SD}{1} = 0.213 \text{ in.} \times 25.4 \text{ conversion factor for millimetres}$$

$$\frac{\text{Screen}}{\text{distance}} = 5.410 \text{ mm}$$

In practice, a slight correction must be made to allow for the thickness and refraction of the back cover glass.

Controlled exposure

Example. A good quality paper is being used so a fine screen of 60 lines per cm can be employed. Exposure times have been determined on the v/64 ratio, so all that remains is the calculation of the screen distance.

$$\frac{1}{\text{Ratio}} = \frac{SA}{SD}$$

$$\frac{1}{64} = \frac{\frac{1}{120}\text{ cm}}{SD}$$

$$\frac{SD}{1} = \frac{1}{120} \times \frac{64}{1}$$

$$\frac{SD}{1} = 0.53 \text{ cm} \quad \text{(To convert to millimetres move the decimal point back one place)}$$

$$\frac{\text{Screen}}{\text{distance}} = 5.3 \text{ mm}$$

In practice, a slight correction must be made to allow for the thickness and refraction of the back cover glass.

The screen distance will have to be altered if a new v/ratio number is used. Generally speaking, the variations in screen distance are similar to the following examples.

v/32

$$\frac{1}{\text{ratio}} = \frac{\text{Screen Aperture}}{\text{Screen Distance}}$$

$$\frac{1}{32} = \frac{\frac{1}{120}\text{ cm}}{SD}$$

$$\frac{SD}{1} = \frac{1}{120} \times \frac{32}{1}$$

$$\frac{SD}{1} = 0.266 \text{ cm}$$

$$SD = 2.66 \text{ mm}$$

v/48

$$\frac{1}{\text{ratio}} = \frac{\text{Screen Aperture}}{\text{Screen Distance}}$$

$$\frac{1}{48} = \frac{\frac{1}{120}\text{ cm}}{SD}$$

$$\frac{SD}{1} = \frac{1}{120} \times \frac{48}{1}$$

$$\frac{SD}{1} = 0.40 \text{ cm}$$

$$SD = 4.0 \text{ mm}$$

Convert to millimetres by moving the decimal point back one place.

v/64

$$\frac{1}{\text{ratio}} = \frac{\text{Screen Aperture}}{\text{Screen Distance}}$$

$$\frac{1}{64} = \frac{\frac{1}{120}\text{ cm}}{SD}$$

v/96

$$\frac{1}{\text{ratio}} = \frac{\text{Screen Aperture}}{\text{Screen Distance}}$$

$$\frac{1}{96} = \frac{\frac{1}{120}\text{ cm}}{SD}$$

v/128

$$\frac{1}{\text{ratio}} = \frac{\text{Screen Aperture}}{\text{Screen Distance}}$$

$$\frac{1}{128} = \frac{\frac{1}{120}\text{ cm}}{SD}$$

Systematic working

$$\frac{SD}{1} = \frac{1}{120} \times \frac{64}{1} \qquad \frac{SD}{1} = \frac{1}{120} \times \frac{96}{1} \qquad \frac{SD}{1} = \frac{1}{120} \times \frac{128}{1}$$

$$\frac{SD}{1} = 0.533 \text{ cm} \qquad \frac{SD}{1} = 0.80 \text{ cm} \qquad \frac{SD}{1} = 1.06 \text{ cm}$$

$$SD = 5.33 \text{ mm} \qquad SD = 8.0 \text{ mm} \qquad SD = 10.6 \text{ mm}$$

In practice, settings have the same ratio relationship as the lens apertures; e.g., in the case of the $v/64$ ratio, the lens aperture is $\frac{1}{64}$ of the camera extension and the screen distance is 64 times the screen aperture.

	$v/32$	$v/48$	$v/64$	$v/96$	$v/128$
Screen distance	2.7 mm	4.0 mm	5.4 mm	8.0 mm	10.8 mm
Lens aperture	25.4 mm	16.76 mm	12.7 mm	8.38 mm	6.35 mm

If the method of printing and quality of paper demand a change of screen ruling, then the screen distance must be altered to suit the new screen ruling. If one v/ratio number is being used, say $v/64$, the degree of alteration in the screen distance to suit a selection of different screen rulings would adhere closely to the calculations below.

Systematic working—Screen distances for glass halftone screens

Lines per centimetre

20   30

$$\frac{1}{\text{ratio}} = \frac{\text{Screen Aperture}}{\text{Screen Distance}} \qquad \frac{1}{\text{ratio}} = \frac{\text{Screen Aperture}}{\text{Screen Distance}}$$

$$\frac{1}{64} = \frac{\frac{1}{40} \text{ cm}}{SD} \qquad \frac{1}{64} = \frac{\frac{1}{60} \text{ cm}}{SD}$$

$$\frac{SD}{1} = \frac{1}{40} \times \frac{64}{1} \qquad \frac{SD}{1} = \frac{1}{60} \times \frac{64}{1}$$

$$\frac{SD}{1} = 1.6 \text{ cm} \qquad \frac{SD}{1} = 16.07 \text{ cm}$$

Convert to millimetres by moving the decimal point back one place.

$$SD = 16.0 \text{ mm} \qquad\qquad SD = 10.7 \text{ mm}$$

Controlled exposure

40

$$\frac{1}{\text{ratio}} = \frac{\text{Screen Aperture}}{\text{Screen Distance}}$$

$$\frac{1}{64} = \frac{\frac{1}{80}\,\text{cm}}{SD}$$

$$\frac{SD}{1} = \frac{1}{80} \times \frac{64}{1}$$

$$\frac{SD}{1} = 0.80\,\text{cm}$$

$$SD = 8.0\,\text{mm}$$

48

$$\frac{1}{\text{ratio}} = \frac{\text{Screen Aperture}}{\text{Screen Distance}}$$

$$\frac{1}{64} = \frac{\frac{1}{96}\,\text{cm}}{SD}$$

$$\frac{SD}{1} = \frac{1}{96} \times \frac{64}{1}$$

$$\frac{SD}{1} = 0.66\,\text{cm}$$

$$SD = 6.6\,\text{mm}$$

60

$$\frac{1}{\text{ratio}} = \frac{\text{Screen Aperture}}{\text{Screen Distance}}$$

$$\frac{1}{64} = \frac{\frac{1}{120}\,\text{cm}}{SD}$$

$$\frac{SD}{1} = \frac{1}{120} \times \frac{64}{1}$$

$$\frac{SD}{1} = 0.53\,\text{mm}$$

$$SD = 5.3\,\text{mm}$$

80

$$\frac{1}{\text{ratio}} = \frac{\text{Screen Aperture}}{\text{Screen Distance}}$$

$$\frac{1}{64} = \frac{\frac{1}{160}\,\text{cm}}{SD}$$

$$\frac{SD}{1} = \frac{1}{160} \times \frac{64}{1}$$

$$\frac{SD}{1} = 0.40\,\text{cm}$$

$$SD = 4.0\,\text{mm}$$

Lines per inch

50

$$\frac{1}{\text{ratio}} = \frac{\text{Screen Aperture}}{\text{Screen Distance}}$$

$$\frac{1}{64} = \frac{\frac{1}{100}\,\text{in}}{SD}$$

$$\frac{SD}{1} = \frac{1}{100} \times \frac{64}{1}$$

$$\frac{SD}{1} = 0.64\,\text{in} \times 25.4$$

$$SD = 16.25\,\text{mm}$$

75

$$\frac{1}{\text{ratio}} = \frac{\text{Screen Aperture}}{\text{Screen Distance}}$$

$$\frac{1}{64} = \frac{\frac{1}{150}\,\text{in}}{SD}$$

$$\frac{SD}{1} = \frac{1}{150} \times \frac{64}{1}$$

$$\frac{SD}{1} = 0.426\,\text{in} \times 25.4$$

$$SD = 10.82\,\text{mm}$$

Systematic working

100

$$\frac{1}{\text{ratio}} = \frac{\text{Screen Aperture}}{\text{Screen Distance}}$$

$$\frac{1}{64} = \frac{\frac{1}{200}\text{ in}}{SD}$$

$$\frac{SD}{1} = \frac{1}{200} \times \frac{64}{1}$$

$$\frac{SD}{1} = 0.32 \text{ in} \times 25.4$$

$$SD = 8.12 \text{ mm}$$

120

$$\frac{1}{\text{ratio}} = \frac{\text{Screen Aperture}}{\text{Screen Distance}}$$

$$\frac{1}{64} = \frac{\frac{1}{240}\text{ in}}{SD}$$

$$\frac{SD}{1} = \frac{1}{1} \times \frac{64}{1}$$

$$\frac{SD}{1} = 0.266 \text{ in} \times 25.4$$

$$SD = 6.75 \text{ mm}$$

150

$$\frac{1}{\text{ratio}} = \frac{\text{Screen Aperture}}{\text{Screen Distance}}$$

$$\frac{1}{64} = \frac{\frac{1}{300}\text{ in}}{SD}$$

$$\frac{SD}{1} = \frac{1}{300} \times \frac{64}{1}$$

$$\frac{SD}{1} = 0.213 \text{ in} \times 25.4$$

$$SD = 5.40 \text{ mm}$$

200

$$\frac{1}{\text{ratio}} = \frac{\text{Screen Aperture}}{\text{Screen Distance}}$$

$$\frac{1}{64} = \frac{\frac{1}{300}\text{ in}}{SD}$$

$$\frac{SD}{1} = \frac{1}{400} \times \frac{64}{1}$$

$$\frac{SD}{1} = 0.16 \text{ in} \times 25.4$$

$$SD = 4.06 \text{ mm}$$

These equations will give a good idea of how screen distances are varied in accordance with the particular situation. There are other important considerations, such as the interpretation of the halftone theory which is being used as a basis for the dot formation. This and other contributory factors are covered in the chapter on tone reproduction.

Exposing through coloured filters

A colour filter is a piece of transparent material (dyed glass, plastic or gelatine) that transmits light of its own colour and absorbs light of other wavelengths. High quality colour printing is the collective result of at least three separate printing plates. Their effectiveness and faithfulness to the original depends largely upon the accuracy of the initial photographic colour separations. These photographs, which are representations of the original divided into three basic colours, yellow, magenta and cyan, are produced separately

Controlled exposure

by exposing three panchromatic films through three coloured filters, which transmit blue, green and red light respectively. Exposures through coloured filters are also required when monochrome reproductions are being made from coloured, toned or stained originals.

As soon as a filter is introduced into the path of light travelling from the original to a light-sensitive emulsion a considerable amount of the light is lost through the filter's selective absorption. This means that the exposure time which was correct for white light will be hopelessly too short with the filter. The amount of increase in exposure time with the filter depends upon the filter's transmission, the emulsion's spectral sensitivity and the colour temperature of the light source. There are two systems which indicate the required increase in exposure time when filters are being used, firstly filter factors numbers and secondly filter ratio numbers.

Filter factor

This is a number which indicates the amount an exposure should be increased when using a particular filter. This factor number is employed as a multiplier on the white light exposure that would be correct under the same photographic conditions.

Example.
White light: factor No. 1;
 Exposure time: $1 \times 10^* = 10$ Light counts
Red filter: factor No. 5;
 Exposure time: $5 \times 10^* = 50$ Light counts
Green filter: factor No. 7;
 Exposure time: $7 \times 10^* = 70$ Light counts
Blue filter: factor No. 10;
 Exposure time: $10 \times 10^* = 100$ Light counts

* Found by experiment

Filter ratio

This is a number which indicates the increase in exposure time needed for the various filters. Filter ratios take the correct exposure time through a tri-colour red filter as their base. Once this correct red filter exposure has been found, the exposure time for the other filters can be determined by multiplying the correct red filter exposure time by the appropriate filter ratio number.

Example.
Red filter: ratio No. 1;
 Exposure time: $1 \times 25^* = 25$ Light counts

Systematic working

Green filter: ratio No. 1·4;
Exposure time: 25 × 1·4* = 35 Light counts
Blue filter: ratio No. 0·9;
Exposure time: 25 × 0·9* = 22 Light counts

* Found by experiment

Filter factor and ratio numbers are listed for all emulsion types by the manufacturer, but their photographic conditions may differ from the users; therefore it is a sensible act to find 'house' factors using the manufacturer's numbers as starting points. It is also good practice to expose colour separation negatives in the same order every time; red filter, green filter and blue filter.

Controlled development

The amount of silver blackening obtained in a light-sensitive emulsion after exposure depends largely upon the developer's consistency, agitation rate and temperature throughout the required development time—in other words a factor formula of dilution × agitation × temperature × time.

If developing solutions are purchased in powder form, careful attention must be given to the developer's formula, the mixing sequence and ensuring that all the chemicals are completely dissolved. Once in liquid form, developing solutions should be stored in large plastic vats fitted with tap pourers and air-tight lids. The effectiveness of the developing solution decreases with the number of films passed through it; therefore fresh developer needs to be added with each additional film. With dish development the effectiveness of the developing solution is kept more uniform if half of the old developer is removed and replaced with fresh developing solution. This must be accompanied by another important condition, one that is often forgotten—the same quantity of developer should always be used for a particular film size.

To achieve complete standardization in the development stage the answer is a developing machine, either a nitrogen burst tank installation or an automatic roller-fed processor. If these machines are employed, replacement of the developer becomes unnecessary. Controlled replenishment is carried out, related to either the film size or film size plus the area of developed silver. These machines hold the development factors of dilution, agitation rate and temperature in a tight grip of constancy, while the gamma value and maximum density results are controlled by the development time.

Developing continuous-tone emulsions

The make-up of a continuous-tone photographic image

An organized glance

may be dissected into two major sections—the light-densities in a negative or positive result which are mainly produced by the exposure time, and the heavy densities which are controlled by the development time. The transmission density or contrast range recorded on the continuous-tone negative compared to the reflection or transmission density range of the original is termed the gamma value (see page 334). Once the required gamma value has been found the correct development time can be determined by consulting a time gamma curve which has been constructed and tested under the prevailing conditions.

Developing lith-type emulsions

In graphic reproduction the halftone process—the translation of continuous-tone originals into dots of varying size—is required for most printing processes. The appropriate dot formation is exposed on to a lith-type emulsion. The gamma value is not normally altered a great deal because the dots have to reach a maximum density and optimum quality. Once again effective development follows the factor formula of dilution × agitation × temperature × development time. The precise development time producing optimum dot density and quality can be determined by carrying out a series of test halftone negatives.

Example. Seven lith halftone negatives with the addition of a continuous-tone stepwedge are exposed under identical conditions. Development is carried out varying the time factor for each negative.

1	2	3	4	5	6	7
2 min	$2\frac{1}{4}$ min	$2\frac{1}{2}$ min	$2\frac{3}{4}$ min	3 min	$3\frac{1}{4}$ min	$3\frac{1}{2}$ min

After processing, each test negative is carefully examined with a microscope for dot quality. Once the optimum negative has been selected the lowest density recorded on the continuous-tone stepwedge is noted. This density figure becomes the standard evaluation for all subsequent results. If this density is achieved from the same step on the continuous-tone stepwedge on every exposed and developed lith emulsion, optimum dot quality will be achieved. This system can also be used to evaluate line negatives and positives on lith emulsions.

An organized glance

Let us attempt to organize most of these procedures, systems and equations into a comprehensible whole by summarizing them as they are applied to graphic reproduction photography.

Systematic working

Table 12. Systematic working

Stage	Method	Organized steps		
		1	2	3
Original	Procedure	Classification	Grouping	Densities readings
	System	Period of familiarization, record aim points, compare with appropriate printing standard	In accordance with magnification factor, check this and final size, work-out camera extension	Find the density range of the original, mask out the quality control-strip to the same values
	Equation	Original's characteristics = Process' capabilities	$M = \dfrac{I}{O}$ or $\dfrac{V}{U}$ $V = f(1 + M)$	Original's densities = Exposure times
Illumination	Procedure	Lamp distance and angle	Uniformity	Fluctuations
	System	Use angled rules to set the lamps, these settings can be related to the magnification factor	Check this by taking a series of readings with a luxmeter	Use voltage control and a light integrating meter
	Equation	$R.I. = \left(\dfrac{\text{Old dist}}{\text{New dist}}\right)^2 \times 100\%$	$\dfrac{\text{Total}}{\text{area}} = \dfrac{\text{Division into nine lux readings}}{}$	Constant amount of light = One light count
Exposure	Procedure	Exposure	v/Ratio system	Inverse system
	System	Intensity and Time used as exposure factors	A constant exposure time related to a varying lens aperture	A constant lens aperture related to a varying exposure time
	Equation	Intensity × Time	$\dfrac{1}{\text{Ratio}} = \dfrac{\text{Lens aperture}}{\text{Camera extension}}$	$\dfrac{\text{New exposure time}}{\text{Old exposure time}} = \left[\dfrac{(M_N + 1)^2}{(M_O + 1)^2}\right]$

An organized glance

		Line repro	Halftone repro	Con/tone repro
Exposure classified	Procedure			
	System	Original's density range recorded and compared to previous results	Original's density range and maximum density compared and used to calculate the new exposure times	Original's density range plus the required gamma value used to find the new exposure time
	Equation	$\text{Log of New } E = \text{Log of Old } E + Dr_N - Dr_O$	Reproducible density range of the screen = Main exposure $\text{Log of New } E = \text{Log of Old } E + \text{Highlight density}_N - \text{Highlight density}_O$ Flash exposure: $F = F_B[1 - \text{antilog}(D_S - D_O)]$	$\text{Log of New } E = \text{Log of Old } E + D_N^{Shadow} - D_O^{Shadow}$ N.B. Take into account Lens flare factor and Film speed correction
Exposure classified	Procedure	Contact work	Direct colour repro	Indirect colour repro
	System	Transposing a negative image to a positive image without undue image variation	The halftone repro system is used applied to the red filter exposure, once this has been established, filter ratios are used	The continuous tone system is employed with the red filter exposure as the base, then filter ratios
	Equation	$\text{Log of New } E = \text{Log of Old } E + \text{Max } D_N - \text{Max } D_O$	The same equation as above*	The same equation as above*
Processing development	Procedure	Development	Line and H/t emulsions	Con/tone emulsions
	System	Dilution, agitation rate, temperature and development time are the important factors	The amount of variation in development time is limited and needs to be calculated with the use of a continuous tone stepwedge	Development time mainly determines the gamma value. This is shown on a time/gamma curve
	Equation	$D \times A \times T \times \text{Time}$ constant and standardized × varying factor	$\text{Development} = \dfrac{\text{Density}}{\text{time}}$ correlated with optimum dot and line quality	$\text{gamma value } \gamma = \dfrac{\text{Required negative density range}}{\text{Original's density range}}$ N.B. Flare factor = Camera density range

Systematic working

A matter of economics

For economic reasons the printing of this book has been confined to one process—lithography, so all the illustrations and diagrams have been reproduced by the photo-lithographic method. The following experiments and results are for each of the major printing processes. Unfortunately this means that the appropriate line, tone and colour printing standards for gravure, letterpress and screen process have been represented by simulated results. Of course this cannot be compared to viewing, measuring and analysing the actual result produced on a selection of papers by the correct printing method with the precise ink formulation under the prevailing conditions, but it is hoped that these simulated results and the methods of obtaining information from them will spur the reader to produce his own printing standards for the process in which he is involved.

Quality control strip

We have now reached the point where the application of systematic working should be extended to the remaining stages of the reproduction process. A photographer must be aware of the capabilities and limitations of the printing process if he is to produce photographic images which are really suitable and utilize the attributes of the process to the fullest extent. One way of obtaining this information is to carry out a series of experimental printings using a control-strip to test the capabilities of the process.

This control-strip should contain images which will prove to be exacting tests for line, tone and colour reproduction. A comprehensive control-strip can be used as a resolution-tester, exposure-calculator and ink-film thickness indicator. It needs to be reasonable small so that it can print in the trim areas of all subsequent reproductions, providing a means of checking, comparing and controlling quality. There are many control strips currently available and the one employed in the following experiments embraces most of the popular features.

A line standard

Produce a final line photograph for the individual processes retaining the two extreme ends of the control strip, namely,

(1) fine black lines surrounded by white;
(2) minute white areas surrounded by black; and
(3) a line stepwedge—black lines becoming progressively finer.

Quality control strip

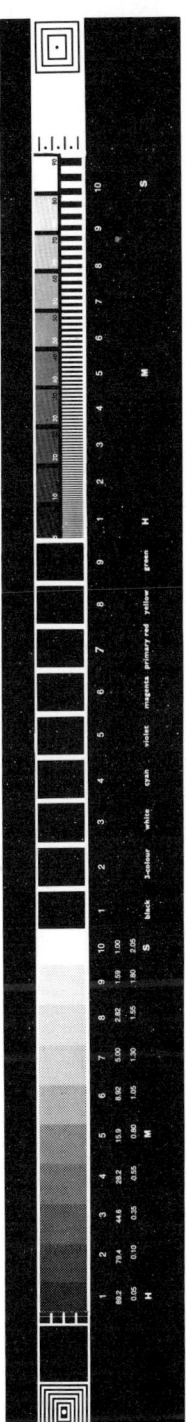

Negative: Transmission control strip

Positive: Reflection control strip

Shadow control

Fine tonal control wedge

Fine line resolution wedge

Solid colour patches

Continuous tone step wedge 'masked-off to correspond with the original picture'

Highlight control

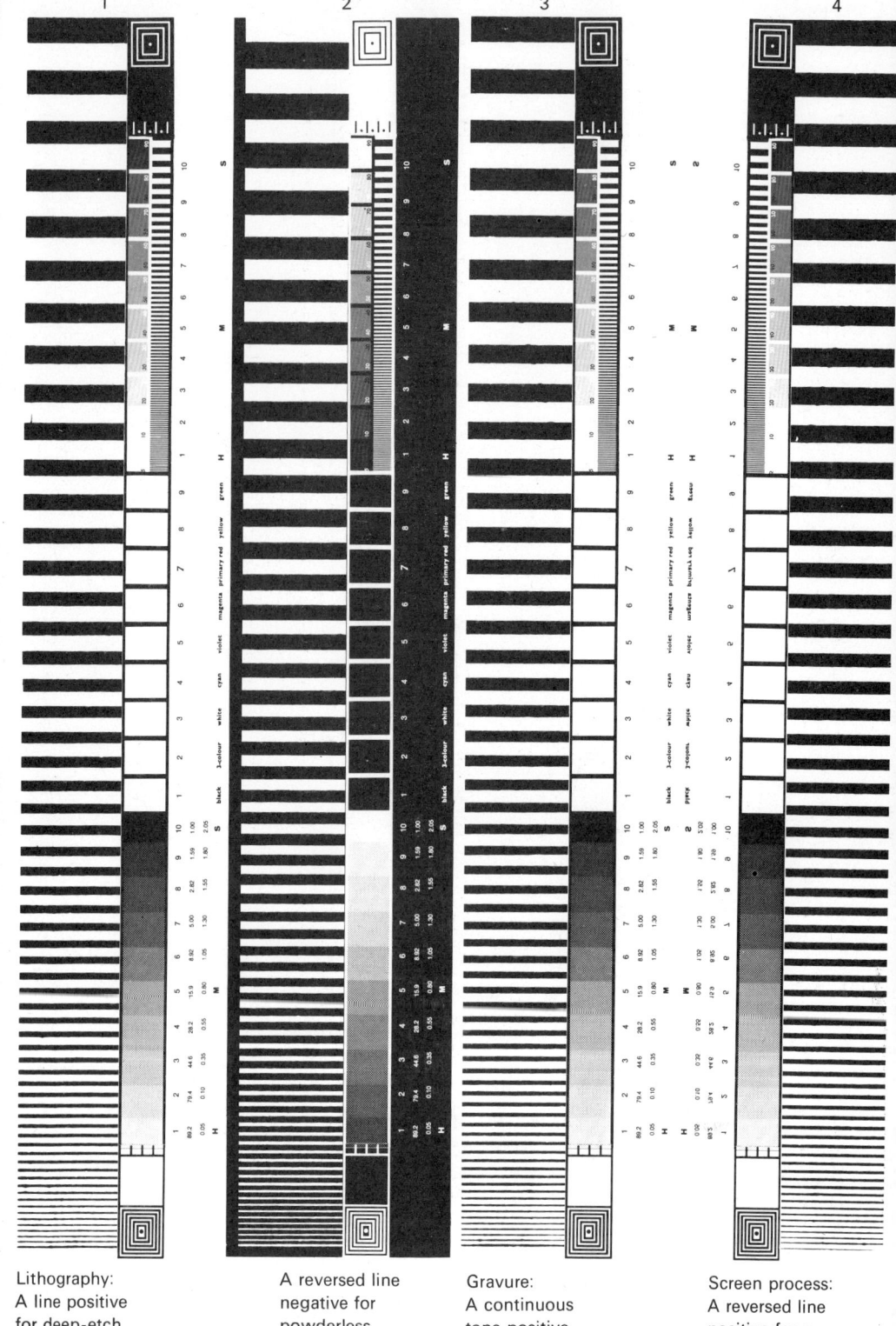

1. Lithography: A line positive for deep-etch plate making
2. A reversed line negative for powderless etching on zinc
3. Gravure: A continuous tone positive for conventional etching on copper
4. Screen process: A reversed line positive for a photographic stencil on a nylon mesh

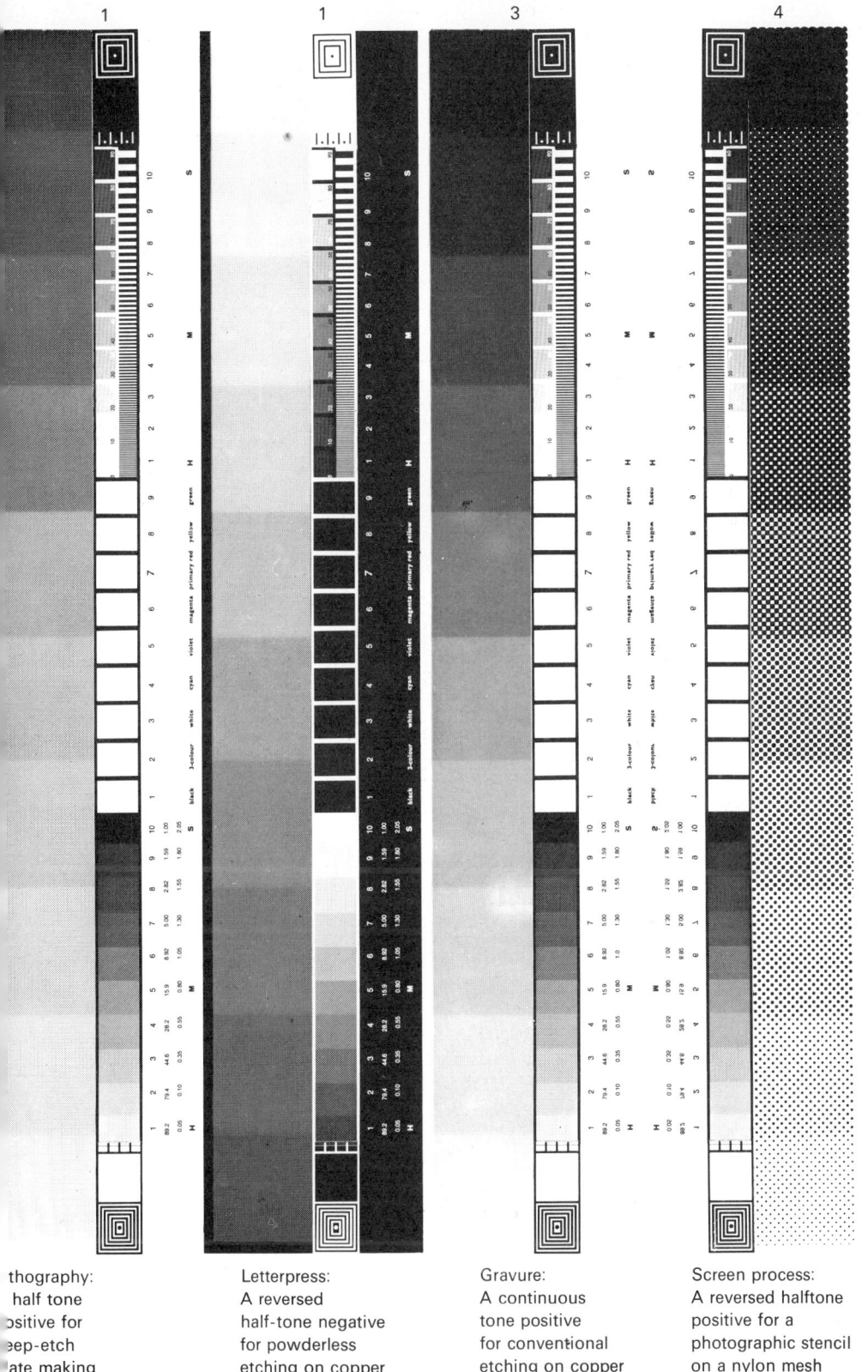

Lithography:
A half tone
positive for
deep-etch
plate making

Letterpress:
A reversed
half-tone negative
for powderless
etching on copper

Gravure:
A continuous
tone positive
for conventional
etching on copper

Screen process:
A reversed halftone
positive for a
photographic stencil
on a nylon mesh

Systematic working

The photographic images should be similar to those on the previous two pages.

After controlled plate and stencil-making procedures have been executed using continuous-tone sensitivity guides, printing may commence using three grades of paper representing the range usually encountered, e.g. machine finished (MF), supercalendered (SC) and coated paper (art). Ink film thickness needs to be recorded by taking reflection density readings from a solid black patch. In every case maximum ink film thickness should be aimed for, but the degree achieved must be conducive to easy and straightforward printing.

From these experiments it can be established what is the minimum width of a single black line and area of white surrounded by black possible on a particular paper or material, once the printing process has been selected. This is extremely helpful and applicable when large line originals are reduced to a small printing size, a necessary step in map-production, technical illustration and printed circuitry.

A tone standard

Once the line standards have been produced measurement of the maximum ink density capability of the process being investigated is possible. As previously stated from current experiments and recorded data, under favourable conditions and on good quality paper the maximum reflection densities of the individual processes appear to set at average figures of: lithography 1·4; letterpress 1·6; gravure 1·8; and screen process 2·0. To extend the individual processes use a control strip containing a tonal stepwedge ranging from the maximum density capability of the process down to white, plus fine lines to check resolution and ink film thickness. Such control strips will appear as follows:

Printing methods employing the halftone process need a final photographic image which presents the maximum density as a solid area and all the successive tonal movements by graduated dot sizes until the white step is represented by a very small dot or no dot at all. One screen ruling can be selected and printed on various grades of paper, or a number of control strips may be produced, each one containing a different screen ruling which will eventually, after printing, associate itself with one grade of paper to produce the optimum tonal reproduction (see page 323).

It will be noticed that the reproductions are not exact renditions of the original stepwedge. This is mainly caused by

Quality control strip

the characteristic screening curve of the halftone screen, the number of times the image is duplicated from original to printed copy and the method of ink transference during the printing operation. All these deviations add up to considerable tonal distortion. The degree of distortion in these examples has been kept to a minimum. This was made possible by counteracting it at the negative stage, mainly by utilizing the toe of the characteristic curve. Printed results such as these on different grades of paper give the photographer an immediate visual guide and measurable comparison for any similar work in the future.

Tracing tonal distortion

True appreciation of this work begins by making an elementary inspection of the tonal and image duplication taking place throughout one particular process. By recording density and dot size readings at each tonal movement or image duplication the amount of tonal distortion can be represented by a curve related to the original reflection density readings, whose relationship and movement we wish to reproduce faithfully on the printed sheet. These curves are drawn in quadrant diagrams made famous by L. Jones in his classic work on tone reproduction. One quadrant is used to represent each movement throughout a process.

We will now look at a simple quadrant diagram illustrating three basic movements in graphic reproduction.

$$\text{Original} \rightarrow 1, \frac{\text{Continuous-tone}}{\text{negative image}} \rightarrow 2, \frac{\text{Halftone}}{\text{positive image}} \rightarrow 3, \text{Printed image}$$

In the first diagrams the tonal densities of the original were exposed and developed so that they appeared exactly on the straightline portion of the negative emulsion's characteristic curve—the region of correct tonal reproduction, but because of the halftone screen's characteristic curve and the method of ink transference a considerable amount of tonal distortion has taken place when the actual result is compared with the ideal reproduction curve—a straight line, in the last quadrant.

Once the degree of distortion is known, the next step is to minimize it. The characteristic curves of the halftone screen, positive–negative emulsions and method of ink transference

Systematic working

Quadrant diagram 1:
Tonal distortion

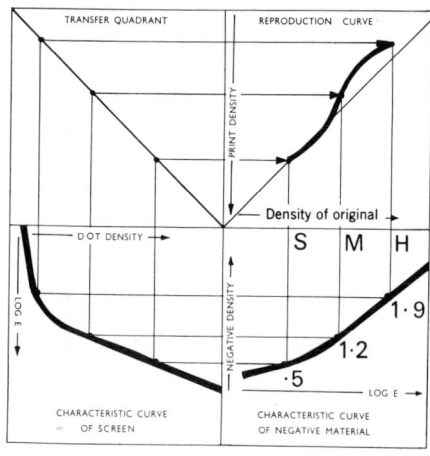

Pictorial quadrant 1:
Copy ⟷ Original
↑   ↓
Positive ← Negative

are set, so we have to make the best of them. Take the reference points on the ideal reproduction curve and track them back. This will give a very good indication of where the original's densities should appear on the negative emulsion to counteract or minimize the subsequent distortions.

A colour standard

In colour printing a precise combination of dots of varying sizes printed in basic colours is needed to reproduce the coloured hues, tints, and shades apparent in the original. It is fairly obvious that before work of this nature can be carried out with certainty, the printing capabilities and limitations of

Quadrant diagram 2:
Corrected distortion

A colour standard

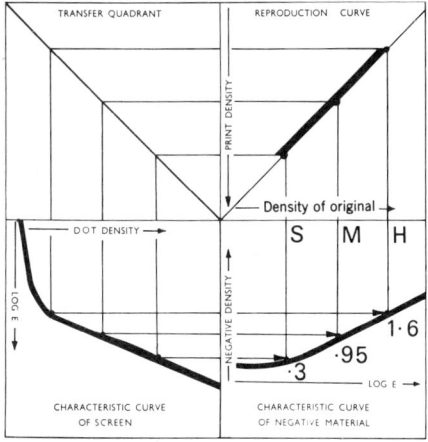

Pictorial Quadrant 2:
Copy ⟷ Original
↑   ↓
Positive ← Negative

the selected printing process need to be ascertained. To do this we must produce a number of tonal stepwedges which will show us the capabilities of the process in the form of an ink chart. These stepwedges should possess similar tonal ranges to those used to produce a tone standard for the process, the only difference being that we need two identical halftone stepwedges made at different screen angles superimposed on a third printing which is one of the steps (a particular dot size or tonal value) appearing as a flat tint. For an appropriate printing example we are using halftone stepwedges suitable for lithographic printing.

By printing these stepwedges in a suitable fashion any combination of the basic inks can be seen using different

Systematic working

Cyan stepwedge 75°

Magenta stepwedge 15°

Yellow tints 90°, black tints 45°

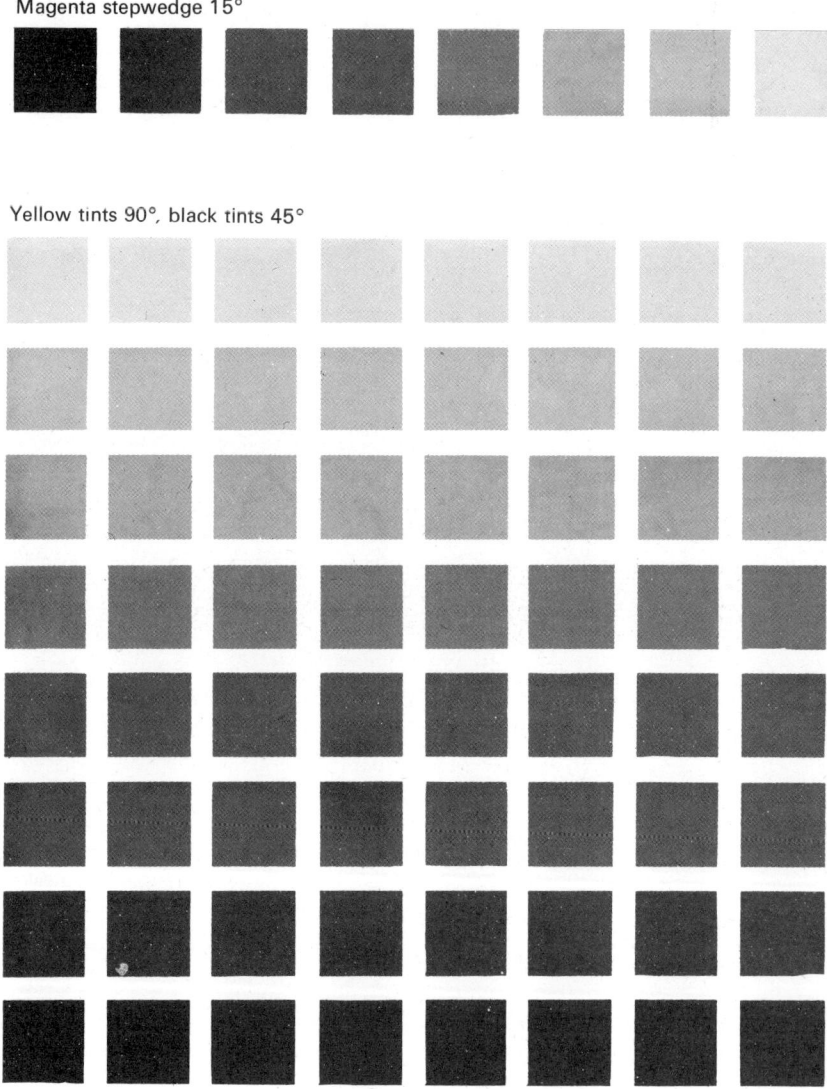

A colour standard

DUPLICATION OF IMAGES
e.g. CYAN stepwedge, MAGENTA stepwedge, YELLOW 50% tint dot size, BLACK 10% tint

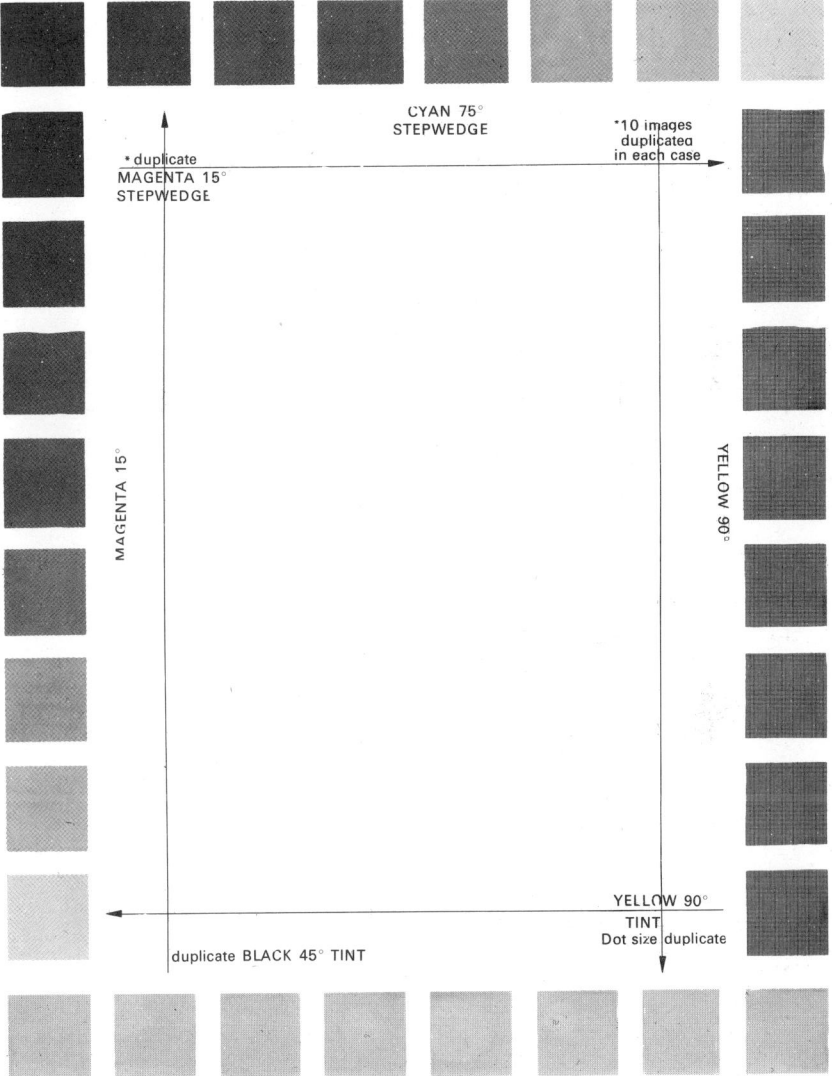

RESULT: 10 pages each one a different dot size of YELLOW (with or without black printing)

Systematic working

dot sizes or tonal values (conventional gravure). This work begins by using a basic block format and stripping the appropriate and corresponding stepwedges into their printing positions. See previous page.

Extra stepwedges, representing black variations, may be printed on top of the colour combinations at a screen angle of 45°, or a set of film positives with dot sizes similar to the set of cyan flat tints can be made and the appropriate dot size selected and just placed on top of the colour combination. The printing test should be as comprehensive as possible, providing colour charts which represent results on different grades of paper, using different printing sequences, wet on wet, dry on dry, and with the range of tri-chromatic inks used in the particular company. The illustrated example will give a good idea of the appearance of such ink charts.(*right*).

Charts such as these are invaluable. They allow colour keys to be drawn from individual coloured originals which at the very initiation of the reproduction indicate the continuous-tone densities and halftone dot sizes required in the subsequent negatives and positives, so that the printing operation can reproduce colours which are as near facsimile as the capabilities of the process allow.

Predicting the colour separations

The separation negatives need two types of correction; colour correction, because the tri-chromatic inks carry out unwanted absorptions, and tonal correction, which is caused by the amount of image duplication and characteristic deviations of the process. Once a colour key has been produced the colour charts can be compared with the original and the continuous-tone densities and halftone dot sizes required to reproduce the important areas of the original recorded on the colour key. This information can be used to indicate the degree of colour correction needed to compensate for the deficiencies of the inks. The colour correction is usually carried out by making photographic masks—images which increase the amount of colour separation. For example, if we consider a particular coloured area of an original, a green area, by comparing it against the ink charts, we find a close colour match at the following figures:

	Yellow printer	*Magenta printer*	*Cyan printer*
Green area negative, continuous-tone densities	0·3	1·6	0·6
Positive halftone dot sizes which will combine to produce the required green	90%	10%	75%

A colour standard

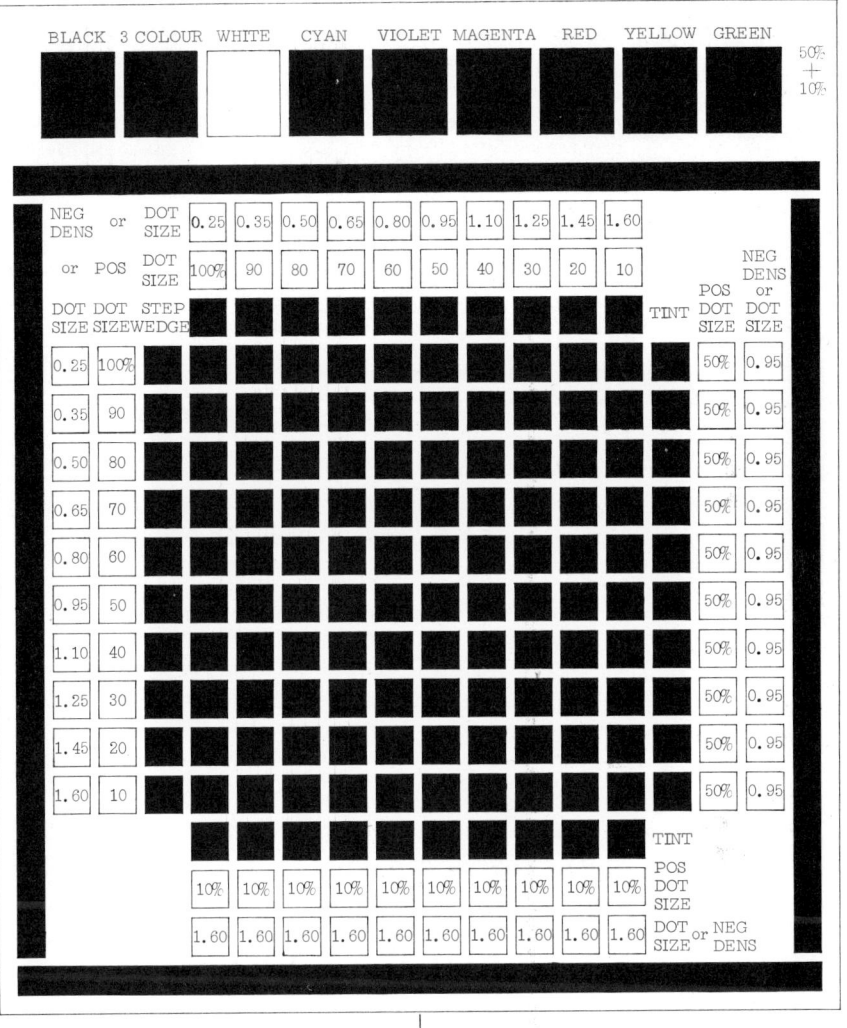

Systematic working

After suitable colour separation photography negatives are produced, which for colour and tonal rendering are as near ideal as is possible with available materials, we can now record the continuous-tone densities and compare them with the required figures. This will indicate the colour correction mask strength in this specific green area for the three separations.

	Yellow printer	Magenta printer	Cyan printer
Green area separation negatives, continuous-tone densities	0·3	1·3	0·35
For the magenta printer a mask strength of	—	0·3	—
For the cyan printer a mask strength of	—	—	0·25
Mask + separation negative	0·3	1·6	0·60

Of course this is a simple example, but it does indicate the value of the colour charts. If we know what the required densities and dot sizes are before we start, then action can be taken to ensure they are achieved before printing commences. Colour correction is rather a complex subject so it is important that these elementary steps are taken before embarking on the theory and practical systems described in the chapter on colour reproduction.

Predicting the tonal separations

Once the printing and halftone screening characteristics

Quadrant diagram:
1. Transfer quadrant
   The highlight (H), middletone (M) and shadow (S) areas of the original are transferred to printing densities. In practice, the limitations of the printing process must be taken into consideration. This usually means a compression of the original's tonal range

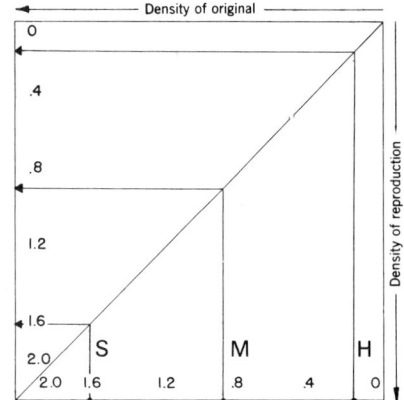

A colour standard

2. Quadrant—required dot sizes:
   The H, M and S areas are compared to the ink charts. Each area is divided into its appropriate dot size for yellow, magenta and cyan inks. These required dot percentages are plotted

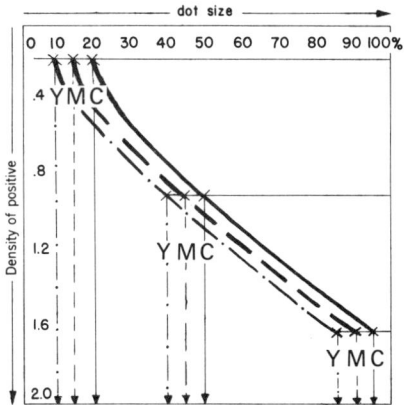

are established they can be used along with the original's tonal densities to predict the required separation negatives. Again, similar to tone reproduction, quadrant diagrams are employed to link the major stages of the process, starting with with the original. Reflection or transmission readings are taken from the highlight, middletone and shadow areas of the original. These readings are transferred to a grey step-wedge positioned alongside the original and are referred to as the H, M and S reference points. These neutral areas are compared against the colour charts until a suitable match is found and the required dot sizes are noted. From this information the first quadrant may be constructed.

The next step is to take the required dot sizes indicated by the colour charts for each printing plate and plot them against

3. Quadrant—characteristic screening curve:
   The individual dot values are deflected by the screen's curve

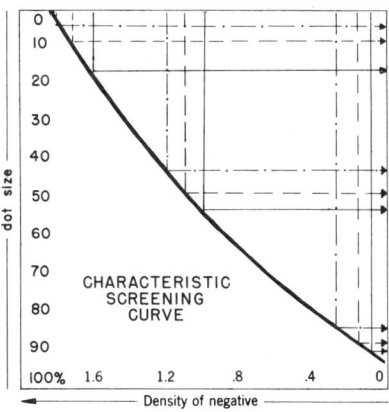

Systematic working

4. Quadrant—predicted negative curves:
The lines representing the printing dot values are linked with the H, M and S areas of the original to show the required negative curves

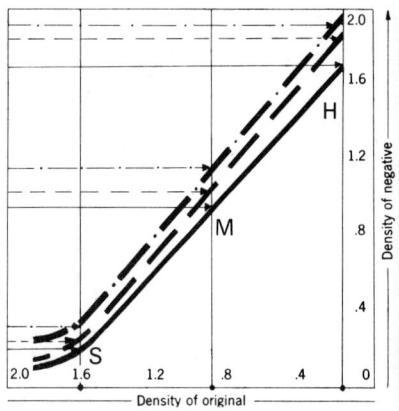

the original's densities to produce the second quadrant.

This is followed by the characteristic curve of the halftone screen employed.

Finally, they can all be joined together and by plotting the points of intersection created by the lines emanating from the characteristic screening curve and the H, M and S points on the original the density ranges and tonal curves of the separation negatives can be seen.

To produce separation negatives possessing such curves and densities a non-image (flash) exposure may be required. The aim in every case is to produce colour separated negatives and positives which possess densities and dot sizes which are as near the required values indicated by the colour charts as it is photographically possible.

Record of work

An efficient method of recording work and applying quality control to graphic reproduction photography is to employ computer programming of mathematical equations for:
1. Calculating the precise alteration in exposure for line, tone and colour reproduction, when increases and decreases in the density range of the original illustration or photographic image are encountered.
2. Scaling a graphic reproduction camera in terms of conjugate foci measurements related to lenses possessing different focal lengths. In each case maximum and minimum magnifications are obtained.

Record of work

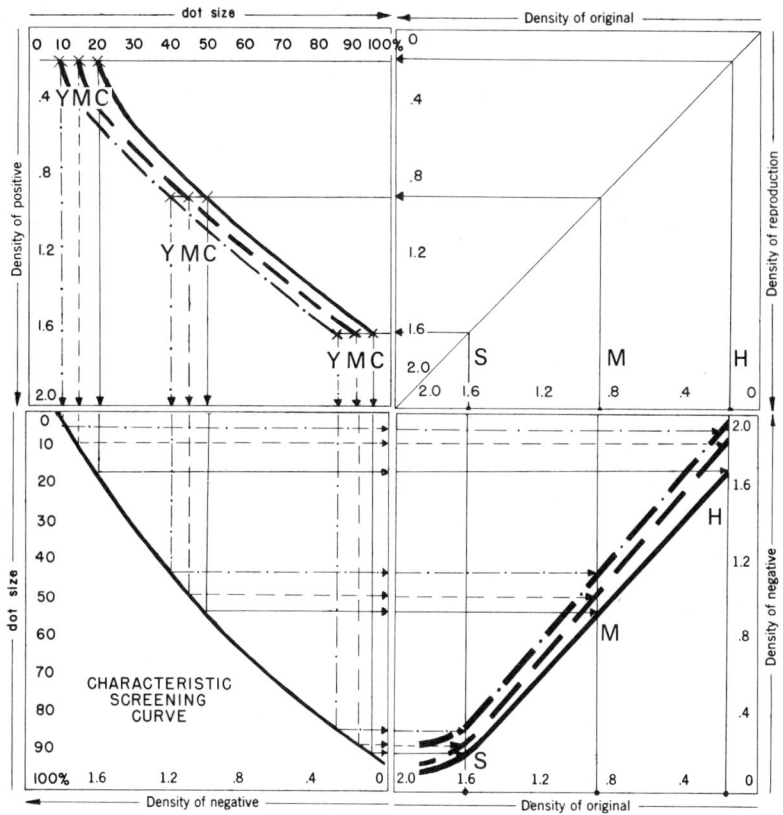

Dot sizes ← Original
↓   ↑
Screen curve → Negatives

Complete Quadrant diagram:
1. Transfer quadrant. 2. Required dot sizes.
3. Characteristic screening curve. 4. Predicted negative curve.

Systematic working

OPERATOR	J.W. Burden			NUMBER	82487			
DESCRIPTION OF ORIGINAL				DATE	1·1·1970		DENSITY RANGE	
Transparency of table top. Focal points – flowers, fruit & bottle of wine. See colour key N° 82487 for dot sizes & tonal ranges.							·2 2·0 1·8	
RESULTS	APPARATUS	SENSITIVE MATERIAL	EXPOSURE TIME AND EXPOSURE CONDITIONS	DEVELOPER	DEVELOPMENT TIME	DENSITY RANGE	NOTES
MASKS	frame 1 light 8	Pan Lith	10 secs Highlight mask register pins	ID2 1-6	2 mins	·4- 0·0	four steps to view
MASKS	frame 1 light 10	M13p	45 secs M2f Multimask	G24	4 mins	·6- ·2 ·4	reduces trans range to 1·4
CYAN	camera 4	TYPE 1	RED f 30 l.cs	ID2 1-3	2½ mins	1·65- ·25 1·4	correct
	frame 1 L 10	Lith 81p	75° 100 l.cs	Lith A+B	3 mins	15% 95%	H/L sl'FLAT
MAGENTA	camera 4	TYPE 1	GREEN f 45 l.cs	ID2 1-3	3 mins	1·8- 1·35 1·45	H/L = 1·85
	frame 1 L 10	81p	15° 150 l.cs	Lith A+B	2½ mins	10% 85%	✓
YELLOW	camera 4	TYPE 1	BLUE f 65 l.cs	ID2 1-3	3½ mins	1·9- ·5 1·4	
	frame 1 L 10	81p	90° 175 l.cs	Lith A+B	2½ mins	5% 80%	Shadow heavy
BLACK	camera 4	TYPE 1	SPLIT f 10,15&30	ID2 1-3	4 mins	2·0- ·55 1·45	✓
	frame 1 L 10	81p	45° 200 l.cs	Lith A+B	2½ mins	0% 60%	✓

Record of work for quality control. Note stages and items listed.
Work number and date, Description of original
Density range of original. Exposure times
Development times. Apparatus
Materials. Quality of result

3. Applying the two basic exposing systems, i.e.
   INVERSE SYSTEM—exposure time related to the magnification factor.
   V/RATIO SYSTEM—exposure time related to the lens aperture size.
4. Comparison for penumbral and diffraction screen distances for use with a ruled glass screen.
5. Calibration of a magenta contact half-tone screen with regard to its reproducible density range, correlation to colour compensating filters and the influence of a "no-screen" exposure.

These equations would be programmed for computer use in an international computer language such as ALGOL and presented in table, chart and graph form so that problems may be solved by consulting the most appropriate format. These computer generated tables will last as long as twelve months, becoming invalid only when a major variant in the process alters, for instance film speed or the type of illumination. A change of this nature would necessitate running the computer programme once more with new data.

A competent Algol programmer would be able to produce completed tables for the five areas of work in one week. The cost of this work at a computer centre would be—manpower: £25, computer time: £20, overheads: £10, a total cost of £55. To use programmes to produce further tables incorporating new data would cost only £1 per table.

All those responsible for camera operations should investigate this valuable and inexpensive way of employing the accuracy and speed of a computer to extend the utilisation of electronic exposure meters, automatic processing and systematic working.

The moment of truth

We have now travelled halfway. The five basic factors of an original, the camera, light, emulsions and processing should be the firm foundations of photographic knowledge spliced together with systematic working. If this is so and *only if it is,* the camera technician is ready to embark on line, tone and colour reproduction—the ingredients of graphic reproduction.

9. LINE REPRODUCTION

Historical introduction

During the early years of printing, line reproductions from autographically produced metal and wooden printing plates were taken to new pinnacles of excellence by the finesse and skill of Dürer (1471–1528) and Thomas Bewick (1753–1829). Although this type of line reproduction possessed artistic and altruistic qualities the art of printing was rapidly expanding and demanded a faster method of graphic reproduction. With the birth of photography and its application in a photo-mechanical manner these beautiful, but slow autographic methods gradually gave way to photogravure, photo-engraving, photolithography and photostencil processes.

As the nineteenth century began to unfold Joseph Nicéphore Niépce became very interested in lithography, the newly invented art of printing from limestone surfaces, and endeavoured to use it, but his local limestone lacked the required fine and regular grain structure. This led Niépce to replace the stones with pewter plates which he coated with a layer of light-sensitive bitumen of Judea (a type of asphaltum). Niépce completed his printing paraphernalia with what he called 'a kind of an artificial eye; which is nothing but a camera.'

An early line reproduction:
Cardinal d'Ainboise
(by Niépce)

Historical introduction

With these materials in the year 1826 Niépce not only recorded the first permanent view from nature, but produced a heliogravure printing plate. He took a line reproduction of Cardinal d'Amboise, Minister of Louis XII, printed from a hand engraved copper plate and made it transparent by soaking it in oil. Niépce then positioned this positive image on to the surface of the sensitive pewter plate. After an exposure to sunlight for approximately three hours the bitumen under the transparent non-image areas became light-hardened while the bitumen below the image (opaque printing ink) areas remained soluble and was removed by applying a solvent of oil of lavender and white petroleum. The picture now appeared on the surface depicted in lines of bare pewter on a background of hardened bitumen. Niépce then carefully etched these image lines with a solution of acetic acid. Further etching, hand correction and intaglio printing was carried out professionally by Lemaître, a skilful and intelligent engraver.

As Niépce's heliographs were improved and changed to Daguerreotypes, so the early heliogravure printing plates became etched daguerreotype plates, and new metals, etching techniques and electroplating methods were employed to produce metal intaglio printing plates capable of longer production runs and a more faithful retention of fine lines. Also at this time other contemporary photographic pioneers entered the competition of trying to capture the camera's image permanently in printing ink. William Henry Fox Talbot, upset and annoyed about the fading dilemma of his silver Calotype prints, evolved the Photoglyphy process for etching steel plates, while a Viennese, Paul Pretsch (1808–1873), presented an electro-platemaking process which used electrolysis to form an intaglio printing surface in copper from a relief mould which was a representation of the picture in lines of hardened gelatine.

There can be no doubt that the link between photography and the printed page which established the photo-mechanical processes was brought about by the application of the wet collodion photographic process and the sensitizing properties of potassium bichromate. The discovery of these properties was the result of astute observations made by a Scot, Mungo Ponton (1802–1880), while exposing photogenic drawings. Ponton related,

'When paper was immersed in the bichromate of potash alone, it was powerfully and rapidly acted on by the sun's rays. It accordingly occurred to me to try paper so prepared to obtain drawings, though I did not at first see how they were to be fixed. The results exceeded my expectations.'

Line reproduction

Until this time the majority of line impressions from metal printing plates were produced by the intaglio method. This made line illustrations difficult to correlate with text printed from moveable relief typefaces. This situation motivated Firmin Gillot (1802–72), a Frenchman, and later his son. Charles Gillot, to invent and develop 'Paniconography', a method of etching in relief line illustrations transferred to a sheet of zinc.

This relief etching process was very successful and in 1872 Charles Gillot, emulating his father's example of hard work and intelligent thought, became aware of, and conversant with Frederick Scott Archer's wet collodion photographic process. Charles applied this knowledge and exposed wet collodion negatives of line engravings, drawings and lithographs directly on to zinc plates sensitized with a solution of bichromated albumen. The resultant positive image was rolled up with a greasy ink, developed in running water and dusted with an acid-resisting resin. Etching followed in a bath of nitric acid. This simple and efficient method of producing a relief printing line plate was widely used and named the 'Gillotage'.

Together with the intaglio and relief processes planographic printing in the form of lithography was developing at a fast pace. Alphonse Louis Poitevin (1819–1882) employed bichromated albumen as a light-sensitive coating for lithographic stones. After exposure under a wet collodion line negative the exposed image areas became insoluble and water repellent. The greasy lithographic ink was readily accepted while non-image areas remained moist, the ideal conditions for lithographic printing. Eventually the stone support was replaced with grained zinc plates.

The use of photography linked with light-sensitive bichromated colloids became widespread and was adopted in the early nineteen hundreds to produce on a paper base the first photographic stencils for screen-process printing.

Assessment of line originals

We will begin by defining the term 'line original'. Line originals are pictures delineated in solid black images on a white background, e.g. pen and ink drawings, type pulls, scraperboard illustrations, stipple and tint work. Efficient reproduction photography commences with a period of familiarization—a detailed study of the original's qualities, peculiarities and method of delineation.

A line image

When we see a line original our visual senses are stimulated

Assessment of line originals

by the light energy emanating from the white areas and the absence of light occurring in the black regions. The mental picture which we form of the original is built up from the amount of difference between these two diverse visual sensations and can be simplified into the difference between the width of a black line and the area of white surrounding it.

When man invented photography he was really trying to emulate himself because a human being is equipped with the best twin lens camera in the world. Our eyes are superb lenses converging light energy on to a sensitive emulsion of rods and cones. This emulsion generates impulses just like a photocell. These impulses activate our brain into creating a coloured positive image of the original. This mental picture may be appreciated, minutely investigated, interpreted and then stored for future reference.

This is all applicable to line reproduction photography because we must, with a glass lens and a light-sensitive silver emulsion, capture exactly the visual differences of the line original. The subsequent printed copy must compare favourably with the original; the maximum density or absence of light of the copy should be similar to that of the original. The width of black lines and the area of white surrounding them needs to be as near facsimile as is possible. If this is so when our eyes traverse from original to copy there will be no immediate differences.

The photographer must become fully aware of the line original's maximum density by taking a reflection density reading of the largest opaque area. This is followed by a reflection density reading of a suitable area in the white background. By subtracting these two figures the original's density range becomes apparent and comparable to ideal 'white' and 'black' reflection density readings.

The original's attributes are now becoming apparent and may be evaluated against the previously mentioned line standards for the different processes printed on a variety of papers. This comparison is best completed by following a number of steps.

1. Compare the original with the line standard printed on a similar paper to the stock stipulated for the subsequent printing.
2. Check the maximum ink density of the standard with the original's darkest area, e.g.:

Process	Paper	Printing max. density	Original's max. density
Lithography	Machine finished	1·45	1·65

Line reproduction

3. Note the loss or retention of the finest black lines surrounded by white paper on this line standard, e.g.:

Process	Paper	Size of finest line	Original's finest line
Lithography	Machine finished	0·25 mm	0·20 mm

4. Record the 'filling-in' of the minute white areas surrounded by black ink, e.g.:

Process	Paper	Size of Smallest white area	Original's smallest white area
Lithography	Machine finished	0·75 mm	0·50 mm

Line control strip

From the information obtained the line control strip can be masked-off to indicate the extremities of the original, but due note must be taken of the limitations predicted by the appropriate line standard. It can be seen from the example figures that the line reproduction in question will, on machine-finished paper using the lithographic printing process, look slightly less opaque than the original accompanied by a loss or 'breaking-up' of the finest black lines while the minute white areas will tend to 'fill-in'.

Using this information, especially the density range and maximum density values of the original now clearly indicated on the control strip, calculate the required exposure time using the system which the manufacturer recommends. This will be based on the previously mentioned basic equation of:

$$\text{Log of the New exposure time} = \text{Log of the Old exposure time} + \text{New density range} - \text{Old density range}$$

The manufacturers' systems take into account such important factors as reciprocity failure, the exposing system (v/ratio or Inverse system), minimum and maximum exposure times. These factors which influence the reaction of the light-sensitive emulsion are correlated by the manufacturers to their range of emulsions. Factors such as these also play an important role in camera procedure.

Camera procedure

This work commences with a series of checks on the original's dimensions compared to the required image size and confirmation of the magnification factor. These figures,

Camera procedure

plus the focal length of lens, will provide the correct information for focusing (U and V distances) and finally the size of the required lens aperture. In order to follow an organized and economical work-pattern, line originals should be grouped into batches of the same magnification factor and, if possible, similar density ranges.

Exposure data

At this point it must be made quite clear that any one camera procedure or recorded exposure data is only suitable for the camera on which the tests were carried out. All cameras have their own peculiar working conditions related to their design, method and efficiency of illumination and amount of lens flare.

The majority of line photography for the printing processes, with the exception of gravure, is carried out on 'lith' type emulsions which give a high degree of resolution together with extreme contrast. So we will look at this work first.

Working with lith

Lith emulsions are prepared from a basic silver chloride mixture with other halides and sensitizers added later. Critically sharp images are produced possessing maximum density right up to the edge of clear lines, the result of a fine grain emulsion having a characteristic curve with a short toe and an extremely steep straight-line portion producing high gamma values of 10 or more.

Correct exposure times for line reproduction photography can only be established after a series of tests have been conducted to determine the correct exposure time under the prevailing conditions and with standardized materials and equipment from a representative original.

1. *Exposure tests*

A line original possessing an average density range together with a line and continuous-tone control strip is given a series of progressively increasing exposures producing a number of small steps, resulting from exposures of 6, 7, 8, 9, 10, 11, 12, 13, 14, 15, 16, 17, 18, 19, 20, 21, 22, 23, 24, 25, 26, 27, 28, 29 and 30 light counts. The exposed series of tests are then cut into a number of strips so that each strip contains all the exposure times. Processing is carried out, developing each strip for a different time period, e.g. $1\frac{1}{2}$, $1\frac{3}{4}$, 2, $2\frac{1}{4}$, $2\frac{1}{2}$, $2\frac{3}{4}$, 3, $3\frac{1}{4}$ and $3\frac{1}{2}$ min. The final test strips are compared and carefully inspected for image resolution, width of clear lines and optimum image density using a microscope and transmission densitometer. Exposure times producing acceptable image

Line reproduction

Exposure tests:
1. Under-exposed negative—heavy print
2. Correctly-exposed negative—acceptable print
3. Over-exposed negative—thin print

quality can be related to development time and if each acceptable step contains a continuous-tone stepwedge the number and density of the last opaque step should be noted.

Now each development time will produce not one, but a number of steps of acceptable quality. The exposure range of these steps needs to be carefully recorded. A development time of $2\frac{1}{2}$ min may have up to 10 steps containing images of acceptable quality ranging from say, 10 light counts to 20 light counts with an optimum exposure time of 15 light counts. From these figures the exposure latitude for a development time of $2\frac{1}{2}$ min may be ascertained in the following manner. The logarithm of 10 is 1·00 and that of 20 is 1·30; subtraction of these values gives an exposure latitude of 0·30. Exposure latitude depends on the interpretation of the words 'acceptable quality' and a figure of 0·30 would tend to be too great for

Exposure latitude:
10 to 20 lc's = 0·3 latitude

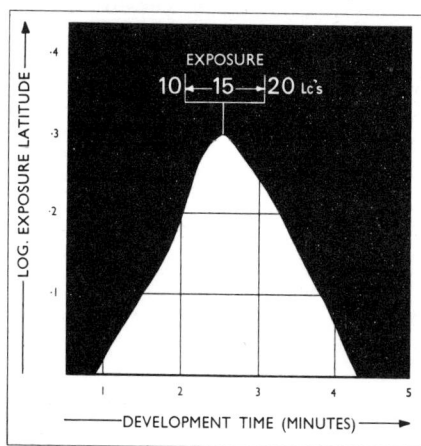

Camera procedure

extremely fine line work. Once this information has been compiled a suitable graph may be produced.

2. **Application of lith exposure methods**

Exposure latitude values are used when development times and development techniques are limited as is found when using automatic processing machines. With most machines, development times are interpreted into speeds—the rate at which the film is passed through the developing solution. Using the example graph the exposure latitude values would be applied to original reflection densities and calculated exposure times in this way.

Table 13. Lith exposure altitude

Original's reflection density range	Calculated exposure time (light counts)	Exposure latitude value	Development time Machine mm (ft) per min	Dish (min)
1·00	6			
1·20	6	0·15	450 (1·5)	$3\frac{1}{2}$
1·30	8			
1·40	10			
1·50	12	0·30	600 (2·0)	$2\frac{1}{2}$
1·60	15			
1·70	19			
1·80	24			
1·90	30	0·10	760 (2·5)	$1\frac{3}{4}$
2·00	30			

When dish development is employed exposure times must be correct within a fairly limited tolerance. Appreciation of this fact can be gained once the original reflection densities and their calculated exposure times are related to specific development times, as shown in this chart.

Table 14. Lith exposure and development

Original's reflection density range	Calculated exposure time (light counts)	Related development time (min)
1·00	6	$3\frac{1}{2}$
1·20	6	$3\frac{1}{2}$
1·30	8	$3\frac{1}{4}$
1·40	10	3
1·50	12	$2\frac{3}{4}$
1·60	15	$2\frac{1}{2}$
1·70	19	$2\frac{1}{4}$
1·80	24	2
1·90	30	$1\frac{3}{4}$
2·00	30	$1\frac{3}{4}$

Line reproduction

(3) *Lith development*

Lith development begins where maximum exposure has been received and spreads to the silver halide grains which have accepted the minimum exposure. The developing solution is basically a formaldehyde-hydroquinone relationship with a minimal amount of sodium sulphite. This formula produces the characteristic 'lith' development of a slow induction period while the exposed silver halide grains reach a density of approximately 0·2. Blackening then occurs followed by rapid acceleration until the exposed image areas reach gamma infinity and a density of at least 3·0.

A reduction process of this type is termed 'infectious' development because after the silver halides in the areas of maximum exposure have been reduced the quantity of developer responsible becomes partially oxidized and by-products (namely semiquinone) of the reducing agent hydroquinone are formed. These by-products 'infect' the neighbouring silver grains and accelerate their reduction. Escalation of development continues producing extremely high densities even in areas which have received minimum exposure. This reaction is continuous because if development is not curtailed at the appropriate time the clear areas of the emulsion (unexposed silver grains), which up until this time have remained at fog level, will be subjected to the developer's 'infectious' nature and the image areas will spread into the clear areas, narrowing them and finally producing a veil across them.

With dish development good results can be readily achieved by combining time-control with intelligent inspection. This counteracts minor inaccuracies in exposure.

Development is carried out normally (rocking or brushing at a steady rate) for two-thirds of the required development time. Then agitation is stopped and a red safelight under the transparent developing dish is switched on. Using a magnifying glass careful inspection of the fine clear lines and the minute opaque areas is carried out. Their relationship determines the development technique for the remaining period of time. For example, if the minute opaque areas have appeared but the clear lines are still too wide, rapid agitation is continued. If the reverse is seen, with the clear lines beginning to close, but with the minute opaque areas still to appear, then 'still bathing' is employed. The lack of motion means that the developing solution reducing the image areas which have received maximum exposure becomes exhausted and is not replaced by fresh solution. The accelerating by-products are therefore not so readily created and development in these areas

Camera procedure

is thus retarded, while the minute opaque areas can be sufficiently reduced by the motionless developing solution surrounding them. Still bath techniques become advantageous when very fine line originals are encountered.

Working with continuous-tone emulsions

Photogravure, unlike letterpress, lithography and screen process, is an intaglio process using minute ink cells etched to varying depths below the surface of the plate to hold and impart the printing ink. To produce these ink cells an intermediary is employed to carry the image from the photographic positive stage to the metal cylinder. This is a layer of light-sensitive pigmented gelatine on a paper base, termed 'pigment paper' or 'carbon tissue'.

Following an exposure to a gravure 'fish-net' screen and a suitable continuous-tone positive of the line original the sensitive gelatine becomes hardened. After viewing a continuous-tone positive stepwedge, it is quite easy to translate the thin highlights and the deep shadow tones into a 'wedge-shape', that is a side-view, representing the density of silver present. The depth of the hardened gelatine image, referred to as the resist, is directly proportional to the density of the continuous-tone positive. This hardened gelatine layer is transferred to the copper surface of the printing cylinder, where it controls the penetration time of the etching acid.

The etching of the copper surface into a myriad of square ink cells, the result of the 'fish-net' screen exposure, begins when the etching acid has percolated through the resist and reaches the copper; but because the resist is a *negative* image of the original in hardened gelatine the acid etches the shadow ink cells first and the highlight cells last. Therefore the shadow ink cells will be etched to a greater depth than the highlight and will consequently transfer more printing ink on to the paper than the shallower highlight cells. All this information may be condensed into the statement, 'the depth of the etched ink cells is directly proportional to the thickness of the resist, this thickness being dependent upon the densities of the continuous-tone positive, which in turn should be a faithful reproduction of the original'.

The density range of the continuous-tone positive is mainly determined by the hardening properties of the pigment paper. Ideally the highlight depth of the resist should not exceed a maximum of 0·01 mm, while the shadows are retained at a minimum of 0·001 mm. If this ideal is realized a full range of tones or a line situation can be successfully etched with a 'one-bath' technique (an acid solution mixed

Line reproduction

to one specific gravity reading). It has been generally found that a continuous-tone positive possessing a density range of 1·3 (highlight density 0·3—shadow density 1·6) produces an ideal resist as long as the continuous-tone emulsion used for the positive provides a characteristic curve which will accommodate densities of 0·3 and 1·6 on the straight-line portion. Compression of tones on the curve's toe is not required.

1. *Continuous-tone development*

To form a basis for line reproduction on continuous tone emulsions we may accept a positive density range of 1·3 (white areas 0·3—opaque lines 1·6). The white areas may, if required record as clear film 0·00, but it is essential that the density of opaque lines compares very favourably with the maximum density used in tone reproduction because in order to provide ideal printing conditions the etched depths in both cases need to be similar.

Therefore as far as the camera operator is concerned standardized continuous-tone positives will be his aim, bearing in mind the diverse nature of originals, which may range from reflection densities of 1·0 to 2·0, and the necessity of producing a suitable continuous-tone negative first. Once this is understood it is usual practice to use the negative stage to equalize the density ranges in order to produce standardized positives by the simple and straight-forward contact method. This work revolves round the following formulae:

$$\text{Negative gamma} = \frac{\text{Required density range of the negative}}{\text{Reflection density range of the original}}$$

$$\text{Positive gamma} = \frac{\text{Required density range of the positive}}{\text{Transmission density range of the negative}}$$

For example, two line originals are supplied, A with a reflection density of 0·9, and B with a range of 1·8. A continuous-tone positive possessing a standardized range of 1·5 is required from both originals. To obtain this we must develop the two negatives in such a manner that they both acquire a similar density range. This allows a simple positive stage, and each positive will acquire the standard range of 1·3 after development has been carried out to the same gamma value.

The next important consideration is the detrimental effect of lens flare. As has been shown previously a normal flare factor of 1% reduces the original's density range considerably. Using our examples they would appear as

Camera procedure

Table 15. Original density and camera density

	Original density range			Camera density range		
	White area	Opaque line		White area		Opaque line
A	0·10	0·90	1·00 →	0·09	0·87	0·96
B	0·10	1·80	1·90 →	0·09	1·56	1·65

A careful study must be made of the original density ranges usually encountered together with the required negative and positive density ranges and their gamma values. This information needs to coincide with the maximum density, gradation and gamma capabilities of the emulsions chosen

Negative curves:
A.1.24 vigorous developer
B.0.62 low contrast developer
Positive curves:
A and B 1.2 standard developer

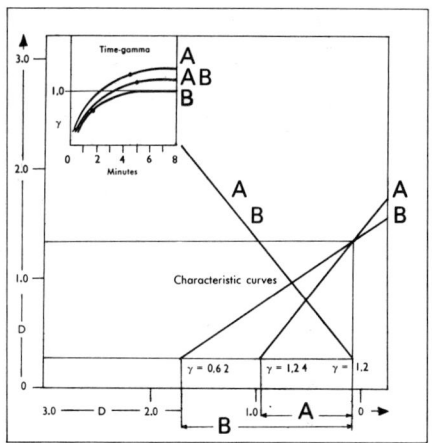

for the reproduction. If the emulsions satisfy all requirements then development and time/gamma curves need to be constructed for the negative and positive emulsions. It goes without saying that all other contributory factors, such as illumination, exposure, developer concentration, temperature, agitation, etc., must be maintained in a rigid state of constancy.

Unlike lith materials, continuous-tone emulsions and developers can produce acceptable results when a wide variation of exposures and development times are used.

(2) *Determining the development time*

First take the range of the standardized positive and apply this formula, but note that it has been found in practice that improved gradation is achieved if a positive gamma of between 1·2 and 1·3 is used.

Line reproduction

$$\text{Required density range of the negative} = \frac{\text{Required density range of the positive}}{\text{Required positive gamma}}$$

$$\text{Negative density range} = \frac{1 \cdot 3}{1 \cdot 2}$$

Negative density range = 1·08

Also consult the appropriate time/gamma curve to find the development time for a gamma value of 1·2, e.g. 1·2 = 5 min.

The next step is the negative making when this formula is employed:

$$\text{Negative gamma} = \frac{\text{Required density range of the negative}}{\text{Reflection density range of the original}}$$

Let us apply this to our example originals.

(A) Original density range 0·90 = Camera density range 0·87

$$\text{Negative } \gamma = \frac{\text{Density range of negative}}{\text{Density range of original}}$$

$$\text{Negative } \gamma = \frac{1 \cdot 08}{0 \cdot 87}$$

Negative γ = 1·24

Consult the appropriate time/gamma curve—

γ 1·24 = 4½ min development

(B) Original density range 1·80 = Camera density range 1·56

$$\text{Negative } \gamma = \frac{\text{Density range of negative}}{\text{Density range of original}}$$

$$\text{Negative } \gamma = \frac{1 \cdot 08}{1 \cdot 56}$$

Negative γ = 0·62

Consult the appropriate time/gamma curve

γ 0·62 = 1¾ min development

In order to see our overall aim these photographic steps may be represented as a block-diagram

Camera procedure

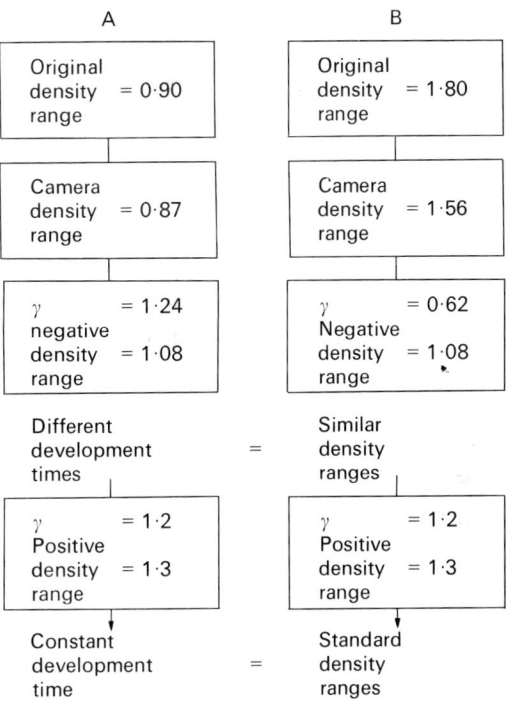

Exposure tests and systems

As in lith work the only way to become fully conversant with the way that the various density ranges of originals affect the exposure time on continuous-tone emulsions is to conduct careful tests using a line original of average qualities. It is usual to conduct these camera tests at same size. After a base exposure has been established this can be altered to suit a new original density range and if the inverse exposing system is chosen, magnification factors may be employed to multiply this exposure time so that it is correct for a particular reduction or enlargement.

Most of the systems for predicting exposure times on continuous-tone emulsions are based on the following formula.

$$\text{Log of the New exposure time} = \text{Log of the Old exposure time} + \text{New shadow density} - \text{Old shadow density}$$

Although this equation provides a very good guide to the increase or decrease required to relate an exposure time to a new original's shadow density, it does not take into account such important factors as emulsion characteristics, reciprocity failure, lens flare and the exposing system employed. If all

Line reproduction

the basic concepts which have been used until now are quite clear the Agfa-Gevaert *Repro Slide Rule* and *Gevarex* system is recommended. This is a unique method of pre-assessing exposure and developing times based on practical tests and uses all the important factors to solve practical problems.

Line results for each reproductive process

The photographic requirements for making line blocks, plates and stencils for the letterpress, lithographic and screen process printing methods are basically the same and are all admirably satisfied by using lith-type emulsions, although small differences do occur in the need for image reversal and positive images. Gravure printing, on the other hand, requires negative and positive images captured on continuous-tone emulsions. This is because of its printing method of imparting ink from ink cells. Individual printing companies do have their own specifications as far as maximum densities, line widths and density ranges are concerned, but generally speaking line results for each reproductive process follow the requirements listed below.

Photo-engraving (letterpress)

A reversed line negative on lith material is required. This means that the image must read correctly on the emulsion side of the film and appear as an exact black 'stencil' of the line original. To obtain this a prism or mirror is used to turn the image through 90° while lith emulsions, developers and contrast techniques are exploited to the full.

Photography proceeds only to the negative stage because a positive image is created on the light-sensitive, acid-resisting coating employed during the plate-making process to transfer the photographic image on to the metal surface.

Photolithography

Photography suitable for this process must be divided into two sections.

1 *Surface printing plates*

Plates of this type require non-reversed line negatives on lith materials. The image must read incorrectly on the emulsion side of the film. To achieve this an exposure is given through a lens on to a lith emulsion. Appropriate processing follows producing a characteristic lith-type line negative.

Negatives are supplied because the positive image (as in the photo-engraving process) is formed on the metal surface

Line results for each reproductive process

of the printing plate—the result of a light-sensitive colloidal coating becoming light-hardened in the image areas.

2. Deep-etch printing plates

Because of the plate-making procedure a reversed line negative on lith film is made, its image reading correctly on the emulsion side of the film. From this a non-reversed line positive is produced by contacting, i.e. placing the emulsion side of a second sheet of lith film in contact with the emulsion side of the negative. After suitable exposure and processing a line positive is obtained. This positive will read incorrectly on the emulsion side and needs to appear as an exact replica of the line original.

The plate-making procedure is a positive to positive process. The light-sensitive colloidal coating produces a hardened negative stencil of the image. After careful processing the metal surface of the plate appears in the image areas. This allows microscopic etching to take place in these areas, which is responsible for the longer printing life of this type of lithographic plate.

Screen process

Once again negative-positive stages are employed to produce a non-reversed line negative with its image reading incorrectly on the emulsion side, the result of a lens-only exposure. The positive, after contacting, appears reversed, that is the image reads correctly on the emulsion side of the lith film. Because of the high degree of contrast required lith materials are used extensively.

This type of positive is needed for the stencil-making process. The back of a light-sensitive stencil film is exposed under the emulsion of the reversed positive. After development in warm water a hardened negative stencil is produced, which reads correctly on the emulsion side. The emulsion side of the stencil is then firmly adhered to the underside of the printing mesh and the film base carefully removed. The ink may now be forced through the clear image areas on to the paper underneath.

Photogravure

In this process the negative and positive photographic images produce the gelatine resist for etching the ink cells to different depths. The image needs to be reversed by a prism or mirror to produce a correct-reading negative image on the emulsion side of a continuous-tone film. Using a contact method a non-reversed positive (image reads incorrectly on

Line reproduction

the emulsion side) is made to a standardized density range on a suitable continuous-tone emulsion.

In order to produce positives directly for cylinder-making many gravure printing houses have adopted the 'Rinco' monochrome reproduction method. This is a method of commencing the reproduction process with a negative image of all type-matter, line drawings and monochrome tone illustrations. The monochrome tone illustrations are photographed on to rinco bromide paper to give a negative image possessing approximate end densities of 0·2 (shadows), 1·15 (highlights), a range of 0·95. Line drawings are copied on to a specially prepared photogravure line bromide paper which appears slightly yellow because of its ultra-violet absorbing properties. The density range of this paper line negative must compare favourably with the negative of the tonal illustration. Finally, the type-matter is printed on black paper using an off-white ink. The whiteness or density of ink is adjusted to match the density ranges obtained on the paper negatives. This is essential because all the type-matter and illustrations are planned together to make up a complete page or imposition of pages. This enables the cameraman to take one photograph which translates all these negative line originals into positive form. If due care is taken in matching the negative density ranges, followed by efficient photography, this positive result will possess in every opaque area a maximum density of 1·6 which fulfils the specific requirements for cylinder-making.

Cylinder-making begins with light-sensitive pigment paper being exposed under the positive, emulsion to emulsion. The exposed emulsion of the pigment paper is adhered to the metal surface of the cylinder. After development in warm water the paper base and unexposed areas are removed. This allows the incorrect-reading image to be etched into the metal surface.

Short-cut methods

It can be seen that a positive image on lith material is needed frequently for the different printing processes. This involves another sheet of film accompanied by extra processing time and as printing is an industry obsessed with time rates and economics it was not long before methods were introduced to produce a positive image directly from a positive original. The two major methods are:

1. *Chemical reversal*

A lith negative is produced with an increased exposure

time (approximately 25%+). Development is carried out normally and when the time period ends the image should appear opaque right through to the film base as a result of the increased exposure time. The negative is then immersed in a fresh acid stop bath where development should be completely arrested. Once this has been achieved the negative is transferred to a specially prepared etch–bleach bath—the negative *is not* fixed in the normal manner.

This etch–bleach bath, consisting of cupric chloride, acetic acid and hydrogen peroxide, dissolves away the exposed and developed areas leaving the unexposed silver halides. The dissolving action is helped by delicately swabbing the emulsion with cotton wool. After the majority of the blackened negative image is removed the white lights in the room may be switched on and this completely exposes or fogs the previously unexposed silver halides. An exposure of this kind is quite suitable for linework, but if a halftone image is being chemically reversed, then a controlled exposure must be given.

If after careful inspection it is seen that the original negative image has been effectively removed even in the finest lines, then washing follows for 5 min in a copious supply of running water. The film, now containing a fogged or exposed positive image is redeveloped in a lith or high contrast M/Q developer until the positive image is opaque right through to the film base. Washing and drying conclude the chemical reversal process. Suitable formulae for the etch–bleach bath can be obtained from the emulsion manufacturer.

The second method of obtaining a direct positive is to use autopositive, autoreversal or duplicating film.

2. *Autopositive films*

These emulsions are a modern extension of the 'Herschel' effect, i.e. latent images formed on silver halide grains may be destroyed by an exposure to wavelengths to which the emulsion is not sensitive. This phenomenon is exploited to the full.

If on receiving this type of film from the manufacturer, development is carried out prior to exposure, an overall blackening will occur. By exposing to either a filtered light source emitting light to which the emulsion is not sensitive, or to a lamp rich in these wavelengths, a positive image may be produced from a positive original. For instance, an emulsion sensitive to blue light would be exposed to a light source covered with a yellow filter or an intense tungsten lamp would be used.

With careful thought quite complicated results containing

Line reproduction

both positive and negative images can be achieved, as long as these rules are remembered:

(a) Exposure to yellow light before development produces clear film—positive to positive.

(b) If an exposure to white light (*blue* wavelengths) is given after the yellow light exposure, development produces a black image—positive to negative.

These emulsions are usually capable of being handled in normal room lighting and provide an easy way of producing direct positives, outlines, block-outs, line, halftone and tint montages.

Line and tint combinations

Frequently a camera operator is asked to produce a line and tint reproduction. The inclusion of tints, whether negative (over 50%) or positive (under 50%), can on certain line illustrations increase the visual impact and three-dimensional appearance. Although there are many techniques of producing a final line photograph with the required sized tint appearing in the appropriate areas, this work can be simplified to four steps.

Line and tint negatives:
1. Original
2. Negative
3. Tint

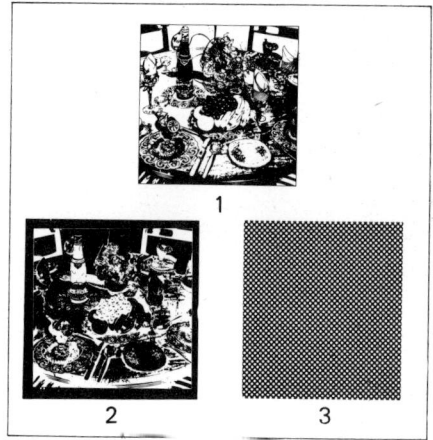

1. *Line and tint negatives*

A line negative is made to the required size and specifications for the printing process in question, followed by a tint negative possessing the correct negative dot size. A tint negative can be produced from a cross-line or contact halftone screen in the camera by reflecting light from a sheet of clean white paper. It helps the uniformity of the tint if the

Line and tint combinations

Guide positive:
Positive in register

Blocking out:
Unwanted tint areas opaqued

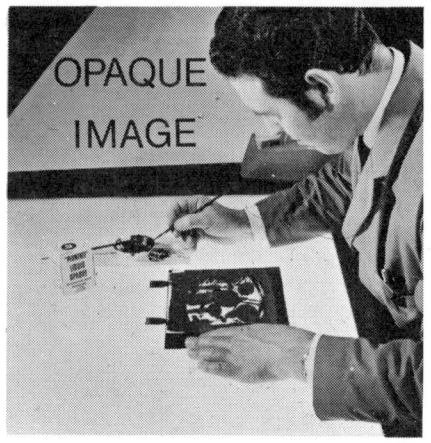

camera is not in focus. The size of dot produced is controlled by the exposure time. Some companies (map-printers in particular) who use tints a great deal buy or make a range of master tint negatives. Producing a suitable tint is just a matter of contacting.

2. *Making the guide positive*

The line and tint negatives have register holes punched into them. Using these holes and register pins a positive is made on to a similar punched sheet of lith film from the line negative using the contact method.

3. *Producing the tint block-out*

When dry the guide positive is laid underneath the tint negative, register being maintained by the use of pins. By

Line reproduction

Complete positive:
1. Line negative
2. Tint block-out
3. Complete positive

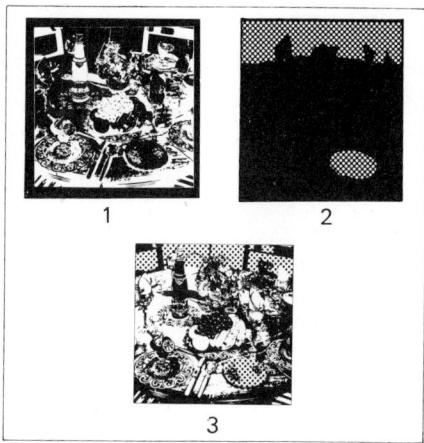

carefully applying opaque or cutting a rubylith mask all the areas where the tint *is not* required are blocked-out. The precise outline of the tint area is indicated by the guide positive underneath.

4. *Complete combination positive*

Once the block-out has been completed a line negative and a tint negative have been produced whose individual images can be brought into accurate register by superimposing them on the pins. The next step is contacting. A sheet of lith film is punched so that it contains the same register holes as the two negatives. By laying each negative on the pins in turn and exposing on to the punched sheet of lith film, the individual line and tint images are accurately superimposed on the one sheet of film.

Producing good line reproductions can bring immense pleasure along with a sense of fulfilment when, for instance, numerous printed copies of a complicated technical illustration present to everyone the intricacies of the motor car engine with its wealth of pistons, gears and differentials; in which the finest line has been captured and the smallest shadow detail retained.

10. TONE REPRODUCTION

Historical introduction

Apart from hand-painted illustrations the first pictures to appear in a printed publication as true tonal renditions of the original scene were photographic prints produced by William Henry Fox Talbot in the year 1844 for his book *The Pencil of Nature*. Issued in six parts, this labour of love contained a total of twenty-four photographs presenting the magic of the camera in landscapes, still-life, architecture, human studies and reproductions of sculpture, engravings and botanical specimens—all readily available for the princely sum of three guineas.

Unfortunately, Talbot was daunted by the persistent fading of his prints and related:

'I have met with difficulties innumerable in this first attempt at photographic publication, and I therefore hope all imperfections will be candidly allowed for and excused. I have every reason to hope the work will improve greatly as it proceeds, and that British talent will come forward and assist the enterprise.'

Talbot was still determined to publish photographs and turned to the use of printing ink, but at this time aquatint, mezzotint, collotype, woodburytype and hand-drawn lithography were the only process capable of printing tonal images, and these had to be proofed separately from the typematter. This led Talbot to investigate the possibility of printing a tonal picture with a uniform thickness of printing ink so that it could be printed together with type. He deduced that the only way to print 'halftones' (medium greys) is to break up the photographic image into fine dots of varying size. Talbot said:

'This break-up may be obtained by placing a piece of folded gauze, or alternatively a glass plate covered with an innumerable quantity of fine lines, or else with dots and specks which must be opaque and distinct from each other, between the photograph or object to be copied, and the sensitive plate.'

The largest dots would represent the shadows, the smallest the highlights, while intermediate size dots would depict the halftones. The eye thinks it is seeing tones when in fact

Tone reproduction

Halftone image:
1. Continuous-tone, shadow, middle-tone, highlight
2. Negative halftone, 5%, 49%, 95%
3. Positive halftone, 95%, 50%, 5%
4. Dot patterns

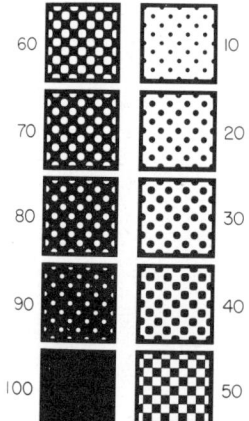

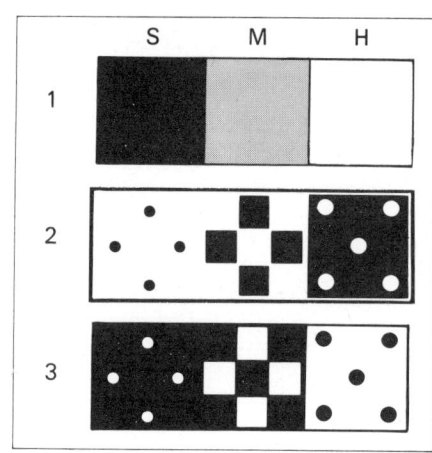

Dot patterns.

each dot carries the same thickness of ink. This idea of printing an illusion of tone became popular and evoked the interest and inventive powers of a number of men in different countries.

To begin with a slight deviation from Talbot's original conception took place. Charles Petit, a Frenchman, and Frederic Eugene Ives, an American, quite independently in the year 1878, presented physical systems of producing halftone dots. Although including some variations, both systems worked on a similar basis. A gelatine relief image was produced from the original picture, and from this a white plaster cast was made. Its surface was blackened and lines ruled through it using a V-shaped stylus. This ruling operation produced small dots in the highlight areas, represented by the heights of the cast, and large shadow dots in the low regions. The surface of this halftone cast was then photographed on to a wet-collodion emulsion, facilitating its journey to the printing machine.

One year later in England Joseph Wilson Swan, a prolific inventor, propounded the idea of a single line glass screen containing approximately 40 lines per cm (100 lines per in.). He explained that the screen should be placed in front of the emulsion during exposure and suggested turning the screen through 90° halfway through the exposure time in order to produce dots. This idea led Georg Meisenbach, a Munich engraver, to make just such a screen by photographing finely-ruled lines produced from an engraved copper plate. The halfway turn was employed and the resultant negative was used to form an etched halftone zinc block for letterpress printing. The printed copies were so successful that later,

Historical introduction

in the year 1884, Meisenbach established profitable block-making establishments in London and Munich.

This advancement made the American, Frederic Eugene Ives, dissatisfied with his physical system and spurred him to improve the single-line screen. Ives thought the halfway turn was cumbersome and this motivated him to produce the first cross-line screen, formed by cementing two single-line screens together so that their lines crossed at right-angles. He very cleverly produced fine single lines on photographically blackened wet-collodion plates using a ruling machine. After completing the screen's construction Ives went on to investigate such important topics as screen distance, lens aperture size and other contributory factors.

The final touch was made when Max Levy, a fellow American, skilled in draughtsmanship and engineering, was asked to improve the uniformity of the screen's lines with regard to width, opaqueness and edge sharpness. Perfection was gained by engraving the fine lines through an acid-resist coated over the glass plates. After etching the lines appeared as minute recesses which Levy filled with an opaque pigment. The halftone screen was finalized by cementing the two plates together with canada balsam. Fine screens with 40 or more lines per cm (100 or more lines per in.) and coarse screens below 40 cm (100) were commercially manufactured. The combination of Frederic Eugene Ives's ingenuity and Max Levy's prowess firmly established the halftone process in all fields of printing.

The halftone process is certainly a practical method of printing tonal pictures on letterpress, lithographic and screen process machines which impart one weight of ink; but these reproductions tend to lack depth, detail and gradation when compared with tonal reproductions printed in the same manner as the original was created, that is with different thicknesses of pigments or dyes. The collotype and gravure methods of graphic reproduction are worthy examples of printing tone for tone or thickness for thickness. It is sad to relate that for economic reasons the use of collotype has declined during recent years, but gravure, being more adaptable to rotary printing, has progressed rapidly.

It is fitting that gravure printing should stem from heliogravure plates produced by the world's first photographer, Joseph Nicéphore Niépce, whose ambition was to recreate all the living scenes surrounding him in pictures of tone and colour. In 1879 another man of vision, Karl Klič (pronounced Klitch), a Czech painter and engraver, realized the importance of true tonal reproduction. These thoughts motivated him to

Tone reproduction

expose a continuous-tone positive on top of a sheet of carbon tissue. Carbon tissue consisted of a layer of gelatine impregnated with finely powdered carbon, backed with a sheet of paper. The gelatine layer was sensitized in a bath of potassium bichromate. This photographic medium was the cumulative result of individual research work carried out by Alphonse Louis Poitevin, Mungo Ponton and Joseph Wilson Swan. After exposure the sheet of carbon tissue was transferred to the surface of a copper plate. This surface contained a grain structure of fine resin dust fused on by the application of heat. The carbon tissue was flooded with warm water which removed the paper base and unexposed areas of gelatine, leaving a negative image of the original in hardened gelatine. This acted as a resist for the successive etching baths of ferric chloride at different strengths. Once the resist was removed the copper plate presented its etched surface, the shadow areas of the original picture being etched the deepest, while all the other tones gradually climbed to the surface. Tonal printing is the result because the deeply-etched shadow areas impart more ink than the shallower highlight regions, while the overall detail is retained in a delicate grain structure.

At this time most gravure printing was carried out from flat plates, but Klič could see the importance and economics of rotary printing, and in 1890 he replaced the aquatint grain with a cross-line screen in order to achieve this end. Although a loan from the halftone process the lines of this screen were slightly modified in their ratio to the clear space dividing them (1:1 becoming 3:1). This was necessary because this screen *was not* producing dots. Its function was to form the walls of minute square ink cells, which varied in depth and made up the image. Its use also provided supports for the 'doctor' blade (which wiped away excess ink from the surface of the plate) and made the inclusion of typematter on the same plate a practical proposition.

In recent years there has been even more interplay between the different printing methods. Processes using the cross-line ruled screen have gradually adopted the vignetted dyed-dot contact screen. These screens, made photographically on a film base, allow the halftone dot to change its shape as well as its size, thus improving the resolution of detail. Photogravure has combined the halftone dot shape with the continuous-tone wedge in the Invert Halftone methods which produce etched ink-cells varying in depth and width. Slowly but surely new technological advances are merging the individual processes. The innate advantages of each process are being retained without creating too many new dis-

Types of halftone screens

advantages. What people like Niépce, Talbot, Gillot, Senefelder, Poitevin, Klič and Swan would have to say about modern graphic reproduction with its influx of electronics is hard to guess, but one thing is absolutely certain, they would be both excited and stimulated by the challenge.

Types of halftone screens

The majority of tone originals are the visual result of a pigment or dye being skilfully used in different thicknesses to create the shadow tones, middle tones, highlights and details of a picture. The graphic reproduction processes have the task of retaining these pictorial attributes and presenting them all over again in printing ink. This task becomes more realistic for the gravure process because its printing action can deliver different thicknesses of ink, while the letterpress, lithographic and screen process methods have to rely on a tonal illusion created by the halftone process.

Generally speaking, the reproduction processes in printing lower the tonal movement between the lightest tones and subdue shadow tones. Each printing process has a maximum density capability. This is shown as a reflection density value after reading the darkest and lightest areas of tone standards (stepwedges containing maximum and minimum continuous-tone densities or halftone dot sizes printed on a variety of papers) with a reflection densitometer. On a coated art paper using a single black ink impression the individual processes record average maximum ink density ranges of lithography 1·4, letterpress 1·6, gravure 1·8 and screen process 2·0. Tone originals containing greater densities than these values will be printed with black tones flattened to dark greys, or both highlight and shadow areas in a visually lower condition.

For the next page or so let us concentrate on the wide range of halftone screens available for the reproduction of tonal pictures by the 'one weight of ink' printing processes.

Glass ruled halftone screen

The screen consists of two sheets of polished glass, each possessing a set of opaque lines. These two sets of opaque lines cross each other at right angles and are ruled and etched to such a width that they are equal to the clear spaces dividing them. Screens with this ratio of 1:1 obstruct approximately 75% of the light reflecting from the original and it is the remaining 25% passing through the square transparent openings that produces the dot formation.

Screens may be ruled in such a way that a selection of screen rulings (number of lines per cm) become available,

Tone reproduction

Glass ruled screen:
1. Two sheets of glass with single lines
2. Glass sheets superimposed
3. Ratio, space 1: line 1
4. Ratio, transparent 25%, opaque 75%

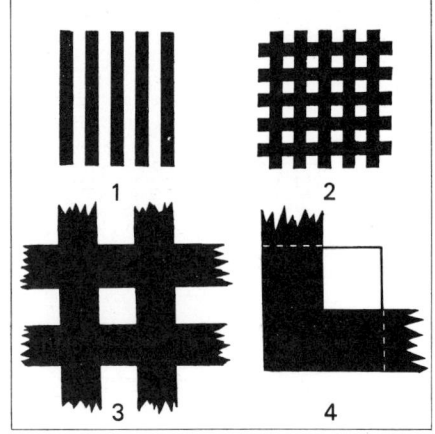

providing the printer with a coarse or fine tonal illusion. The primary consideration here is the surface of the material, usually paper, on to which the picture is to be printed. The selection of the appropriate screen ruling is an important decision which is dealt with in the following section on assessment of the original. Glass screens may be purchased in rectangular or circular shapes. For monochrome reproductions the lines of the screen are set at 45°, because when the dots are printed across the paper this is the angle which is least apparent to the eye. When coloured reproductions are required each printing plate responsible for the build-up of colours must contain dots which run at different angles, in order to prevent unpleasant moiré patterns being set up by dots running along the same angle. With circular screens this angle change may be achieved by a turn of the hand, while

Halftone translation:
1. Continuous-tone original
2. Ruled screen (side view)
3. Light focused into dots—negative image
4. Final positive halftone image

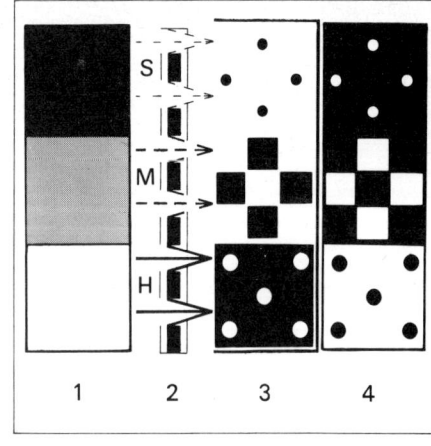

Types of halftone screens

a rectangular shape necessitates the purchase of at least three screens with their lines set at different angles. As the question of screen angle mainly occurs in colour reproduction we will reserve this topic for that chapter.

The glass rules screen is positioned behind the lens of the camera in close proximity to the light-sensitive emulsion. The screen is parallel to the surface of the emulsion, but held a short distance away by the screen gear. Each minute screen opening becomes a pin-hole camera focusing and photographing a small part of the original picture in accordance with its reflectance or transmission. It translates the continuous-tone picture into thousands of minute dots of varying size, but of equal density and gradation.

A screen ratio of 1:1 has the disadvantage of producing an overall transparent to opaque ratio of 1:3. Combine this with the light 'chopping' action caused by a mesh of opaque lines and considerable loss of detail is the result. This ratio also restricts the screen to straightforward reproductions of original density ranges of 1·4 or lower. Screens are made with a ratio of 0·8 (black line):1·2 (clear space) which enables original density ranges up to 1·8 to be photographed with a single exposure. It would be possible to alter the ratio of black line to clear line so that a specific screen would be used for the reproduction of a particular original density range, but this would be both costly and complex. The rather restrictive reproduction range of a screen is overcome by using a suitable 'flash' exposure or applying other contrast control methods.

The loss of detail produced by the opaque line structure led the German company of Klimsch to construct a new glass screen called the Allton Gradar screen. The previously opaque lines have been replaced by vignetted lines produced in a magenta dye. The soft edge of the magenta lines is responsible for the halftone dots changing shape as well as size, thus helping to reconstruct the detail of the original picture. The screen is thinner than normal and less dense, and with the use of contrast control filters produces improved gradation, shorter exposure times and minimal diffraction effects.

Glass ruled screen for rotary gravure

This screen is *not* made up from two glass plates—the fine crosslines are ruled and etched on the surface of a single glass plate. The screen ruling is usually in the order of 70 lines per cm (175 lines per in.) and the ratio between the black lines and clear spaces is 3 or 2·5:1. The function of this screen is *not* to translate continuous-tone into halftone, but to produce ink cells and supports for the doctor blade.

Tone reproduction

Film vignetted-dot contact screen

Many of the early workers in graphic reproduction could see the advantage of using a soft or vignetted screen to produce halftone dots. In the year 1895, only some three years after the perfection of the glass cross-line screen, efforts were being made to produce 'chess-board' screens on a silver emulsion. The idea was to make a screen by exposing through a cross-line screen. This photographic screen would ideally possess 50% dots with central cores of high density which gradually vignette away to fringes of low density. Although proceeding along the correct path these screens were the product of a silver-grain structure with its inherent irradiation effects, which caused the formation of mis-shapen halftone dots. This retarded the use of vignetted screens for many years until the early nineteen-forties when John A. C. Yule, an Englishman working in America at the Kodak research laboratories, overcame the previous stumbling-blocks by making vignetted screens via a cross-line screen on a high-contrast emulsion and using dye-coupled development. The resultant screens possessed magenta-dyed or neutral grey vignetted-dots free from irradiation. During practical tests the emulsion side of these vignetted screens was held, using vacuum pressure, in intimate contact with the emulsion of a sheet of 'lith' type film throughout an image exposure. The contact screen had been invented.

The contact screen, unlike the glass screen, provides a 50/50 relationship between the semi-opaque dots and the transparent openings which divide them. Light rays coming from the original pass through, but are allowed to follow their individual paths set up by the contours of the tone original.

Contact vignetted screen:
1. 50:50 dot structure
2. Enlarged screen dot vignetted dyed image

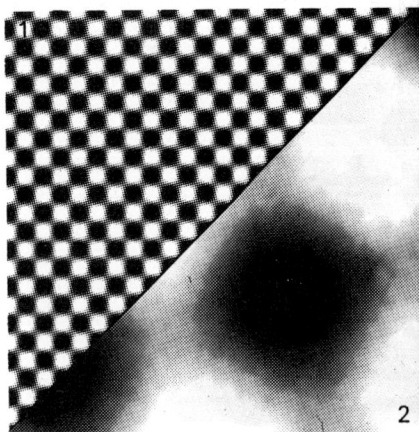

Types of halftone screens

Screen dot structure:
1. Weak light-shadow dot (negative)
2. Medium light-middle-tone dot
3. Intense-light-highlight dot

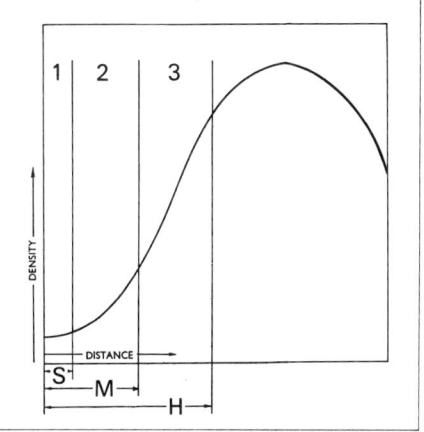

The resultant dots endeavour to redraw the detail of the original by becoming a particular shape. It is the intensity of these rays multiplied by exposure time which determines how much of the screen's vignetted structure will be penetrated, giving rise to dots of varying size.

Contact screens are usually produced from a glass crossline screen. The characteristic light distribution set up behind a cross-line screen is responsible for the vignetted density structure of the dots forming on a new contact screen. It is this density structure, or to be precise the steepness of the gradient which rises from the minimum density on the edge of the fringe to the height of maximum density reached at the apex of the dot's core, that determines the tonal reproduction (degree of tonal separation and graduation) obtainable from any individual contact screen. The difference in density between the minimum and maximum areas of a screen dot, together with its base shape and colour, determines the screen's reproducible density range. A particular screen may only be capable of reproducing originals which have a 1·4 density range or lower, while others under the influence of contrast control techniques or because of their own attributes may have their reproducible density range extended or compressed.

The fact that the density structure of a contact screen controls tonal reproduction was applied to the production of halftone negatives and positives. It was found that 'reasonable' results could be achieved with a gradient of medium steepness, but more 'acceptable' results were obtained when the density structure was changed to a more bulbous slope for negative making and altered to a steeper pinnacle shape for the production of positives. This led to the manufacture of universal, negative-working and positive-working contact screens.

Tone reproduction

Universal contact screen:
1. Density profile
2. Dot structure

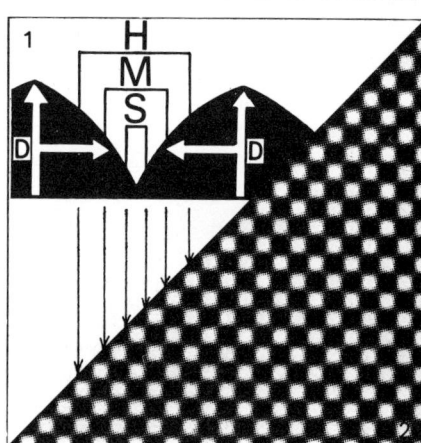

Negative-working contact screen:
1. Density profile
2. Dot structure

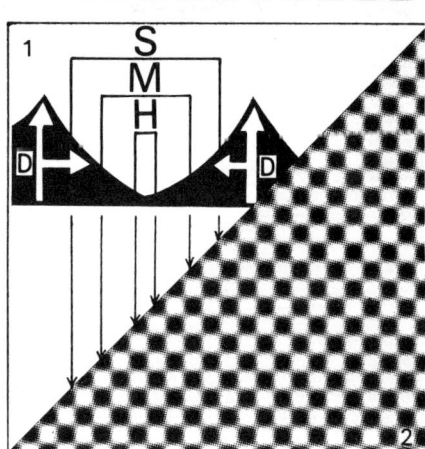

Positive-working screen:
1. Density profile
2. Dot structure

Types of halftone screens

Magenta and grey contact screens

Magenta screens are available for producing negatives from monochrome originals or making positives from continuous-tone negatives. Screens possessing magenta-dyed dots provide the photographer with the facility of being able to alter the screen's reproducible density range by exposing through an appropriate light filter. This system is advantageous when the overall density range of an original needs to be compressed or expanded.

A neutral grey screen is a must for obtaining direct halftone separations from coloured originals, as the magenta screen would act as a second filter. Contrast control is obtained by using 'supplementary' exposure methods, or in work where extreme density changes are encountered by the selection of an appropriate grey screen from a set of screens, each one possessing a different reproducible density range. This type of screen does not provide such flexible contrast control techniques as the magenta screen. The grey screen can with intelligent handling be used for both monochrome and colour reproductions.

Elliptical dot formation

It has become a popular feature for halftone screens to produce an elliptical or chain-dot formation in the middle tones of the reproduction. This was introduced as an answer to the sudden break in tonal gradation which occurs when 50% dots lose their corner link-up. Elliptical dots, even when they are smaller than 50%, remain linked at two diagonal corners. This continuity of dot shape and subsequent ink

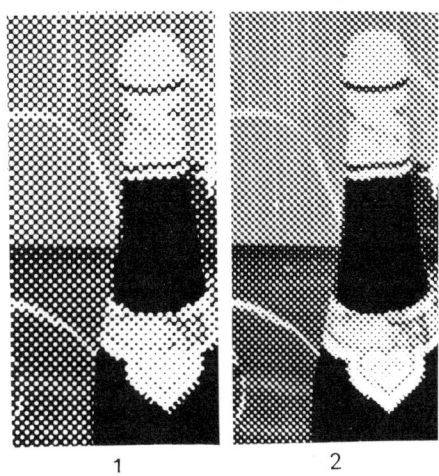

Dot image:
1. Conventional dots
2. Elliptical dots

Tone reproduction

receptance is responsible for the improved smoothness of tone in vignettes and middle tones apparent in most elliptical dot reproductions.

Advantages of the contact screen

It is quite common to be asked the question, 'What are the advantages of using a contact screen in preference to a glass ruled screen?' This is best answered by stating a brief list of both advantages and disadvantages obtained when contact screens are used in the place of glass screens.

Advantages

1. The resolution of critical detail is improved.
2. The screen may have holes punched in it to accommodate register pins. As long as vacuum pressure is provided, one contact screen can be used on a camera-back, enlarger baseboard or in a contact frame.
3. Simplified handling—no screen gear or lens aperture/screen distance calculations. Large lens apertures may be employed to give short exposure times.
4. Contact screens are relatively inexpensive in the first instance.

Disadvantages

1. Contact under vacuum pressure gives rise to Newton's rings, out-of-contact marks and dust spots.
2. Large areas of flat, even tone are difficult to reproduce as a uniform dot-size.
3. Contact screens are handled a great deal and being fragile by nature become easily scratched, stained and kinked. The dye density has a tendency to wear and fade over a period of time.
4. Consistent manufacture with regard to reproducible density range and tonal reproduction is often poor.

Pre-screened emulsions

After seeing the advantages of having a screen in intimate contact with an emulsion, the Eastman Kodak Company of America took this a step further and varied the sensitivity of the emulsion into a dot pattern. Thus the film became a light-sensitive halftone screen. They called this prescreened emulsion Kodak Autoscreen film.

The silver halide emulsion is desensitized in certain areas by applying localized pressure. A revolving cylinder containing a relief dot pattern is used to apply this pressure as the emulsion in a wet state passes between it and a back-pressure

Types of halftone screens

Image resolution:
1. Ruled screen reproduction
2. Contact screen reproduction (view at arm's length and squint)

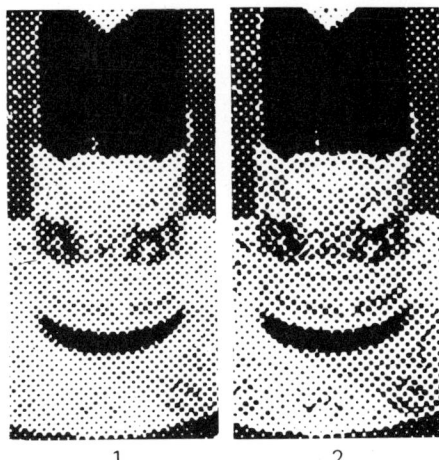

1   2

roller. The areas which receive least pressure become the apexes of maximum sensitivity, while other areas become regions of medium and minimum sensitivity.

Weak reflections from the shadow tones of an original will produce small dots on the apexes of maximum sensitivity, while stronger light rays reflecting from the middle tones and highlights of the original will progressively overcome the less sensitive regions forming larger dots. Contrast control relies mainly on the flash exposure and the film is restricted to 250 mm × 200 mm (10 in. × 8 in.) and 350 mm × 275 mm (14 in. × 11 in.) sizes, patterned with 54 dots per cm (133 dots per in.) running at 45°.

To fully appreciate the densitizing effect of applying localized pressure, keen students will carry out the following experiment. Take a small sheet of lith film from the box and lay a newly-minted penny on the emulsion. Now comes the highly technical part of the experiment—give the penny one jolly good bash with a hammer! Remove the penny and expose the film to the room's white lights for approximately five seconds, finally processing the film in the normal manner. It will be seen that a clear image is formed in areas desensitized by the localized 'pressure'.

Effect screens

The usual halftone result is not always required and many graphic designers often seek other mediums to stimulate the eye. This has resulted in the manufacture of numerous contact screens offering single line, wavy line, target circles, irregular grain and textured effects.

It has been found that making a suitable continuous-tone

Tone reproduction

Effect screens:
1. Regular grain
2. Coarse dot
3. Irregular grain
4. Single line

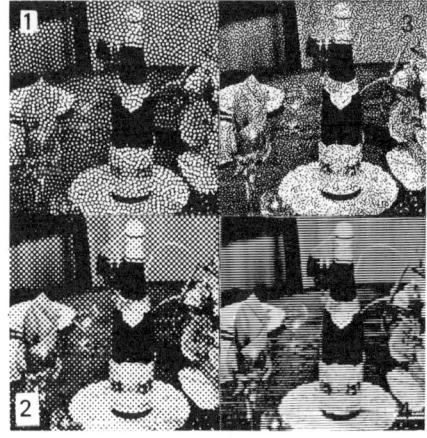

negative first and then introducing the effect screen to produce a positive is one of the best methods, because the photographer has more control over the result, combined with the fact that to achieve another effect it is just a matter of changing the screen and contacting. To give an idea of the appeal of this type of reproduction the following examples have been chosen.

At the time of writing the manufacture and use of contact screens has become widespread. All manner of variations on the basic theme have been introduced, giving the graphic reproduction photographer an immense choice. A screen may be selected to suit a particular process, reproduce a certain density range or to give a specific tonal reproduction. It would be pointless to try to give a list of screen manufacturers and differentiate between their products. It is much better for the camera operator involved to understand the underlying principles, and once it is quite clear what particular screening requirements are needed in his company he should consult the manufacturers who will provide up-to-date information and specific recommendations.

Halftone theories

After using a glass ruled halftone screen it is quite obvious that the screen produces dots, but exactly what influences the dot size, highlight join-up and overall tonal range has been a bone of contention for many years. The answer is very important indeed because once it is known, contributory factors such as screen distance, lens aperture size and shape, may be calculated with certainty.

Halftone theories

Lens-aperture effect:
1. Square aperture
2. Round aperture

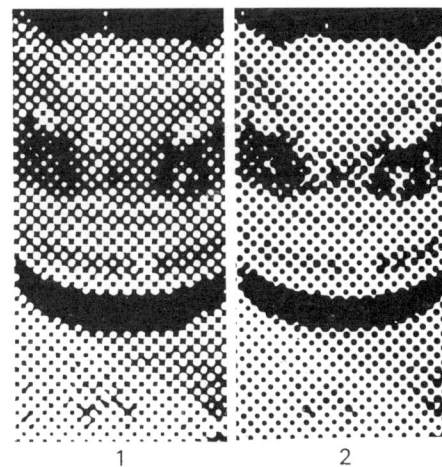

1 2

Pin-hole theory

Owing to the fact that the screen opening acts very much like a pin-hole camera photographing a minute section of the original picture in terms of a dot, it is not unnatural that the first halftone theory, expounded by Frederic E. Ives in 1888, should be the pin-hole halftone theory. This theory proposed that halftone dots forming on the emulsion were in fact pin-hole images of the lens aperture. The central core of high density appearing in each individual dot was explained as a photograph of the lens aperture, while its surrounding fringe, which varied in size, was the product of exposure time. This theory would suggest that very small opaque dots can only be produced by using a small lens aperture, a conclusion which when applied in practice is found to be untrue, but nevertheless the shape if not the size of the lens aperture does have an effect on the dots. This becomes clear after viewing the following examples.

Penumbral theory

Many investigators of the halftone dot phenomenon concentrated on the pin-hole idea and considered the facts. Light passes through the lens aperture and makes its way towards the emulsion, but is obstructed by the opaque mesh of the screen. A portion of the light passes directly through the screen openings, while the remainder is absorbed by the opaque lines. The direct rays form conical light beams whose apexes are brought to a definite focus—represented by a small intense spot of light which may be referred to as the point of maximum illumination. This is surrounded by partial illumina-

Tone reproduction

Distribution of light:
1. Maximum illumination
2. Partial illumination
3. Complete shadow
A and B forming dots

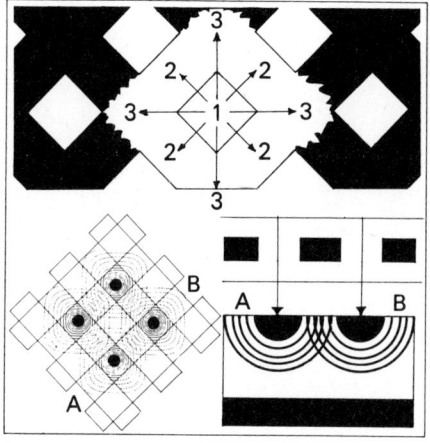

tion termed the penumbral regions. Behind each intersection of opaque lines is an area of umbra—complete shadow.

This distribution of light could be interpreted into an explanation for the dot growth. The opaque lines of the screen cause shadows to be cast upon the emulsion. The shadows vary in intensity from complete shadow to no shadow at all. According to the intensity of light reflected from the original a greater or lesser area of shadow will be sufficiently intense to affect the emulsion. In a highlight area a large area of emulsion is affected and conversely, with a shadow area only, a small area is affected, resulting in the formation of large, well-jointed highlight dots and small, sharp shadow dots.

All these factors may be related to geometrical optics and

Geometrical lines:
1. Maximum illumination
2. Partial (penumbral) illumination
3. Complete (umbra) shadow

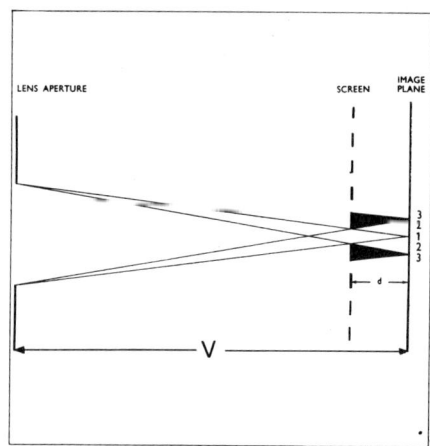

Halftone theories

are well presented by drawing geometrical lines radiating from the lens aperture.

This explanation, known as the penumbral theory, was first developed between the years 1895 and 1896 by A. K. Tallent, A. W. Dolland and E. Deville. The theory advocates that the point of focus is determined by geometrical optics based on the assumption that light always travels in straight lines. In later years it proved popular because it provided a practical equation for camera operators to use when selecting a lens aperture and setting a screen distance, the basic rule being that after a suitable ratio number has been chosen, e.g. 64, then the lens aperture should be $\frac{1}{64}$ of the camera extension and the screen distance 64 times the screen aperture. In maintaining this relationship at all magnification factors similar halftone negatives could be produced from comparable tone originals; in short, the application of the following equation.

$$\frac{\text{Lens Aperture}}{\text{Camera Extension}} = \frac{\text{Screen Aperture}}{\text{Screen Distance}}$$

The lens aperture is generally the diameter of a circular aperture produced by an iris diaphragm or the length of one side of a square slip-in stop. The screen aperture in this case is the length of one side of the screen opening, e.g. a 54 lines per cm (133 lines per in.) screen would have a screen aperture of 1/108 cm (1/266 in.) because the screen has a line ratio of 1 to 1 and therefore in any one centimetre (inch) there are 54 (133) opaque lines and 54 (133) clear spaces. The side of one screen opening would be a 108th (266th) part of a centimetre (inch). The equation using the 64 ratio would appear in practice as:

$$\frac{1}{\text{Ratio}} = \frac{\text{Lens Aperture}}{\text{Camera Extension}}$$

$$\frac{\text{Lens Aperture}}{1} = \frac{\text{Camera Extension}}{64}$$

$$\frac{1}{\text{Ratio}} = \frac{\text{Screen Aperture}}{\text{Screen Distance}}$$

$$\frac{\text{Screen Distance}}{1} = \frac{1}{\text{Twice the Screen Ruling}} \times \frac{64}{1}$$

Diffraction theory

The fact that the penumbral theory produced acceptable

Tone reproduction

halftone images when used to calculate the screen distance and lens aperture for medium screen rulings of 40–60 lines per cm (100–150 lines per in.) but gave screen distances which were too short for coarse screens and too long for very fine screens led research workers to believe that this theory was incomplete. So they turned to the diffraction theory propounded by Max Levy in 1894.

Diffraction is the bending of light at the edge of an opaque object. When a beam of light passes through a small aperture or past the edge of an opaque body (e.g. a black screen line) and then falls upon a focusing screen, patterns of light and dark bands or circles are observed near the edges of the beam, and extend into the geometrical shadow. These patterns, known as diffraction effects, can be given as an example of interference and are the result of the wave nature of light.

A wavefront is formed by the combination of all the wave-

Diffraction:
Wavefronts → wavelets
1. Primary wavelets
2. Secondary wavelets
3. Tertiary wavelets

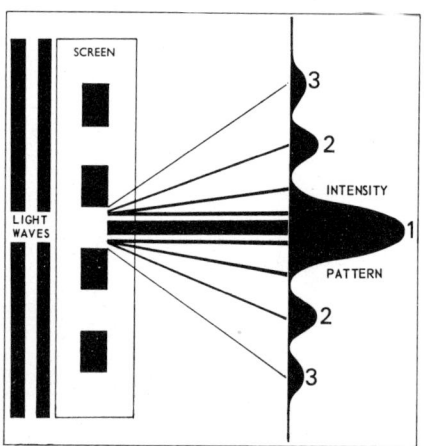

lets. On the side of the obstruction no wavelets can be formed, but wavelets do originate from the non-obstructed side. The wavefront in the neighbourhood of the obstruction is thus made up of wavelets from the unobstructed side only. This gives a weak wavefront passing round the obstruction and a corresponding deficiency in the remainder of the front.

To use an analogy, light is similar to the wave-pattern set up by the sea. Standing on a high cliff and watching the waves rolling in to a beach, the formation of definite wavefronts can be seen, but if there happens to be a rock rising out of the sea a few yards away from the beach this will interrupt the wavefronts. Waves striking the centre of the rock roll backwards, waves impinging at the edges are caused to bend round

Halftone theories

forming secondary wavelets which follow a different path to the un-impeded waves. In halftone photography the rock becomes a single opaque line in the screen.

The original idea of the screen openings acting as pin-hole cameras is understandable. If the light rays emerging from the openings are all sharply focused to one apex of maximum illumination then the halftone dots would certainly vary in intensity owing to the diverse reflecting surfaces of an original picture, but little or no variation in size would take place. The penumbral theory states that the shadows caused by geometrical optics are the influencing factor in the dot's growth. This is by no means conclusive because if the lens aperture is reduced to its smallest diameter the penumbral regions should disappear, producing a negative image of the screen lines. However, if a reduction is carried out it will be seen that this is not the case, diffraction effects preventing it happening.

In fact, V. G. W. Harrison in his comments on the diffraction theory states that the halftone dot is *not* an image of the lens aperture. He states:

'The shapes of the halftone dots are fundamentally the same whether the diaphragm be round or square. The middle tone dots remain unaffected whatever the orientation of a square lens aperture. Moreover, by suitable adjustment of screen distance, round stops may be made to give square dots and square stops round dots.'

Some researchers believe that the dot growth is caused entirely by diffraction—the interference of secondary and tertiary wavelets bending round the edges of the screen lines. The more intense the light, the stronger the wavelets and the greater the diffraction effect which increases the area of light exposing the emulsion, therefore building up a larger dot and vice versa.

Diffraction effect:
1. Maxima, plus adjacent 2nd max
2. Secondary maxima
3. Tertiary maxima

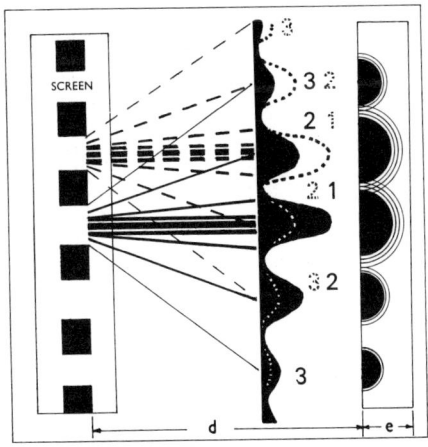

Tone reproduction

This build-up of light may be appreciated if intense white light emerging from a medium size aperture is projected through a fine screen (e.g. 60 lines per cm (150 lines per in.). Using a magnifying glass a definite light pattern will be seen on the focusing screen. Adjust the screen distance until the central area of the light pattern emanating from one screen opening becomes a small apex of maximum illumination surrounded by circles of light which gradually diminish in intensity. These circles are caused by the secondary and tertiary wavelets (diffraction) followed different paths to the central wave. The diffraction theory is based on the idea that if the light of these surrounding circles overlaps the apexes of adjacent dots their individual size will increase and they will become firmly joined together at the corners.

One conclusion established by the diffraction theory is that every screen has its own focal length—the precise screen distance at which small sharp shadow dots are formed. This distance is controlled mainly by the screen ruling. Each screen can be given a speed ratio by dividing the diameter of the screen opening into the screen's focal length. Taking the 54 lines per cm (133 lines per in.) screen as a base of 1, fine screens usually need shorter exposure times to produce a similar dot range, while coarse screens need a substantial increase. Finally, the screen distance must be proportional to the area of a screen opening and *not* to the length of one side as recommended in the penumbral theory. If the area of the screen opening is to be used, the wavelength of the light exposing the emulsion must also be taken into account, together with the fact that the ratio of screen aperture to screen distance needs to be the same as the ratio between the lens aperture and the camera extension.

Determined researchers into halftone dot formation, such as J. A. C. Yule, V. G. W. Harrison and W. B. Hislop, have gone to great lengths in their investigations, producing photo-micrographs, isophot diagrams and experimental evidence to explain the light distribution and intensity variations set up behind a screen opening. The result of these individual approaches is an agreement that an optimum screen distance would be found by the formula:

$$\frac{d^2}{3\lambda}$$

This is obtained from calculating the maximum sharpness of a pin-hole image including diffraction effects. The diameter of the pin-hole in our case is the length of one side of the screen opening squared, (d^2) and the wavelength of the light

Halftone theories

being used becomes λ (the Greek letter lambda). It has been found that taking the wavelength of blue light, 460 nm, as an average figure, gives satisfactory results.

$$\frac{d^2}{3\lambda}$$

Example. Using a 54 lines per cm glass ruled screen the screen opening would be termed d.

$$d = \frac{1}{108} \text{ cm}$$

As the screen distance is usually set in millimetres convert

$$d = \frac{1}{10 \cdot 8} \text{ mm}$$

A wavelength of 460 nm (blue light) is normally taken as λ

$$\text{Screen distance} = \frac{\left(\frac{1}{10 \cdot 8} \text{ mm}\right)^2}{3 \times \frac{460}{1{,}000{,}000} \text{ mm}}$$

$$\text{Screen distance} = \frac{1{,}000{,}000}{10 \cdot 8^2 \times 3 \times 460} \text{ mm}$$

$$\begin{array}{r} 92592 \cdot 92 \\ 108 \overline{\smash{\big)}\ 10000000} \end{array}$$

$$\begin{array}{r} 8573 \cdot 39 \\ 108 \overline{\smash{\big)}\ 925929 \cdot 20} \end{array}$$

$$\begin{array}{r} 2857 \cdot 80 \\ 3 \overline{\smash{\big)}\ 8573 \cdot 39} \end{array}$$

$$\begin{array}{r} 6 \cdot 21 \\ 460 \overline{\smash{\big)}\ 2857 \cdot 80} \end{array}$$

Screen distance = 6·21 mm

It appears from experimental research and practical experience that no one theory can provide complete explanations for all conditions taken up by the dot as it grows in size. The individual theories seem to become more relevant as the dot grow in size. Perhaps the pin-hole theory accounts for the change of shape which occurs in small shadow dots differently

Tone reproduction

shaped lens stops are used. Then as the intensity of the reflected light increases the penumbral shadows are cast, influencing the formation of middle tones. The progressive increase of intensity produced by the highlights of the original sets up strong diffraction effects round the edges of the screen openings, the secondary wavelets increasing the size of the highlight dots and forming strong 'join-ups' at their corners. Whatever the complete answer proves to be, it is quite clear now that when glass ruled screens are being used great care must be taken in finding the precise screen distance and corresponding lens aperture size if optimum halftone photographs are to be achieved.

Application of halftone theories

Halftone photography using glass ruled-line screens is controlled by the following variants. A complete understanding of their individual effects upon the resultant negative combined with a knowledge of their inter-relationships with one another is essential.

1. SCREEN DISTANCE. The precise focal length of the screen which becomes an actual air space between the screen's lines and the emulsion. The screen distance is reliant upon the screen ruling and the halftone theory being used as a base.

2. LENS APERTURE. The diameter of a circular aperture or the length of one side of a square stop. The size of the lens aperture and the length of the screen distance are related, but to provide contrast control over the resultant negative these two important factors may be altered individually, leaving the other as a constant, or both varied at the same time.

3. IMAGE EXPOSURE. The transmission of light reflecting or transmitting from an original through the screen multiplied by time.

4. SUPPLEMENTARY EXPOSURES. These are given to improve or control the contrast range of the result.
(a) The 'flash' exposure, a short exposure to white light which produces denser shadow dots.
(b) The 'no-screen' exposure, an additional exposure without the screen which accentuates the highlight tones.
(c) The 'filter' exposure. If the screen contains magenta-dyes lines, then an exposure through a light-magenta filter will increases the contrast range of the result, while the use of a yellow filter will decrease the contrast.

To obtain some order in the use of these variables it is

Halftone theories

recommended that a fixed screen distance is used for a particular screen ruling because its calculation and setting is necessarily precise. The lens aperture needs to be a constant proportion of the camera extension so that exposure times remain basically the same at different magnification factors—$v/64$ is generally a good basic ratio to use. For reproducing originals possessing average density ranges of 1·0–1·5 it is normally found that a single image exposure is sufficient. Originals with greater density ranges, 1·5–2·0, will need a 'flash' exposure which will increase progressively in proportion to the increase in original density. Originals which require more contrast in their highlight regions can be given a short 'no-screen' exposure. There are other contrast control systems which move away from the simple v/ratio system. Some use different exposure times through varying lens aperture sizes, and these are known as multiple-stop systems. Others employ small alterations in the screen distance. We will now look at the precise settings of screen distances and lens apertures along with these rather more complex contrast control systems.

Screen distance settings

From the previous work it can be seen that there are two major ways of calculating the screen distance—the penumbral equation and the diffraction formula. In both cases the measurement obtained is the optical distance between the screen lines and the emulsion. To find the actual 'air-space' between these two planes the thickness and refraction of the back glass must be taken into account and subtracted from the calculated screen distance. The influence of the back glass is found by the following formula.

$$\text{Subtracted figure} = \frac{\text{Thickness of back glass}}{\text{Refractive index of glass}}$$

$$\text{e.g. Subtracted figure} = \frac{0 \cdot 125 \text{ in.}}{1 \cdot 5}$$

convert to millimetres—

$$\text{Subtracted figure} = \frac{0 \cdot 125 \times 25 \cdot 4}{1 \cdot 5}$$

$$\text{Subtracted figure} = \frac{3 \cdot 175 \text{ mm}}{1 \cdot 5}$$

$$\text{Subtracted figure} = 2 \cdot 113 \text{ mm}$$

Tone reproduction

Now that the calculation of this figure is known it is an appropriate time to review the optical screen distances produced by the penumbral and diffraction theories. It has been found in practice that penumbral distances related to the $v/64$ ratio are only suitable for medium screen rulings 40–60 lines per cm (100–150 lines per in.). Diffraction distances have been 'modified' by F. J. Tritton so that two convenient v/ratio numbers, $v/80$ and $v/72$, could be used. These distances corrected by subtracting the 'back glass' influence can be noted and their photographic results recorded. By carrying out purely practical experiments comparison screen distances may be found.

A method introduced by W. B. Hislop recommends the use of a magnifying glass focused on the inner surface of the central transparent area of the focusing screen to find the 'diffraction' screen distance. The camera extension is set at the 'same size' distance (twice the focal length of the lens), while the flash exposure lamp is switched on behind a lens diameter of 2·5 mm. The halftone screen is pulled up to the focusing screen. On slowly moving the halftone screen back and looking through the magnifying glass, a diffraction pattern of intense lines is seen emanating from behind a screen opening. These intense lines move to a point of focus forming a cross of bright light. The screen distance is noted and recorded against the screen ruling being used. The mean of this distance and the calculated diffraction distance will produce a halftone result which embraces diffraction effects.

Another practical method is to position the screen slightly nearer to the emulsion than the penumbral distance. Load a photographic plate (not a lith type—use a process or continuous-tone emulsion) in a sloping position, so that its top edge is against the plate clips, while its bottom edge is held away from the lower clips by approximately 3 mm, using a metal spacer made from a strip of composing furniture. Carry out a photographic test by giving a short exposure using the flash lamp. The photographic result will contain small shadow dots which when closely inspected with a microscope will vary in sharpness. Select the line of sharpest dots and mark this on the back of the plate. Reload this plate into its original sloping position. Using a wedge marked off in millimetres find the precise distance between the selected line of dots and the screen. The mean of this distance and the calculated penumbral distance will give a halftone image which includes the effect of penumbral shadows. After producing all these results tabulation is the best way of comparing them.

Halftone theories

Table 16. Screen distances (Backplate 3·174 mm (0·125 in.))

Screen ruling		The diffraction 'mean'		The penumbral 'mean'	
Lines per in.	Lines per cm	Screen distance mm	Aperture ratio	Screen distance mm	Aperture ratio
50	20	22·0		15·0	
55	22	20·0		14·0	v/64
60	24	18·0	v/80	13·0	using a
65	26	16·0		12·0	square
75	30	12·0		9·5	stop
80	32	11·5		9·0	
100	40	6·5		7·0	
110	44	6·0		6·0	
120	48	5·0		5·0	
133	54	3·5	v/72	4·5	v/64
150	60	3·0		4·0	
175	70	2·0		3·0	
200	80	1·5		2·5	

Each screen has its own individuality and it will be found by practice and experiment which screen distance suits its working condition. A worthwhile aid to camera operating is the Agfa-Gevaert Screen Key device, or an assembly of the relevant information in graph form. Settings for screen distances and lens apertures are found by a trace of the finger.

Screen distance and lens aperture graph

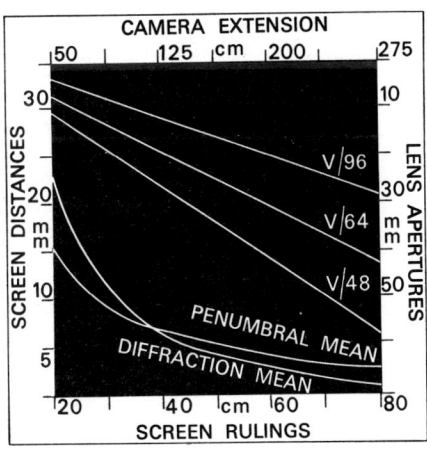

Tone reproduction

Decreased screen distance:
1. Lens aperture V/64
2. Screen distance decreased
3. Halftone image

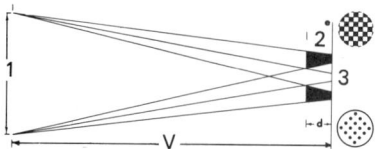

Correct screen distance:
1. Lens aperture V/64
2. Screen distance correct
3. Halftone image

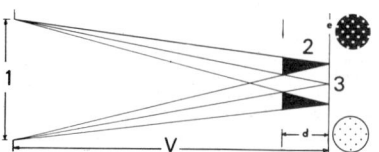

Increased screen distance:
1. Lens aperture V/64
2. Screen distance increased
3. Halftone image

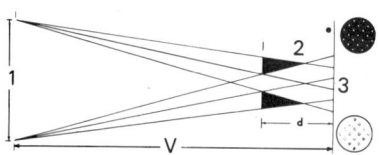

Halftone theories

Reduced lens aperture:
1. Lens aperture V/96
2. × 64 screen distance
3. Halftone image

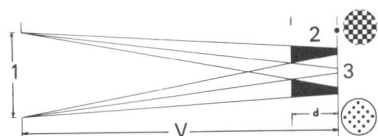

Correct lens aperture:
1. Lens aperture V/64
2. × 64 screen distance
3. Halftone image

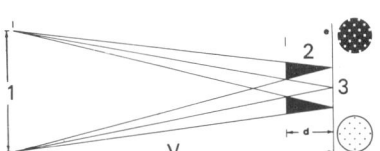

Enlarged lens aperture:
1. Lens aperture V/48
2. × 64 screen image
3. Halftone image

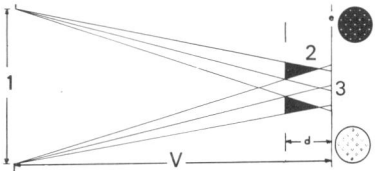

Tone reproduction

Contrast control using the screen distance

Halftone reproductions are usually required from a diverse range of originals. The majority of requests are for comparable negatives, and this means a standard negative must be formulated for the particular printing process and the specific dot sizes obtained by employing contrast control techniques. The first of these is minute movements of the screen distance, keeping the lens aperture ratio constant. Originals may be divided into three main groups.

1. Low contrast—density ranges 1·0, 1·1, 1·2, 1·3.
2. Medium contrast—density ranges 1·4, 1·5, 1·6.
3. High contrast—density ranges 1·7, 1·8, 1·9, 2·0.

By moving a glass-ruled screen and altering the screen distance, that is the actual air-space between the black lines of the screen and the emulsion surface. The originals classified as possessing low or high contrast may be altered during the photographic procedure into negatives of medium contrast and should be comparable to the negatives obtained from the group of medium contrast originals. The final dot sizes of a standard negative and positive should embrace the tonal distortion peculiarities of the printing process together with such important considerations as ink-gain factor and, of course, the printability, reflectance and surface quality of the paper selected for the printing. Once these factors have been investigated a standard negative and positive, in terms of dot sizes in percentage or integrated dot density may be determined.

Standard negative:

	S	M	H
1. Original	1·40	0·70	0·05
2. Negative	10%	50%	95%
3. Positive	90%	50%	5%

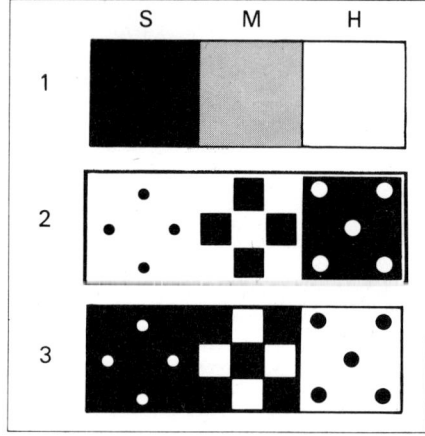

If a standard negative is required the screen distance would be kept at normal (calculated or found by experiment) for medium density originals, increased up to a maximum of

Halftone theories

1·5 mm for originals of low density, and reduced by a maximum of 1·5 mm to accommodate originals possessing high densities. To illustrate the effect of moving the screen distance diagrams and resultant reproductions have been included.

Lens aperture variations for contrast control

Another approach to the problem of producing a standard negative is to maintain a constant screen distance and use different exposure times through lens apertures of varying size. The simplest application of this idea is to use the $v/64$ ratio for originals of medium contrast, $v/48$ ratio (a larger lens aperture) to expose originals of low contrast, and $v/96$ (a smaller lens aperture than normal) to photograph originals possessing high densities. The results of this contrast control technique are illustrated.

An advancement on the varying lens aperture technique is to use a multiple stop system. The total image exposure is divided between two stop sizes—$v/64$ for the 'main' exposure (this ratio should be related to the screen distance) and $v/24$ for the 'highlight' exposure. Contrast control of the original's density range is effected by varying the ratio of highlight exposure to main exposure and when necessary giving a flash exposure.

Generally speaking, the main exposure time ($v/64$) is calculated so that a 75%–80% dot (0·70 integrated dot density) is produced in the highlight areas. The highlight exposure ($v/24$) is now added, being approximately 4% of the main exposure time. Because of the enlarged lens aperture the highlight dots grow in size without increasing the middle-tone dots. The final negative for most processes requires on average a 95% (1·30 integrated dot density) highlight dot, 50% (0·30 integrated dot density) middle-tone dot and a 10% (0·05 integrated dot density) shadow dot. With intelligent application this system can produce halftone negatives containing specific dot sizes.

A further extension of this multiple stop system is to give three different exposures, $v/90$ (shadow dots), $v/64$ (middle-tone dots) and $v/45$ (highlight dots). Medium contrast in originals is reproduced by a ratio of exposure times, namely $v/90$ (2), $v/64$ (4) and $v/45$ (1). Originals of high contrast receive longer exposures through $v/90$, while low contrast originals are given an increased exposure through the $v/45$ aperture. The permutation of exposures in this system can become rather complicated.

Flash exposure

If the original's density range exceeds the reproducible density range of the screen a flash exposure becomes

Tone reproduction

necessary. The precise period of time required may be determined by consulting a 'flash exposure graph' as illustrated in the chapter on systematic working. The flash exposure can be made to a sheet of white paper or better still a piece of white opal plastic, employing the same illumination and lens aperture size as used for the 'main' exposure. Very short exposure times are overcome with the use of neutral density filters as it is advantageous to use the 'main' exposure stop for flashing. Most modern cameras have a flashing lamp—a tungsten lamp behind a piece of opal glass—which may be conveniently positioned in front of the lens. The lamp is switched on and timed accurately by means of a clock and lens-shutter circuit.

When contact screens are used a controlled flash technique is often employed. A bee-hive light source is set up in the darkroom approximately four feet away from the centre of the opened vacuum-back. The film and screen are held on the vacuum-back. Using a precision timer the flash exposure is given first. The vacuum-back is then closed and the camera's image exposure follows. With an electronic timer controlled flash exposures up to $\frac{1}{10}$ of a sec may be given.

No-screen exposure

A short image exposure without the screen can raise the threshold (development capability) of the fringe surrounding highlight dots. This improves the negative's highlight gradation. To facilitate this exposure the glass screen must be moved away into its cassette and a compensating glass (replacing the screen and its refraction effect) introduced into the light path without disturbing the film. The 'main' exposure should produce 80% (0·85 integrated dot density) highlight dot, then a short 'no-screen' exposure at the same v/ratio aperture will increase the highlight dots to the required size. If very short exposure times are encountered introduce appropriate neutral density filters.

Assessment of the original

The definition of a *tone* original is a picture created in black and white, but with intermediate tones of varying density, e.g. photographic prints, chalk drawings, pencil sketches and wash drawings. Before commencing any photographic work the camera operator must study the tonal picture, becoming aware of its origination, feeling of contrast, number of tonal movements, highlights, middle-tone gradation and areas of maximum shadow. Excellent reproductions can only follow complete appreciation of the original.

Halftone theories

A tonal image

The amount of tonal movement in a picture, interpreted by us into greys of varying dilution, is the result of our eyes becoming stimulated by the different intensities of reflected or transmitted light emanating from the object. Our mechanism of vision allows these various light intensities to excite the ultra-sensitive tips of the rods—light-receptor cells housed in the retina. The rods allow complicated patterns of light to be translated by our central nervous system into three-dimensional concepts, each pictorial movement being a translation in terms of grey density. The graphic reproduction photographer has to emulate the way our visual senses record tonal differences by using light-sensitive silver halides and varying amounts of printing ink.

Reproduction of tone originals entails the photographic stages of copying tone for tone, enabling gravure printing to take place or the translation of tone into halftone in the case of letterpress, lithographic and screen process printing. In each method the subsequent printed copy must compare favourably with the original; that is, the maximum shadow area of the copy should be similar to that of the original. The tonal gradation and highlight accentuation needs to be as near facsimile as is possible within the capability of the printing process being employed.

The photographer should commence operations by taking reflection density readings of the original in order to ascertain its density range. The readings obtained may be used for comparison against ideal 'white' and 'black' reflection density readings. The original's attributes are now ready to be evaluated against the previously mentioned tone standards for the different processes printed on a variety of papers. This comparison, as in line reproduction, is best completed by following a number of steps.

1. Compare the original with the tone standard printed on a similar paper to the stock stipulated for the subsequent printing. With gravure printing a particular positive density range and a 'fish-net' screen ruling will associate itself with this type of paper. Acceptable printing on this paper by the other processes will become related to a particular screen ruling, e.g.:

Process	Paper	Most suitable screen ruling	General comparison with original
Lithography	Machine finished	40 lines per cm (100 lines per in.)	Loss of detail higher contrast

Tone reproduction

2. Check the maximum ink density of the tone standard with the original's darkest area, e.g.:

Process	Paper	Maximum density reading	Original's darkest tone
Lithography	Machine finished	1·2 (solid area)	1·60

3. Note the loss or retention of the lightest tones on this paper, e.g.:

Process	Paper	Minimum density reading	Original's lightest tone
Lithography	Machine finished	0·07	0·05

4. Record the maximum and minimum densities or dot sizes together with the number of tonal movements capable of being printed on this paper by the particular printing process, e.g.:

Process	Paper	Largest dot dot size retained	No. of tonal movements	Smallest dot size retained
Lithography	Machine finished	75% (0·60 integrated dot density	8	15% (0·07 integrated dot density

Tone control strip

From the information obtained the tone control strip can be masked-off to indicate the extremities of the original, but due note must be taken of the limitations predicted by the appropriate tone standard. It can be seen from the example figures that the tonal reproduction in question will at the optimum, on machine-finished paper using the lithographic printing process, look less opaque in the shadow areas than the original, and there will be a loss of tonal gradation and detail because of the use of 40 lines per cm (100 line screen ruling), while the highlights will look 'greyer' than the original's.

Using this information, especially the density range and maximum density values of the original now clearly indicated on the control strip, calculate the required image exposure time which will produce the end densities or dot sizes predicted by the tone standards, using the system recommended by the emulsion manufacturer. This will be based on the previously mentioned basic equations of:

Halftone theories

Continuous-tone image

Log of the   Log of the   New   Old
New exposure   =   Old exposure   +   shadow   −   shadow
time   time   density   density

Development = $\dfrac{\text{Negative density range}}{\text{Original density range}}$ Consult time/gamma curve

Halftone image exposure

Log of the   Log of the   New   Old
New image   =   Old image   +   highlight − highlight
exposure time   exposure time   density   density

Flash exposure = $\dfrac{\text{Percentage of basic flash exposure time}}{\text{(Consult 'flash-exposure' graph)}}$

Development γ = Original's density range   Related to the   Lith development 'latitude' graph

The manufacturer's systems embrace all the relevant factors and relate them to their range of emulsions.

Screen ruling and contrast

When exceptionally high quality reproductions are being aimed at, the drawing of a 'tonal key' is a great help. A sheet of tracing paper is laid over the original and a light outline drawn round the important pictorial areas. In these areas the predicted dot sizes or positive densities obtained from the tone standards can be recorded along with related negative

Tonal key:
Negative densities and positive dot sizes plotted

Tone reproduction

densities. The values recorded must always be the optimum figures associated with the paper and printing method. A key of this description gives the camera operator and retoucher an immediate check on the photographic results achieved.

A fine screen will *not* automatically reproduce more detail. Three controlling factors influence the choice of screen rulings.

1. The surface of the paper—a harsh rough surface will break up or distort fine highlight dots and maximum shadow dots.
2. Printing conditions—careless adjustment of inking rollers, plate and impression pressures can lead to fine screen images being damaged by abrasion and excessive pressure.
3. Length of run—difficult conditions may be bearable during a short run of a few hundred sheets, but will become totally intolerable when continuously encountered by the machine operator as he endeavours to run thousands of sheets through the printing press.

It will be noticed that coarser screens produce greater contrast and a more 'pleasing' visual sensation because there is a greater area of white reflecting light energy from between the dots. Although the following tabulation is a general guide to the selection of screen rulings, it is infinitely better to rely on tone standards created in the factory under the prevailing conditions.

Table 17. *Screens**

Printing	Paper	Letter-press	Litho-graphy	Screen process	Gravure
Fine art reproductions	High quality art	60(150)	70(175)	40(100)	79(185)
Book illustrations	Process coated or super calendered	48(120)	54(133)	34(85)	66(165)
Magazines	Machine finished	40(100)	48(120)	30(75)	60(150)
Newspapers	Newsprint	26(65)	40(100)	26(65)	40(100)
Coarse packaging materials	Boards	20(50)	26(65)	20(50)	34(80)

* Inch screens in parentheses

Camera procedure

There are two basic ways of approaching the camera procedure for tone reproduction. The reasons associated with these approaches vary according to the printing process employed, but in general they appear as follows.

Camera procedure

1. *The direct method*

In this approach the light passes directly through the screen producing a halftone negative. From this negative a contact positive is made when required. The direct method gives a 'hard' halftone dot on the contact positive which makes it suitable for all plate-making techniques. Using this method a master negative is obtained from which identical positives with regard to dimensions and dot sizes are made with every contact.

2. *The indirect method*

The first stage of this method is the production of a continuous-tone negative. Using this continuous-tone negative a continuous-tone or halftone positive may be made at same-size in a contact frame or at a different magnification using a camera or enlarger. The indirect method is particularly suitable for reproductions requiring extensive retouching. The continuous-tone negative may be masked or retouched by any of the autographic methods. The halftone positive which may be obtained at the conclusion of the indirect process contains dots built up from a central core allowing extensive dot-reduction to take place.

Printing requiring two or more reproductions at different sizes from the same original lends itself to the indirect approach. One continuous-tone negative is made at any intermediate size on which the required masking or retouching is carried out. This negative is positioned in an enlarger or camera and the final positives produced at the appropriate magnification factors.

The indirect method tends to be rather a long way round for letterpress printing because it involves three stages to produce a halftone negative, but at times the masking and retouching advantages make it worthwhile. For the other three processes the continuous-tone negative stage is an ideal chance to equalize the rather diverse nature of the originals encountered. The production of a standard negative from which a halftone screen positive or continuous-tone positive can be made using a constant procedure in the contact frame, camera or enlarger is an ideal work-pattern.

Relevant data and tests

The conventional glass screen, because of its rigid and opaque structure, offers limited possibilities for tone correction. This, together with the general popularity of contact screens, is the reason for the concentration on the use of contact screens in this section on camera procedure. The problem of

Tone reproduction

tonal reproduction is such that it has proved necessary to make two different contact screens, one providing improved tone rendering in halftone negatives, and one to ensure the same with halftone positives.

If any sort of screen is to be used, before acceptable results can be achieved its individuality and reproducible density range must become quite clear to the camera operator. The way in which a screen reproduces tonal values and encompasses a certain original density range is dependent upon the efficiency of equipment, flare factor and development technique. Therefore tests of this nature must be conducted under normal working conditions. This familiarization begins with the plotting of a characteristic screening curve. Like many operations this is more easily understood when presented in stages.

1(a) *A negative-working screen.* Take a long-range reflection step-wedge (16 steps from 0·00 to at least 1·80) and place it in the camera beside a tonal original possessing an average density range of 1·40 (including the flare factor). Focus to same-size and stop the lens down to the usual v/ratio lens aperture.

1(b) *A positive-working screen.* Use a long-range transmission step-wedge (20 steps ranging from 0·00 to 2·00) and position it in the contact frame (camera or enlarger if used) alongside a continuous-tone negative containing end densities of approximately 0·30 shadow and 1·70 highlight, a density range of 1·40. Turn the light intensity control to full power.

2. In each case, expose a sheet of lith film through the screen in question recording the test images. Adjust the exposure times to ensure that a full range of dots is obtained. The negative screen should record a 95% dot from the 0·00 'white' portion of the reflection step-wedge, while the positive screen needs to produce a 95% dot through a shadow density of 0·30. After processing read and record the integrated dot density of each step.

3. By assessing the density readings and closely inspecting the halftone stepwedge produced by the negative-working screen it will be seen that the dots gradually reduce in size until no dot is formed at all. Find the reflection density which produced a small 10% (0·05 integrated dot density) shadow dot. This density value is equal to the reproducible density range of that negative-working screen.

The results from the positive screen are assessed in a similar fashion. The transmission densities which produced a

Camera procedure

Characteristic screen curve

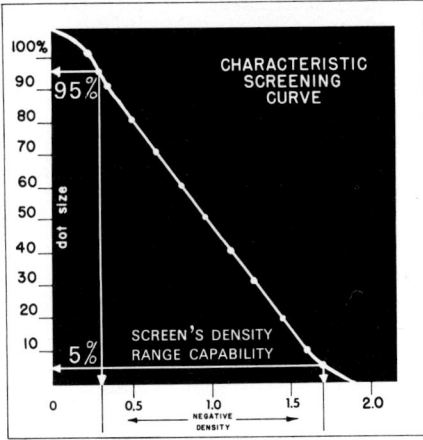

95% (1·30 integrated dot density) shadow dot and a 5% (0·02 integrated dot density) highlight dot, are found and their values subtracted. The answer is equal to the reproducible density range of the particular positive-working screen.

4. Convert all the relevant integrated dot densities into halftone dot percentages and plot these against the appropriate continuous-tone densities obtained from the reflection and transmission step-wedges. The resultant curves predict halftone values from continuous-tone densities and vice versa. The average original and continuous-tone negative will give halftone images from which such important factors as dot sharpness, adjacency (dot drop-out) and tonal gradation can be appreciated. Also different development techniques, such as still-bathing or vigorous agitation, may be carried out to indicate their effect upon the screen's reproducible density range.

Working with lith

When exposing halftone images on to lith-type emulsions the negative and positive contact screens have two specific functions. The negative screen must reproduce original tonal movements as definite halftone steps with special reference to producing firm shadow dots from a wide range of maximum original densities. The positive screen has the same overall task, but should provide increased halftone movement amongst the small highlight dots. This accentuation is necessary because the eye will see these 'white' areas of the reproduction first, because of the light energy reflecting from the greater expanse of white paper surrounding the highlight dots after printing.

Tone reproduction

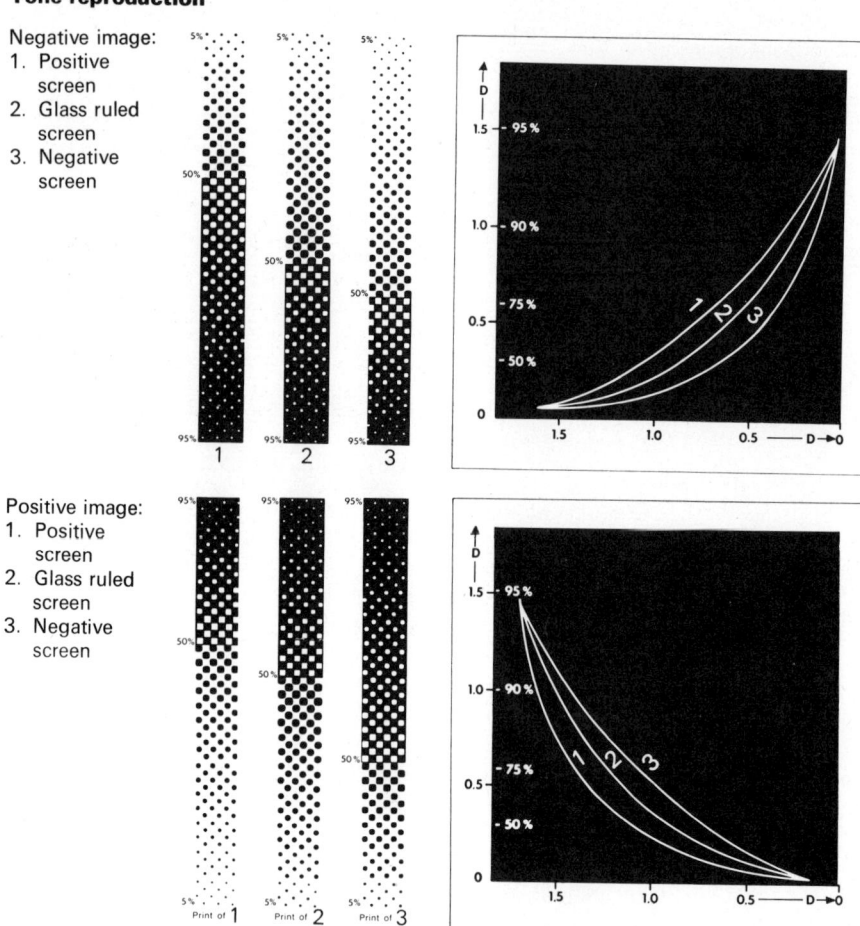

Negative image:
1. Positive screen
2. Glass ruled screen
3. Negative screen

Positive image:
1. Positive screen
2. Glass ruled screen
3. Negative screen

If a single screen made to average specifications (i.e. density gradient, density difference, etc.) was used for both negative and positive work the same portion of the density between the screen's vignetted dots would be used to create both shadow and highlight dots. Neither would approach ideal size or separation from their adjacent tone. This is illustrated by producing halftone step-wedges from the different screens, and comparing them visually and diagrammatically.

Halftone negatives

When the appropriate contact screen is used a simple operation of one 'main' exposure at a large ratio number, e.g. v/22, is sufficient as long as the density range of the original compares favourably with the reproducible range of

Camera procedure

the screen. Therefore it is sensible to purchase a screen which accommodates the majority of the reproduction work encountered in the studio.

Originals with density ranges greater or lower than the screen's reproducible range can be accommodated by using an appropriate flash exposure or no-screen exposure. The no-screen exposure, the requirement for exposing originals of low contrast to a standard negative specification, affects the largest dots which, in the case of a negative, are the highlight dots. Conversely, the flash exposure helps to build the small dots, overcoming the problem in negative production of originals possessing greater tonal ranges than the screen can encompass.

Once the emulsion manufacturer's exposure equation for

Table 18. *Computer programmes (flash exposure)*

$$\text{New flash exposure} = \text{Basic flash exposure} \left[1 - \text{antilog} \left(\text{Screen's reproducible range} - \text{New original range} \right) \right]$$

$$F = F_b \left[1 - \text{antilog} \ (D_S - D_O) \right]$$

Headings	SETV BDFS SETF EXP SETR 9 1) LINE SPACES 5 TITLE TABLE OF FLASH EXPOSURE TIMES LINES 2 SPACES 9 TITLE BASIC DENSITY = 1·3 LINE SPACES 9 TITLE BASIC FLASH EXPOSURE = 60 SECS. LINES 3 SPACES 12 TITLE NEW DENSITY NEW EXPOSURE	TABLE OF FLASH EXPOSURE TIMES BASIC DENSITY = 1·3 BASIC FLASH EXPOSURE = 60 SECS. NEW DENSITY NEW EXPOSURE
Factors	S = 1·3 B = 60	1·3 0·0 1·4 12·3 1·5 22·1 1·6 29·9
Calculation	CYCLE D = 1·3:0·1:3·0 F = S − D F = F = 2·302585 F = EXP F F = 1 − F F = B = F LINE SPACES 15 PRINT D, 1:1 SPACES 10	1·7 36·1 1·8 41·0 1·9 44·9 2·0 48·0 2·1 50·5 2·2 52·4 2·3 54·0 2·4 55·2 2·5 56·2 2·6 57·0
Repeat	PRINT F, 2:1 REPEAT D STOP START 1	2·7 57·6 2·8 58·1 2·9 58·5 3·0 58·8

Tone reproduction

Ideal screen negative:
Integrated dot density plotted against original reflection density

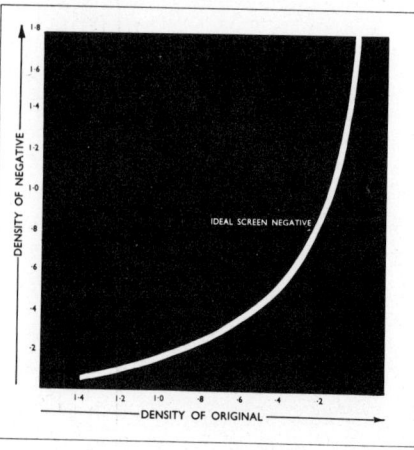

the work in hand is understood and the relevant data collected, complete answers in table-form which may be converted into graphs can be readily obtained at a low price from a company specializing in computer service. The computer programmes in tape-form are kept for producing similar information related to any new problems. An example of this work, taking the flash exposure as a problem, appears in table-form on p. 329.

It is hoped that through applying the previous methods halftone negatives approaching the ideal curve will be achieved.

Halftone positives

When a contact positive is required from a halftone negative the characteristics of the negative are passed on to

Ideal screen positive:
Integrated dot density plotted against negative transmission density

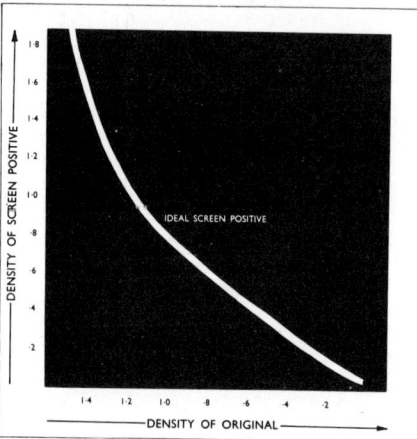

the positive. There is very little chance of altering the dot sizes or tonal gradation, so the direct halftone negative must be photographed with this in mind.

Conversely, the production of an indirect screened positive from a continuous-tone negative lends itself to a wide range of contrast control techniques. It is an advantage to make the continuous-tone negatives to a rigidly controlled standard density range which is the reproducible density range of the halftone screen to be used. This facilitates less complicated work-patterns which consistently produce positives approaching the ideal.

Nevertheless at times it is imperative that the camera operator should be able to alter screened positives to give particular dot sizes or an accurate tonal gradation. There are three basic control methods used in screen positive production.

1. *Flash exposure*

This supplementary exposure controlled with the aid of a density graph can be used with magenta or grey contact screens. The flash exposure builds the small dots; as a positive is being made this means the highlight dots. Due care must be taken as excessive flash exposures replace highlight detail with a flat tint effect. When a contacting procedure is being employed pin-register must be used to enable the flash exposure to be given without disturbing the screen and sensitive film.

2. *No-screen exposure*

This is effective with both grey and magenta screens. The no-screen exposure builds up the large dots, providing a means of accentuating the darker tones of the positive. An exposure of this sort is necessarily short and is best given through a neutral density filter of 1·0. The exposure time is related to the shadow density of the continuous-tone negative. Once an exposure time is established from a standard density, new exposures can be found by the basic exposure equation:

$$\text{Log of the New no-screen exposure time} = \text{Log of the Old no-screen exposure time} + \text{New shadow density} - \text{Old shadow density}$$

3. *Filter control with a magenta screen*

Continuous-tone negatives varying in end densities, but possessing a constant density range related to the screen's reproducible range, can produce acceptable screen positives. Exposure times calculated by the basic exposure equation are associated with the new highlight density in each case.

Tone reproduction

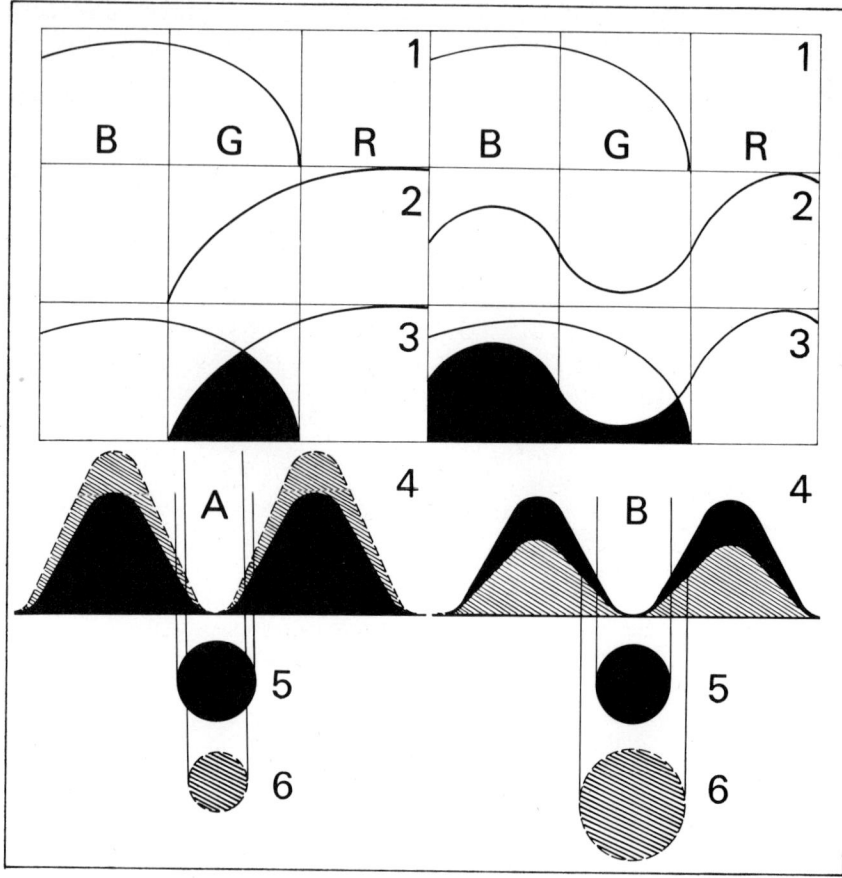

Filter control:
1. Lith emulsion
2A. Yellow filter
2B. Magenta filter
3A. Exposing light
3B. Exposing light
4A. Screen dots increase in size
4B. Screen dots decrease in size
5A. Resultant dot with white light
5B. Resultant dot with white light
6A. Yellow filter dot
6B. Magenta filter dot

Camera procedure

Negatives with density ranges greater or lower than the screen's range can be accommodated by using a magenta screen and changing its apparent dye density with the use of coloured filters. A magenta filter can lower a 1·3 screen to as much as 1·0 or a yellow filter may be used to increase the same basic range to approximately 1·8.

The variable nature of the magenta screen revolves round the following facts. The magenta dye transmits blue light more readily than green light and lith emulsions are only sensitive to blue, green and yellow light. It must be noted that the colour temperature of the light has a drastic effect upon the result. A tungsten lamp with its deficiency in blue wavelengths may produce positives similar to those achieved with an arc lamp and a yellow filter. The effect of introducing a filter upon the light actually exposing the emulsion can be appreciated by drawing wedge spectrograms.

The Kodak company have introduced a series of colour compensating filters, yellow and magenta filters varying in density from 0·05 to 0·50. These filters are coded by their compensating effect, for instance CC 50 *Yellow* or CC 30 *Magenta*. Using a light source producing a fairly even spectrum, the amount by which the CC filter compensates for different density ranges is equal to *half* of the filter's number, e.g.:

Negative range	Screen's range	Difference	Filter
1·15	1·30	− 0·15	CC 30 M
1·50	1·30	+ 0·20	CC 40 Y

The new exposure time for positives from new continuous-tone negatives is readily found on the Kodak exposure computer. This works on the basic exposure equation, e.g.:

Negative highlight density		White light exposures
1·60	=	20 Light counts
1·70	=	CC 20 Y filter

New density is equal to 1·70 plus the effective density of the CC filter, thus 1·70 + 0·10 = 1·80.

Log of the New exposure time		Log of the Old exposure time		New highlight density		Old highlight density
	=		+		−	
		= 1·30		+ 1·80		− 1·60

Consult antilog table—

Tone reproduction

$$\frac{\text{New exposure}}{\text{time}} = \frac{32 \text{ light counts through the CC 20 Yellow}}{\text{filter.}}$$

For camera operators wishing to confine themselves to the use of single yellow and magenta filters, giving a combination exposure of white light and filtered light to obtain the required result, the Ilford screen positive technique is recommended. This is an ideal method of using practical tests to produce workable exposure graphs.

Development

The information regarding lith development in the previous chapter on *line* reproduction is applicable to tonal reproduction. It is a wise step to reduce the number of variables affecting the contrast of a negative or positive to the smallest number possible. Alterations needed to bring about desirable features become clearer this way. After processing, judicious reduction with ferricyanide and hypo produces sharp, clean halftone dots.

Working with continuous-tone emulsions

Conventional gravure uses continuous-tone emulsions throughout the photographic stages of graphic reproduction. The other processes employ continuous-tone photography as part of the indirect method. In every case the production of continuous-tone negatives is an opportune moment to equalize the diverse density ranges of originals in order to obtain a standard negative.

This work revolves round the word 'contrast'. The definition of contrast as applied to graphic reproduction photography is the density range of the photographic image compared to the density range of the original picture. Measurement terms such as gamma, denoted by the symbol γ or bar gee the tangent of an average gradient and its sign $\bar{G}$ are methods of determining a numerical value for contrast.

After exposing and developing emulsions it becomes apparent that image blackening or density rises in accordance with exposure and development times. When a photographic image is presented in graph form it usually follows the characteristic shape of a slope. Look at such a curve and imagine a simple analogy. The flat beginning representing fog level becomes a road, the straight-line portion a hill. The steepness of this hill is the all important factor, because it determines the difference or contrast between travelling along the flat road and climbing to the apex of the hill. The gradient of a hill may be measured

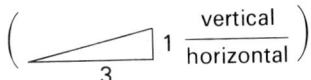

to provide information for the traveller before he attempts the approach, allowing him time to provide favourable conditions for the climb.

This is applicable to photography. The required steepness of a curve's straight-line portion which determines the contrast of the photographic image is found by the equation:

$$\frac{\text{Contrast value}} = \frac{\text{Required density range of the Negative}}{\text{Density range of the Original}}$$

$$\text{Contrast value} = \text{Development time} \quad \text{Consult appropriate } \textit{time/gamma} \text{ or } \overline{G} \text{ curve}$$

The number designated to the steepness of a curve (the contrast value) informs the photographer of the favourable conditions, e.g. suitable emulsion, correct exposure time and related development time, needed to produce such a contrast value.

The numerical value denoting contrast must associate itself with the entire photographic image. Continuous-tone images for gravure printing are controlled by the peculiarities of the pigment paper being employed. This usually results in the use of an emulsion with a long straight-line portion, so that the end density readings of 0·30 and 1·60 fall on this portion of the characteristic curve. Measurement of contrast in this case can be accurately obtained by finding the gamma value γ-tangent of the straight-line portion.

Whenever a continuous-tone emulsion is used on which the lowest image density will leave the straight-line portion and fall on the toe, then contrast measurement resulting from gamma values will give a false idea of the density range of the photographic image compared to the density range of the original, because the image areas on the toe section have been ignored. To overcome this deficiency an average gradient is drawn between the highest and lowest image densities recorded on the photographic image. The tangent of the average gradient termed $\overline{G}$ provides a more accurate measurement of contrast.

Negative emulsions which give the camera operator extra control over contrast are available. These 'variable contrast' films have two emulsion layers producing different degrees of sensitivity and contrast. Although there are a number of types this film usually contains a lower emulsion layer of medium contrast and orthochromatic sensitivity. This layer

Tone reproduction

Finding the gamma value

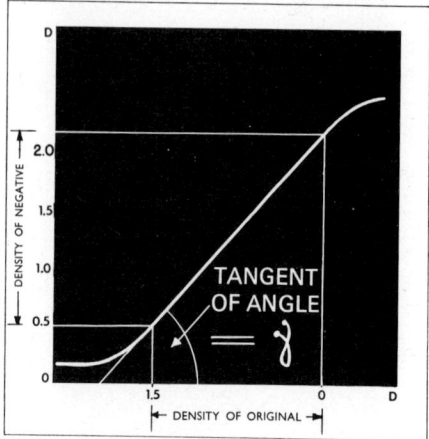

Finding the $\overline{G}$ value

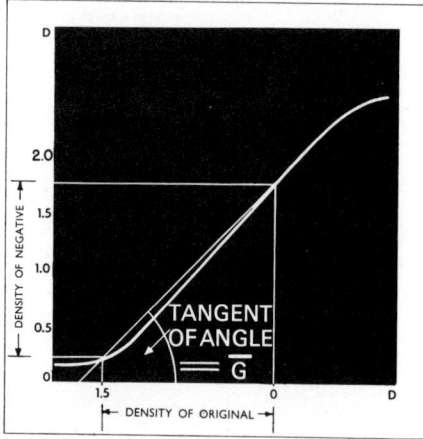

is covered by a high contrast, ordinary emulsion. The full continuous-tone nature of the original is recorded on the lower layer, highlight tones alone affecting the high-contrast emulsion on top. This double-decker result produces acute accentuation of the lighter tones apparent in the original. Exhaustive contrast control effects may be achieved by exposure changes and the use of filters. When employing our example film a blue filter would increase the contrast, while a yellow filter would provide a flattening effect.

Another method of contrast control is the printing of duotones. This is an attempt to overcome the rather restrictive density ranges produced by most printing processes. Two negatives or positives are made. One for a light grey printing of the picture must capture and reproduce all the medium-tones and highlights. The other photograph is for the black printing which should supply the picture's three-quarter tones, dark-tones and shadows. When comparing these final positives or individual printed sheets, the grey needs to be rather full and flat looking, but with definite tonal movements in the lighter tones. The black should give the sensation of contrast and detail. Negatives are made with these end results in mind. The grey may be approached in two ways. Firstly, a thin-looking, low density negative is made, which certainly gives a flat positive, but highlight definition suffers. A second method is to produce a negative to a high gamma with all its densities on the straight-line portion of the curve. By exposing to a strong light-source a suitable grey positive is achieved with good highlight definition. Combining the black and grey printings, which need to be at different screen angles, should extend the visual range of the reproduction.

Continuous-tone development

All the recommendations previously stated in the chapter on *line* reproduction are readily applicable to tone reproduction. The method of development, measurement of image contrast, related exposure and development times only become valid, logical and predictable when consistent work-patterns are carried out in an aura of constancy.

Tone results for each reproductive process

Since the introduction of powderless-etching techniques, presensitized plates and photographic stencils the photographic requirements for tone reproductions by the letterpress, lithographic and screen process printing methods are basically the same and are well met by using contact screens with lith-type emulsions. In many gravure printing houses monochrome tone reproductions are produced by conventional methods employing continuous-tone emulsions throughout the photographic stages, invert halftone methods being reserved for colour reproductions. Individual printing companies tend to have their own specifications as far as densities, dot-sizes and tonal ranges are concerned, but on taking a general view the tone results for each reproductive process follow the requirements listed below.

Photo-engraving (letterpress)

A reversed halftone negative on lith material is required. The highlight dot must not 'drop-out' when printing-down takes place on to the light-sensitive coating, covering the

Tone reproduction

Photo engraving:
Screen negative

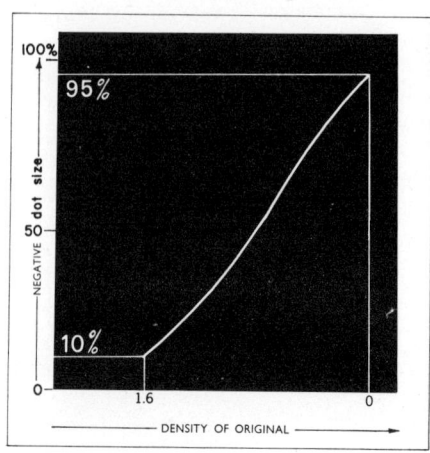

metal surface. This leads to defective etching and troublesome printing conditions. The only time this may happen is when a definite area in terms of a shape is opaqued-out on the negative leaving precise edges. To complement the powder-less, photo-polymer or photo-nylon etching procedures a suitable halftone negative would appear as the following graph.

Photolithography

As in line reproduction this process is divided into two sections.

1. *Surface printing plates*

A non-reversed halftone negative on lith material is required. The highlight dot may 'drop-out' if the original picture contains extreme highlight regions. In graph form the negative follows this curve.

2. *Deep-etch printing plates*

The photographic stage begins with a reversed negative. From this a non-reversed halftone positive is made on lith film. Whether the direct or indirect method is employed the final halftone positive would have a curve-form as that which appears in the diagram.

Screen-process

Negative—positive stages are employed to produce a reversed halftone positive. Lith materials are used with medium or coarse screen rulings and dot 'drop-out' should be avoided. Once again a suitable halftone positive curve would be similar to the one appearing as an example.

Tone results for each reproductive process

Photolithography:
screen negative

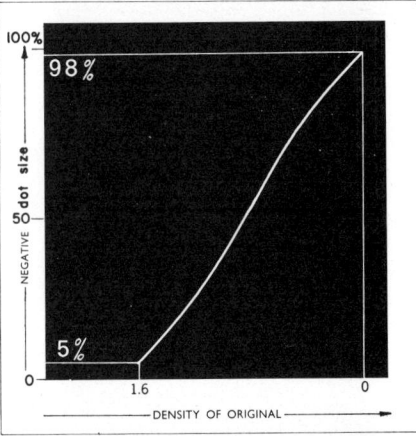

Photolithography:
screen positive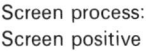

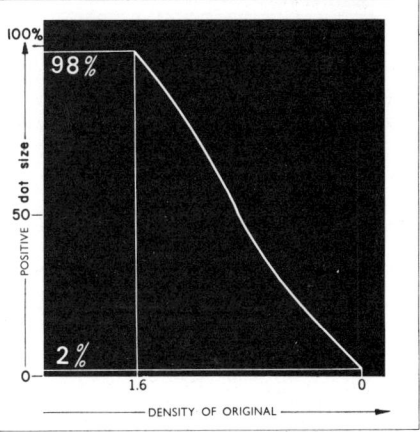

Screen process:
Screen positive

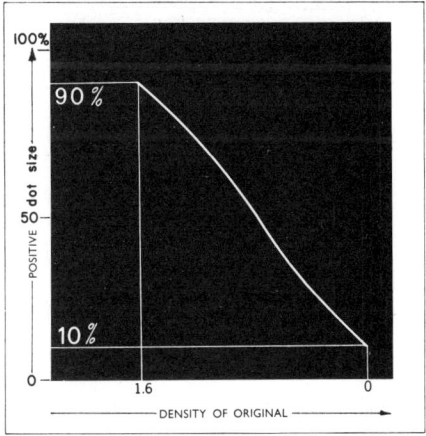

Tone reproduction

Photogravure:
Continuous-tone positive

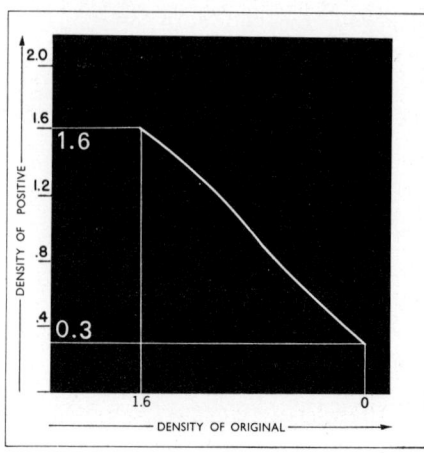

Photogravure

Continuous-tone emulsions are used to produce a reversed negative and a non-reversed positive. The standard negative should produce a continuous-tone positive possessing qualities related to the pigment paper and etching techniques. The following graph illustrates a positive curve suitable for 'one-bath' etching.

Line and tone combination

Tonal reproductions of pictures only are not always required. Quite frequently a camera operator is asked to incorporate line-work with a tone result. The inclusion of black type-matter, in the form of credits or advertising literature running across a tone picture, is a very common practice. There are many ways of producing the required combination of line and tone work, but to begin with it is easier to simplify this work into four steps. In this example a non-reversed halftone positive is the end result.

1. *Preparing the original*

The original must contain suitable central register marks. Each line needs two crosses, the inside one being used on both the line and halftone negatives so that the register of the two images may be assimilated and checked on the combination positive. The outside cross in each case is left on the line negative only. This ensures a fine single register mark on each side of the picture for the plate-maker to use for subsequent positioning operations.

2. *Line and halftone negatives*

A halftone negative is made to the required size and

Line and tone combination

Prepared original

specifications, followed by a line negative. If the line work is drawn on the tone original then the size of the line negative must be exactly the same as the halftone negative, if precise register on the final positive is to be achieved. When using a contact screen for the halftone negative, the line negative should be exposed through a sheet of clear film of the same thickness, and in the same position as the contact screen. With a glass screen the operator has a choice of three substitutes to replace the refraction effect of the glass. A dummy screen of clear glass can be positioned in the screen's clamps, a small compensating glass introduced behind the lens, or, if the camera has a vernier scale, a fine back-adjustment made of the camera extension.

Negatives:
1. Original
2. Halftone negative
3. Line negative

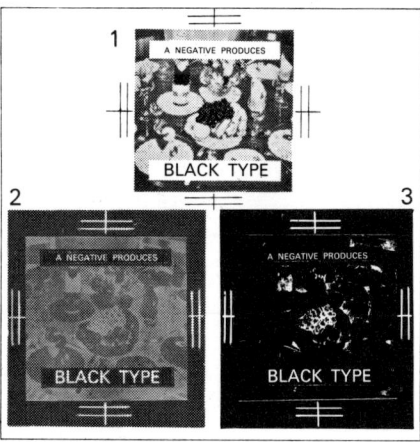

Tone reproduction

Blocking out the negatives

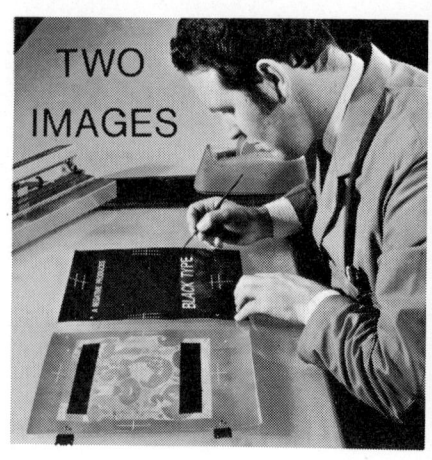

3. *Blocking-out the negatives*

The line and halftone negatives are combined together in precise register and register holes punched into them. In this way size and register may be checked before any further work is carried out. The separate negatives are now inspected and using opaque, red tape or rubylith all tone-work is blocked out of the line negative and all line-work from the halftone negative. The individual negatives should now be separated records of the original in terms of line and tone.

4. *Making the combination positive*

Once the blocking-out has been completed the line negative and a halftone negative can be brought into accurate register by superimposing them on the pins. The next step is contact-

Combination positive:
1. Halftone negative
2. Line negative
3. Combination positive

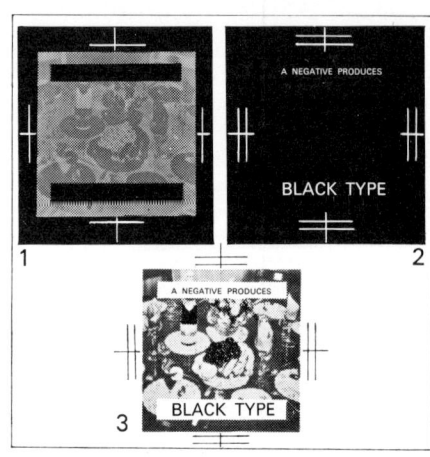

Line and tone combination

ing. A sheet of lith film is punched so that it contains the same register holes as the two negatives. By laying each negative on the pins in turn and exposing on to the punched sheet of lith film, the individual line and tone images are accurately superimposed on the one sheet of film.

The printing of good quality tone reproductions is extremely rewarding. Immense satisfaction is felt by the camera operator when he sees someone appreciating a copy of a publication in which he has recorded the longest tonal range possible, extending the printing process to its maximum capability, reproducing delicate and subtle pencil sketches on one page and bold, bright photographic prints on another.

11. COLOUR REPRODUCTION

Historical introduction

It is surprising how soon people become dissatisfied with a new discovery—the miraculous becomes commonplace. It was not long before the early photographers were looking at their monochrome images and wistfully imagining them to be in colour. 'After all,' they thought, 'we see in colour, so why not photograph and print in colour—capturing the exciting sensations of a rainbow.

It is but a step in logic to understand that colour photography and reproduction techniques aim at two targets. The first is to recreate the amazing colour analysing function of the human eye. This requires the formulation of sensitive emulsions which will analyse the light received from an original into three intensities, i.e. the amount of blue, green and red light. The second aim is to reproduce these intensities and present them to the eye once more as visually identical amounts of blue, green and red light. Because the eye is being stimulated by the same composite collection of light intensities the colour original and coloured reproduction appear the same.

These two targets can be gained in two ways. The first method is additive colour synthesis, and most early colour photographs worked on this principle. The second is subtractive colour synthesis which has become the basis of both modern colour photography and graphic reproduction complemented by colour printing techniques.

The printing of coloured illustrations from wooden relief blocks using coloured inks reaches back to the fifteenth century—the time of Johann Gutenberg, the inventor of moveable type faces. These early printings consisted of numerous blocks, virtually one for every colour contained in the original picture. During the beginning of the eighteenth century a German astronomer and scientist, Johann Tobias Mayer, advocated a three-primary-colour system and wrote lengthy notes on the subject. He proposed the use of three basic colours—yellow, red and blue, the actual red and blue pigments being very similar to magenta and cyan.

The first person to use the tricolour system was a German

Historical introduction

printer, Jakob Christoph Le Blon, born at Frankfurt am Main in 1667. During the year 1722 he published his colour printing process under the title, 'The Harmony of Colouring in Painting, reduced to mechanical practice under easy precepts and infallible rules.' Indeed his process of printing yellow, red and blue inks was an easy precept and colour printing progressed in an elementary form up to the nineteenth century using autographic printing plates suitable for letterpress, lithographic and gravure printing. The psychological result —the composite collection of the three colours becoming a sensation within the human mind—was beyond the imagination of these early colour printers.

Colour as a sensation was not recognized until Thomas Young read a paper to the Royal Society in 1802, explaining that a normal human retina possesses three different kinds of nerves, which when stimulated cause a reflex of the respective nerve elements sensitive to violet, green and red. At a later date, after Niépce had taken the first photograph and the other pioneers had established the early photographic processes, the famous Scottish physicist, James Clerk Maxwell (1831–1879), wrote, 'It is almost a truism to say that colour is a sensation, and yet by recognizing this simple fact Thomas Young clearly realized, nearly forty years ago, that the science of colour must therefore be regarded as essentially a mental science.'

Clerk Maxwell was the first person to employ the newly developed process of photography as a means of determining the relative amounts of blue, green and red light present in any colour. He demonstrated the photographic procedure before the Royal Institution in London on 17th May 1861. The photographic work was carried out by Thomas Sutton, photographer and editor of the magazine *Photographic Notes*. It was in this publication that he wrote a detailed account of the experiment. Three photographs of a coloured ribbon were taken through three coloured solutions. An ammonical solution of copper sulphate became a blue filter, a solution of copper chloride was the green filter and a red filter was produced from a solution of iron sulphocyanide. A fourth plate exposed through a yellow filter was produced as a possible alternative to the red filter separation, but was not used in the actual experiment. From these separation negatives, positive images were made photographically and improved with hand retouching.

The blue filter, green filter and red filter positives were positioned in three separate lanterns and their images projected on to a white screen in register, each positive being

Colour reproduction

illuminated with the same coloured light used to photograph the separation negative. Thomas Sutton described the result in these words. 'And when these different coloured images were superimposed upon the screen, a sort of photograph of of the striped ribbon was produced in natural colours.' The colour photograph was the result of additive synthesis.

Clerk Maxwell's experiment succeeded by accident. At that time the wet-collodion emulsions were not sensitive to every band of the visible spectrum. It was the presence of imperfections in the liquid filters and the unsuspected reflection of ultra-violet light by the red areas of the ribbon which enabled the demonstration to work as well as it did. Nevertheless, Clerk Maxwell's idea of four colour-separated negatives is, in fact, the first stage of multi-colour graphic reproduction, and four contact positives represent the present day colour printing plates.

The idea of photographic colour reproduction also appealed to two Frenchmen, Louis Ducos du Hauron and Charles Cros. Both working independently and unaware of Maxwell's experiment, their work and ideas during the years 1868 and 1869 were very similar. Louis Ducos du Hauron culminated his early experiments with the idea of a screen consisting of primary coloured dots or lines coated with a panchromatic emulsion. The image exposure would be made through the screen's coloured elements. A primary reflection from any area of the object becomes a silver density immediately behind a screen dot of the same colour. After reversal processing the silver deposit behind each coloured dot would represent the original's absorption of that particular primary. A single exposure would be needed to separate the original and only a single projector to effect additive colour synthesis.

During the subsequent years of 1892 and 1893, J. W. McDonough experimented with emulsions coated over minute blue, green and red resin particles, and John Joly produced a grating of coarse blue, green and red lines which was placed over the colour sensitive emulsion during exposure. These coloured lines became the viewing filters after a positive was produced from the separation negative. The employment of additive synthesis was extended once again by the two sons of Antoine Lumière, an emulsion manufacturer from Lyon, France. Auguste and Louis Lumière, in the year 1907, introduced the Autochrome plate—an additive screen plus a colour sensitive emulsion. The screen consisted of a mixture of dyed potato starch grains spread on to a sheet of glass coated with adhesive. The gaps between the grains were then filled with carbon black. This plate was successful, but

Historical introduction

Louis Ducos du Hauron
(1837–1920)

suffered from clumping of the screen's coloured grains. Filter elements were eventually distributed in a uniform manner, when the Finlay plate (1908) and the Dufay film (1930) were introduced. The Dufay additive transparency used fine mechanically-ruled lines which were then dyed to produce a coloured screen or reseau through which the panchromatic emulsion was exposed and subsequently viewed. Unfortunately, with additive colour transparencies there is a considerable loss of light and when reproduced in printed form there is always a danger of the reseau clashing with the halftone screen resulting in moiré patterns.

It is an opportune moment to refer back to the work of Louis Ducos du Hauron, who also invented a three-colour camera in which three colour-separated negatives were made simultaneously. In an application for a French patent he writes:

'We obtain by the aid of the photographic camera three negatives of the same object, the first negative through a green coloured glass, the second negative through a violet coloured glass and the third negative through orange-red glass. Then transparent positives are made by the pigment or a similar process, by the aid of chromolithography, Woodburytype, or by a toning process. From the first negative a red print is made, from the second a yellow print and from the third a blue print. When the three prints are superimposed and thus combined, we obtain a finished print, which is a polychrome reproduction of nature.'

Related here is the foundation of graphic reproduction and trichromatic printing.

Du Hauron, in one of his many disclosures, suggested a triple sandwich—three suitably sensitized photographic emulsions assembled in layers. This arrangement was the fore-

347

Colour reproduction

runner of the present day integral tripack films. These modern colour materials use three emulsion layers coated on a single film or paper support. The layers record separation images without the need for blue, green and red taking filters. This is made possible by making the top layer sensitive only to blue light, the second layer to green light, and the bottom layer to red. The lower emulsions are still blue sensitive as well, but a yellow filter layer positioned under the top layer holds back blue light. Integral tripack materials are manufactured both as reversal (positive) and negative colour films. After exposure the three sensitized layers record their separation images in register. Development produces negative silver images, leaving unexposed silver halide in each layer. With the reversal film this silver halide is converted into positive silver and dye images; yellow dye in the blue-sensitive layer, magenta in the green-recording and cyan in the red-recording layer. On bleaching-out, all the silver images are removed and only the positive dye images are left, which when viewed in combination reproduce the colours and brightness range of the original. A similar process is carried out on a white paper support when positive colour prints are made from colour transparencies. With colour negative film the negative dye images are formed at the first stage of processing together with silver images. Then all the silver is removed, leaving the dyes to view as a negative in tone and colour. This negative is exposed on to a similar colour material coated on to paper or film. Reversal takes place once more producing a coloured positive print or transparency.

Current methods of colour photography, graphic reproduction and colour printing all rely on the marriage of additive analysis with subtractive synthesis.

Colour theory

Colour is light. It is not something that is painted on walls, printed on to paper or incorporated into textiles. This is the application of pigments or the introduction of dyes, whose functions are to absorb specific sections of the visible spectrum and transmit or reflect the remaining light, which is the colour we see.

Daylight consists of the colours of the rainbow—an infinity of colours forming a visible spectrum. To simplify colour reproduction it is sufficient to regard white light as an equal mixture of blue, green and red wavelengths. By mixing these three components in various proportions other colours are formed.

Colour theory

A coloured, transparent object such as a coloured gelatine filter will only transmit its own colour from incident white light, absorbing the rest. Coloured objects which are opaque reflect their colour and absorb the remainder. In printing a thin transparent layer of pigmented ink is rolled on to paper which is a white reflecting surface. The thin layer of printing ink acts as a filter. The white incident light passes through the ink layer which absorbs a section of the spectrum. The remaining incident light strikes the surface of the paper and is reflected back through the ink layer again on its way out, absorption taking place once more. The light that finally leaves the printed paper is the colour we see. When a coloured object is viewed in white light, the colour or light that is finally received by our eyes is always complementary to the light that has been absorbed.

These statements can be appreciated by learning the following table and carefully viewing the diagrams illustrating the action of filters, inks and printed overlaps.

Table 19. *Colour vision*

Coloured object	Light received			Light absorbed		
Black	None			Blue	Green	Red
Blue	Blue				Green	Red
Green		Green		Blue		Red
Red			Red	Blue	Green	
Yellow		Green	Red	Blue		
Magenta	Blue		Red		Green	
Cyan	Blue	Green				Red
White	Blue	Green	Red	None		

Colour measurement

When presented with a large variety of colours we can classify them by using names which refer to their hue, saturation and lightness. This becomes possible if we confine ourselves to the names, violet, blue, cyan, green, yellow, orange, red, purple, magenta, pink and brown, qualifying them with such adjectives as strong, vivid, dark or pale. Intermediate hues are denoted by appropriate combinations, for instance, a dark bluish-green. When dealing with a particular colour in graphic reproduction it is sensible to measure it in terms of spectral composition. Then all the people concerned have a definite target to aim at. There are four basic ways of measuring a colour.

1. Visual comparison with coloured standards. The Munsell, DIN or printed colour chart system are the most popular.

Colour reproduction

2. A comparison with measured amounts of standard coloured lights. Used widely in laboratories for additive colour synthesis.
3. Measurement of transmission and reflection densities through three primary filters. A popular method in printing, but due regard must be taken of the spectral response associated with the instrument's photocell.
4. The use of spectrophotometry. Each wavelength transmitting or reflecting from the object is measured and the result plotted in curve-form.

Spectrophotometry

The spectrophotometer plots a spectrophotometric curve by measuring the amount of light reflected or transmitted, at each wavelength of the spectrum.

A spectrophotometric curve helps us to arrive at three numbers which specify the colour in question. These are termed tristimulus values X, Y and Z representing three hypothetical colour stimuli which would be required in an additive mixture to match the given colour stimulus. This is similar in density measurements made through red, green and blue filters. In fact colorimeters can be purchased which give direct tristimulus values through filtered reflection readings related to the spectral response of the human eye.

Another method of determining the tristimulus values is by recording transmission or reflection readings of the colour at intervals of 10 nm along its spectrophotometric curve. Each reading is multiplied by a factor associated with the spectral composition of the light source employed and the sensitivity of the eye to each wavelength.

Spectrometric curve:

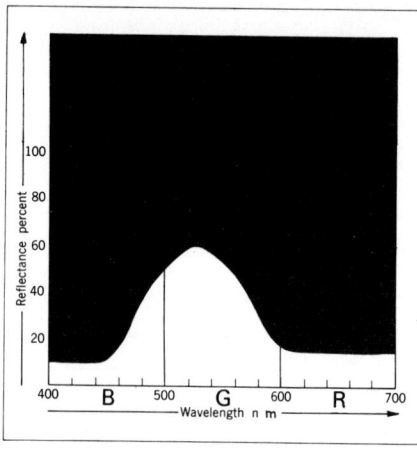

Colour theory

By adding these products the X tristimulus is obtained. This is repeated for values Y and Z using the appropriate factor number. The Y tristimulus value is the luminance of the colour in percentage reflection or transmission, X value the relative amount of *red* light and Z value represents the relative amount of *blue* light. Once the tristimulus values have been found, the trichromatic coefficients x and y are calculated in the following manner.

$$x = \frac{X}{X + Y + Z}$$

$$y = \frac{Y}{X + Y + Z}$$

$$z = \frac{Z}{X + Y + Z}$$

z is not required, but the sum of $x + y + z$ should equal *one*

Tristimulus values allow us to use the CIE (Commission Internationale de l'Eclairage) system of colour measurement. The final trichromatic coefficients x and y define the chromaticity of the colour in question by relating its hue and saturation values. This information is plotted on the CIE chromaticity diagram containing a spectrum locus curve. Saturated blues, greens and reds follow the curve, while **magentas** and purples—the result of mixing the end wavelengths of the spectrum—are indicated on the straight-line region.

Illuminant C represents daylight and the red and green values of the colour are indicated by the coefficients x and y. These two numbers denote the chromaticity of the colour (e.g. when the numbers are low the colour is blue) and are plotted on the curve. Using our example spectrophotometric curve, the x number is 0·20, the y number 0·60. A line is extended from the illuminant C through the plotted mark until it cuts the spectrum locus indicating the dominant wavelength or *hue* of our example colour, e.g. 525 nm. The distance from the plotted mark to the illuminant point, expressed as a percentage of the total length to the spectrum curve, e.g. 57%, is the purity or *Saturation*. To complete our colour specification the luminance Y or *Lightness*, must also be given. Two colours represented by the same tristimulus values will compare visually under the illumination in question. For further investigation into this work and allied subjects the excellent publication, *Principles of Color Reproduction* by John A. C. Yule, Wiley, is a classic textbook for all serious workers in graphic reproduction.

Colour reproduction

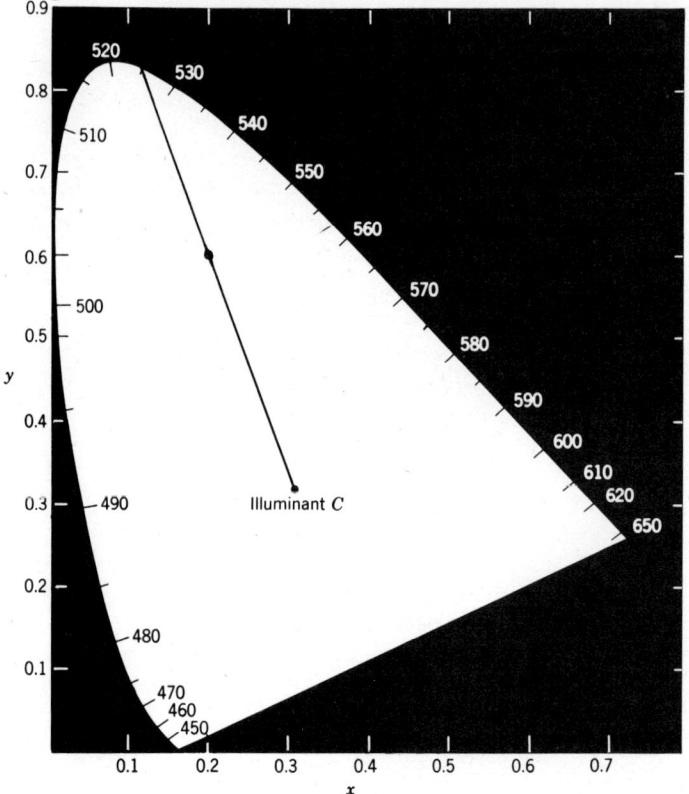

Chromaticity Diagram: Example plotted

Density readings

Every modern printing plant should be well equipped with densitometers. Besides the normal exposure calculations,

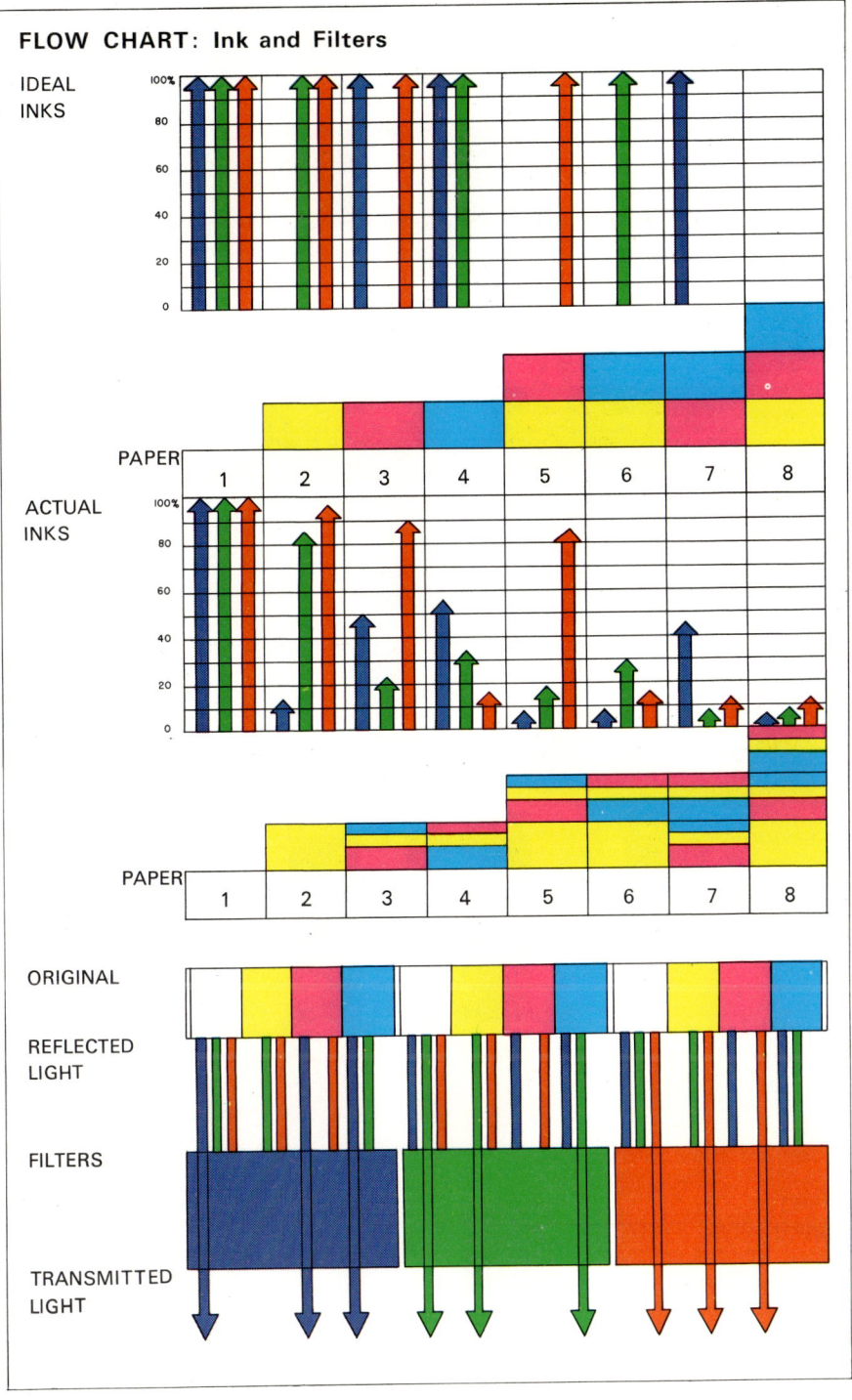

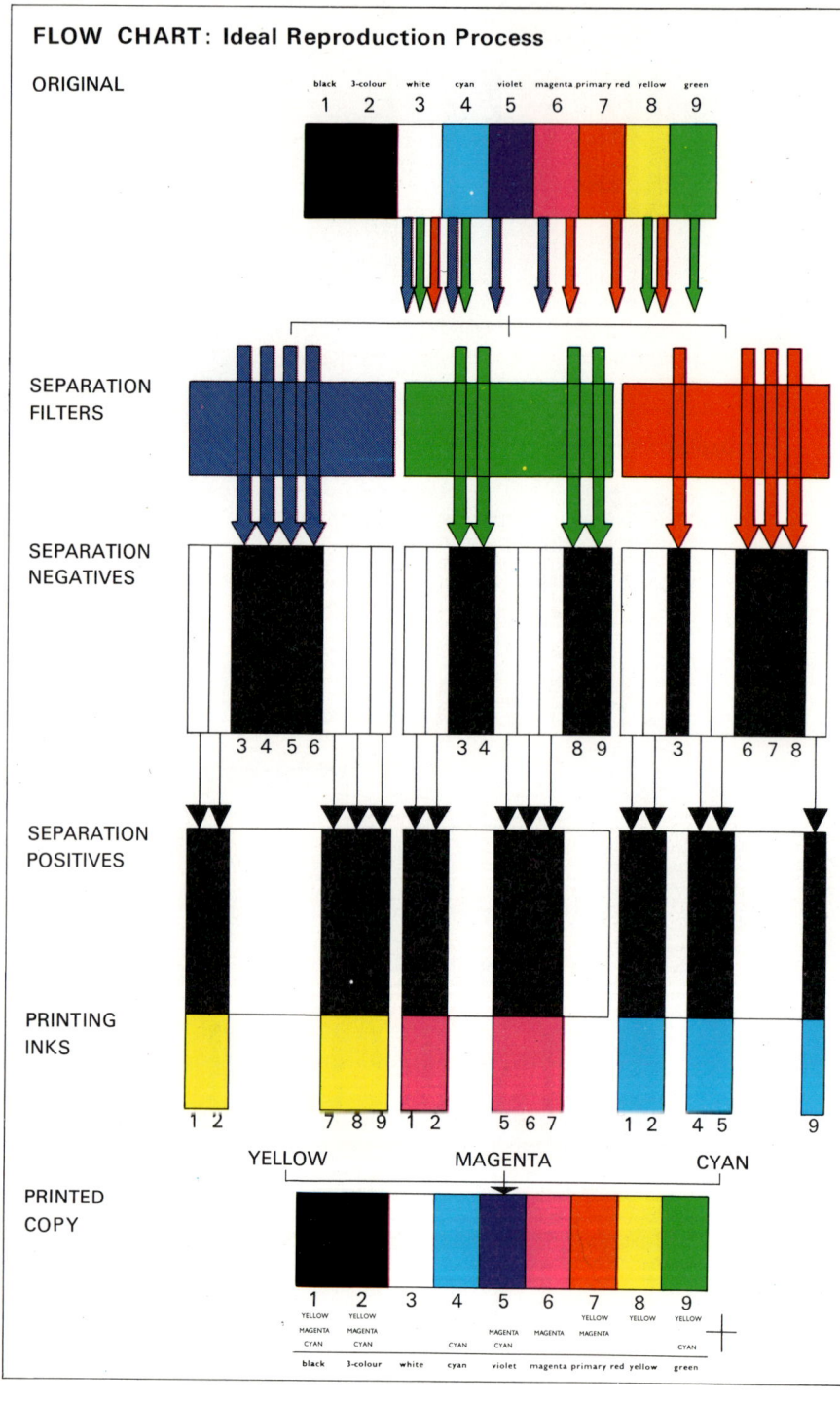

Colour theory

density readings through colour filters can provide additional information. The application of this valuable information may be listed:

1. Corresponding coloured areas in the original picture and printed reproduction can be measured through the primary filters to indicate the cause of poor colour reproduction.
2. Density measurements of trichromatic inks printed separately and overlapped provide an understanding of the problems encountered.
3. Density checks on solid colour-bars printed on the back edge of the paper, using filters complementary to the ink, give the printer a precise reading of the amount of ink present.
4. Colour matching to a particular ink swatch can be assisted by specifying the spectrophotometric curve and the CIE tristimulus values.

A printing dilemma

The foremost dilemma in graphic reproduction occurs when an original picture containing a gamut of colours, probably produced by mixing numerous pigments together, has to be recreated by overprinting three or four coloured inks. When the printing is completed no immediate difference must be seen as a viewer's eyes traverse from reproduction to original.

How is this to be achieved? Well, the viewer's eyes are stimulated by the amounts of blue, green and red light reflecting from the original. Therefore the reproduction must be printed in such a manner that it too reflects the same amount of blue, green and red light. The most practical method of providing these visual stimuli is to start with a material capable of reflecting the visible spectrum, e.g. white paper, and print on to its surface three inks which control by absorption the amount of blue, green and red light that we will eventually see.

The individual printing inks, yellow, magenta and cyan, are complementary to the light absorbed. These inks are printed in various amounts, proportional to the original's absorption of blue, green and red light. The correct proportions, using a halftone dot structure, are related to the original in this fashion. The amount of blue, green and red light absorbed by the original should correspond to the size of the yellow, magenta and cyan positive dots, e.g. 95%, 95% and 10% respectively. The colour we see is the total of the reflected light, e.g. 5% blue + 5% green + 90% red = a strong *red*. Controlling the reflected light in this way makes possible a close visual comparison of the original and reproduction.

Colour reproduction

We wish to print the minus blue, minus green and minus red images of the original, so it is not unnatural that the first stage in colour reproduction is photographing the original picture separately through blue, green and red filters. Ideally, the density of the negative in each case will be proportional to the blue, green and red reflections of the original. The density of the corresponding positives or printing plates will be the opposite—the absorption of blue, green and red light produced in a separated condition, ready for recreation by overprinting with yellow, magenta and cyan ink.

A perfect reproduction process

Before we become engrossed in the problems of colour reproduction and the methods of overcoming them, it is worthwhile to consider the perfect reproduction process. This hypothetical process would commence with exposures through primary filters, each one transmitting exactly one third of the spectrum. The separation negatives would produce continuous-tone positive images—exact records of the original in terms of separated colour and tonal gradation. The printing inks would completely absorb one third of the spectrum and fully transmit the remaining two thirds. After superimposing two of the inks, only the section of the spectrum common to both inks would be seen. Black or the absence of light would be the result of overprinting all three inks.

Tonal gradation from solid areas to highlights would be smooth and linear to the original. Proportionality failure producing distorted hues and shifts in grey balance would not take place. The perfectly transparent inks would print and trap well, overcoming additivity failure, i.e. individual ink densities would equal the total density reading when all three were overprinted. The paper and printing method would not impose any physical restrictions. Long density ranges would be achieved on a background of high gloss and maximum white reflectivity.

Actual conditions

The actual conditions of colour reproduction encountered in practice contain shortcomings and deficiencies. These may be appreciated by dividing them into four basic inadequacies.

1. The transmission of colour separation filters is controlled by the use of different spectral divisions, i.e. medium band and narrow band transmissions. The medium band filters allow an overlap to take place between the three major sections. Certain colours which fall in these overlaps may be recorded on each negative without sufficient separation resulting in

Colour theory

Filters—tricolour:
1. Medium band
2. Narrow band

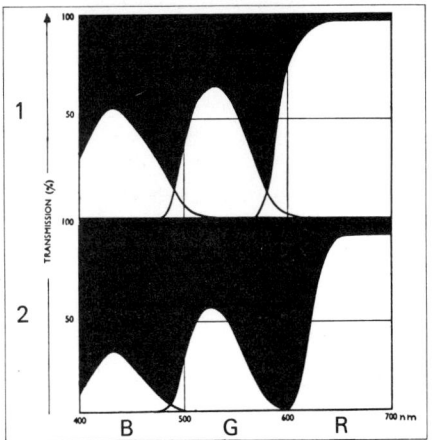

degradation at the printing stage because of an excess of colour. Narrow band filters produce sharper spectral divisions which suit originals containing clean, unmixed pigments, e.g. subtractive colour transparencies. Transmissions such as these prove to be inadequate for originals containing a multitude of mixed pigments, e.g. an oil painting, because intermediate colours will not be recorded in a definite manner, resulting in the printing of a weak, de-saturated colour. These slight inadequacies in filtration are of minor consequence when compared with the other three deficiencies.

2. The use of the halftone dot structure introduces proportionality failure. Small individual dots surrounded by white paper produce an integrated colour which appears weaker and de-saturated when compared with a continuous-tone printing of the same colour. The explanation here is that the continuous-tone layer of ink covers the white paper surface and carries out selective absorption throughout the coloured area. This failure is more prevalent with magenta and cyan inks. Without correction these inks, printed as halftone dots, reproduce the lighter tones of the original as grey, de-saturated colours.

3. The reproduction of colour relies heavily upon the correct overprinting of the three basic inks. This necessity is hindered by additivity failure, i.e. the total density of coloured overprints is less than the sum of their component colours printed separately. The method of printing has an enormous effect upon this failure. Wet-on-wet printing conditions can easily result in component ink layers being under- or over-trapped between the previous and final ink layers.

4. The most serious inadequacy is the hue deficiencies

Colour reproduction

apparent in the standard trichromatic inks available at the present time. The absorption and transmission properties of the yellow ink are nearly ideal. The magenta ink transmits red reasonably well, but is very deficient in transmitting the blue section of the spectrum. Aggravating this deficiency is the fact that magenta ink transmits some green light when it should absorb this section of the spectrum. The appearance of this ink is too reddish because it lacks blue transmissions. The cyan ink contains the worst hue deficiency. It absorbs both blue and green light which it should transmit completely. This is further complicated by the ink's failure to completely absorb the red section of the spectrum. The cyan ink appears too bluish, as a result of the deficiency in green transmissions. It is wise to study the differences between ideal and actual inks (BS 3020) in table and curve-forms.

Table 20. *Colour densities*

Density readings Colour	Blue density		Green density		Red density	
	Ideal	Actual	Ideal	Actual	Ideal	Actual
Yellow	High	1·00	0·00	0·06	0·00	0·01
Magenta	0·00	0·40	High	1·00	0·00	0·10
Cyan	0·00	0·20	0·00	0·40	High	1·30

Tricolour inks:
1. Yellow
   (ideal)
   (transmit)
   ($\frac{2}{3}$G + R)
   (absorb)
   ($\frac{1}{3}$B)
2. Magenta
   (ideal)
   (transmit)
   ($\frac{2}{3}$B + R)
   absorb
   ($\frac{1}{3}$G)
3. Cyan
   (ideal)
   (transmit)
   ($\frac{2}{3}$B + G)
   (absorb)
   ($\frac{1}{3}$R)

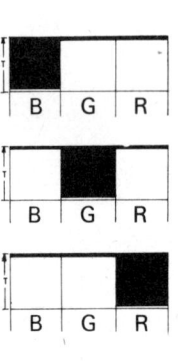
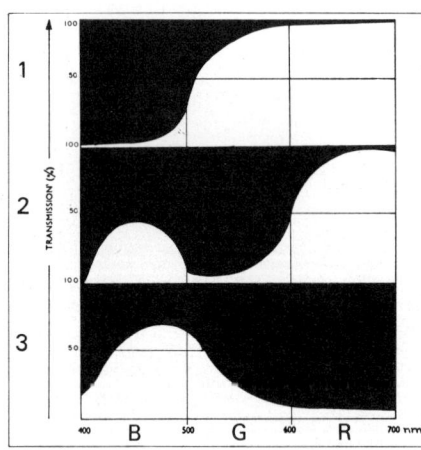

These inadequacies have to be realized and corrected. Taking our control strip containing all the possible solid colours produced by overprinting the trichromatic inks plus a black ink, i.e. 3-colour, violet, cyan, green, yellow, red, magenta, black and white, and imagining that this is an average

Colour theory

coloured original, let us look at a set of typical continuous-tone separation negatives produced from this average original, in terms of density. The colours in each separation negative may be divided into two groups—'wanted' and 'unwanted' colours. The density reading obtained from each *wanted* colour should be comparable to the low density reading obtained from the 'three-colour' or *black* patch. *Unwanted* colours must record at a density level similar to the *white* patch.

Separation table

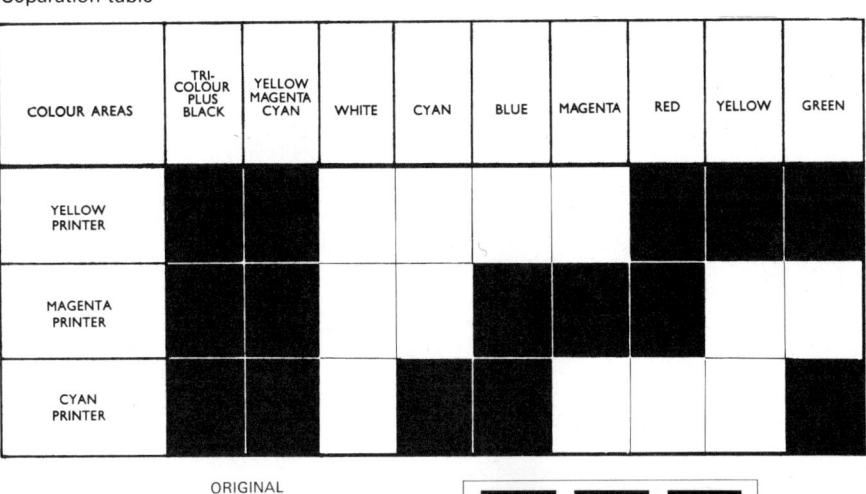

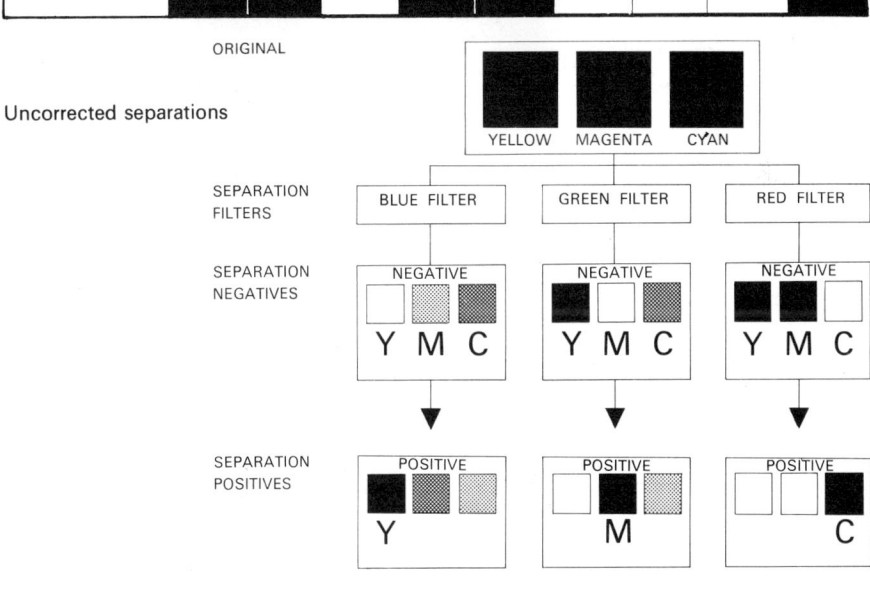

Colour reproduction

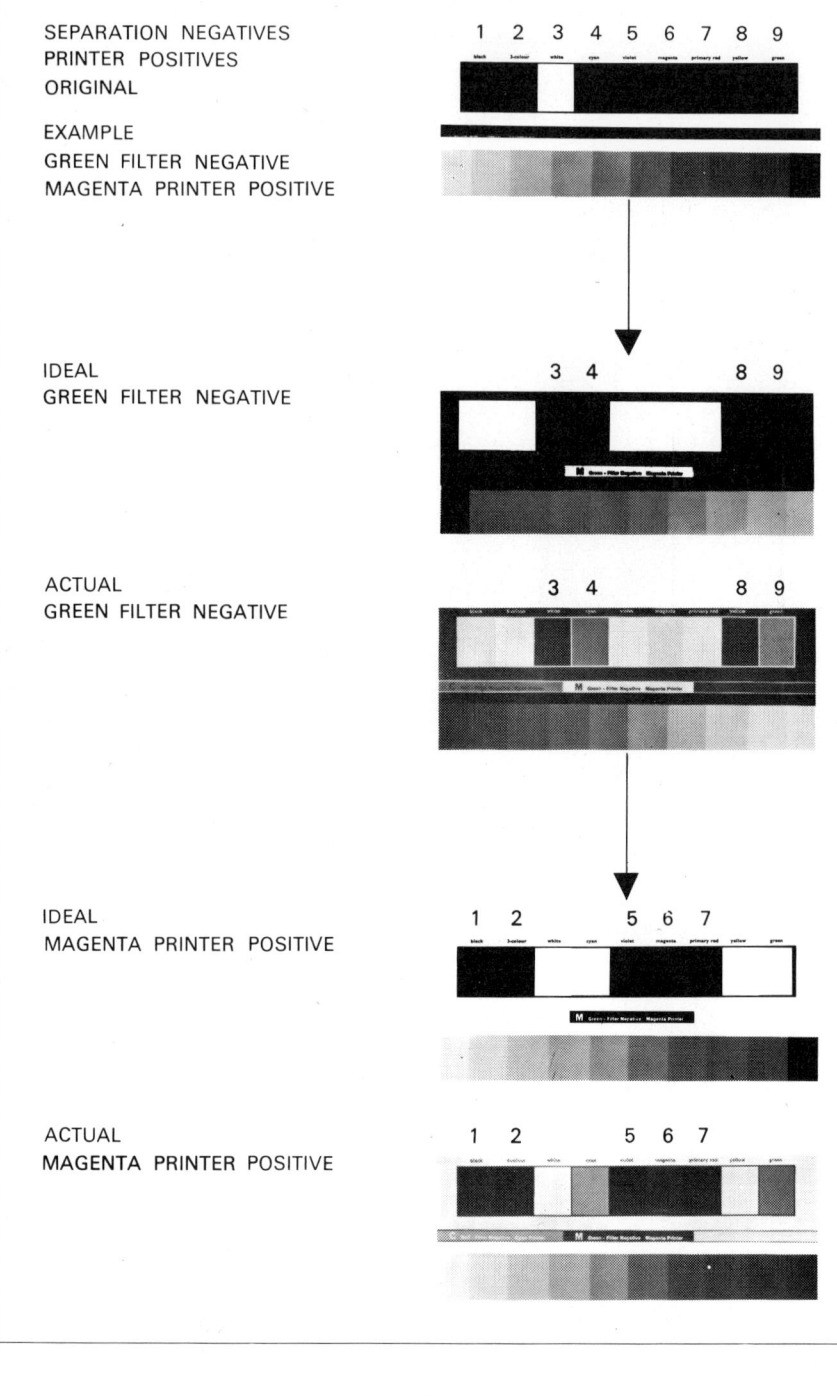

Colour theory

This division can be seen firstly in table-form and secondly as it appears on actual separation negatives compared with ideal separation negatives.

After close examination of the 'unwanted' areas in each actual separation negative the following conclusions can be drawn.

Blue filter negative: yellow printer

Unwanted colours		White	Comment
Cyan	1·20	1·70	These colours are all too
Violet	0·75	1·70	low owing to poor *Blue*
Magenta	1·00	1·70	transmissions.

A high yellow ink density will be printed in the colours classified as 'unwanted'.

Green filter negative: magenta printer

Unwanted colours		White	Comment
Yellow	1·65	1·70	Yellow good because of high green transmissions.
Green	0·90	1·70	Lower than they should be
Cyan	1·00	1·70	owing to poor *Green* transmissions.

The result would be that too much magenta ink will be printed on the green and cyan areas.

Red filter negative: cyan printer

Unwanted colours		White	Comment
Magenta	1·58	1·70	All colours reasonably good
Red	1·55	1·70	because the *Red* trans-
Yellow	1·65	1·70	missions in each case are strong

The red filter negative is always acceptable and needs less colour correction than the rest.

This can be repeated for the 'wanted' colours, comparing them with the 3-colour density. It will be seen, for instance, that the yellow patch, because of its virtually ideal absorption properties, always records at an acceptable low density when exposed through a blue filter for the yellow printer plate.

If these actual separations were printed with the ideal inks previously mentioned then the unwanted densities, which normally degrade a colour reproduction, would be needed to modify the brilliant ideal pigments so that the saturation level of the original could be maintained. This becomes very apparent if the following illustrations are compared. The first picture represents the original, the second an uncorrected reproduction printed in normal trichromatic inks and the third the same uncorrected separations printed with inks of im-

Colour reproduction

proved hue. There is a definite movement towards the original. A reproduction with ideal inks is a matter of taking a hypothetical journey in the same direction.

Colour and tonal correction

It should have become quite obvious by now that with existing inks the major error is the printing of too much ink in certain areas. Local adjustments must be made to the amount of ink which is permitted to print. The deficiencies in the magenta and cyan inks provide the main reasons for this necessity.

The magenta ink does not absorb green light completely and transmits only half the amount of the potential blue light. Therefore the magenta ink acts as if it is ideal magenta ink mixed with unwanted cyan (transmitting green) ink and yellow (absorbing blue) ink. This mixture results in a magenta image being transmitted together with faint cyan and yellow images. It follows that local adjustments must be made to reduce the amounts of cyan and yellow ink which would print in magenta areas, because the magenta ink itself has already printed the required amounts of cyan and yellow ink.

The cyan ink absorbs quite a considerable amount of blue light and even more green light, both of which it should completely transmit. So it could be thought that the cyan ink functions as if it is ideal cyan ink mixed with small quantities of yellow (blue absorbing) ink and magenta (green absorbing) ink. Therefore when we print cyan ink we also deposit faint yellow and magenta images. To overcome this we must

Table 21. *Uncorrected colour*

Section of original	Colour	Result
Flowers	Blue	Darker and greyer
Flowers and glasses	Mauve	Become brown
Leaves	Blue-greens	Lost their greenish hue
Apples	Green	Darker, degraded to grey
Bananas	Yellow	Become lightened
Flowers	Orange	Reasonable
Flowers	Pink	Become yellowish
Flowers	Reds	Darker, lost their bluish tint
Table	Brown	Acceptable

Colour and tonal correction

reduce the amounts of yellow and magenta ink which would normally print in cyan areas.

By looking at the uncorrected illustration (No. 2) it can be seen that degradation is the result of leaving image density to overprint unwanted ink images. The undesirable effect on clean colours appearing in the original may be listed.

To achieve acceptable reproductions we must produce colour separated negatives with *increased* densities in the 'unwanted' areas where less ink is required. This remedial work is called 'colour correction'. While carrying out colour correction we must not forget the equally important problem of tonal correction. The tonal range of the original should be compressed or expanded to the density capability of the printing process. Grey balance and the reproduction of neutral areas must be maintained. The inclusion of a control strip consisting of colour patches for checking colour correction and a grey stepwedge for controlling tonal correction is essential at every stage of the process. As in most aspects of graphic reproduction this work can be broken down into four methods.

1. *Ink improvement*

There is a constant move towards ideal inks. With each improvement over the years the amount of colour correction has decreased. In fact, with the inks used in illustration No. 3, which are specially formulated for a short-run 'pleasing-colour' process, a very simple single-overlay masking system would be used. At the present time these inks do not always give such comprehensive results as standard trichromatic inks combined with a more complex correction system.

2. *Hand retouching*

This entails autographic work using dye, knife and pencil techniques on continuous-tone images, dot etching, staging and opaquing on halftone images, and etching methods on relief or intaglio metal printing images.

3. *Photographic masking*

This is the use of an additional image, usually photographic which will increase the density of the 'unwanted' areas in the separation negative. In other words, this image will 'mask' these areas with additional density. A positive or negative image is made through a filter. This 'masking' image is combined with a second image made through a different filter. The second image may be the original itself, a projected optical image of the original, or a photographic result from it.

Colour reproduction

4. *Electronic scanning*

The original is scanned by a light spot. The light received is divided into its blue, green and red constituent intensities. This separated light energy is converted into electrical energy by photomultipliers. The electrical energy is fed into a computer, where colour and tonal correction is carried out in relationship to the inks selected for the printing. The corrected signals are then passed on to the output, which may be in three forms.

(a) A glow modulator lamp capable of varying in intensity for exposing different densities on a continuous-tone emulsion.

(b) A glow modulator lamp with built-in interference which results in the exposure of halftone dots varying in size.

(c) A pulsating stylus which engraves a sheet of dyed plastic or a metal plate with varying sized halftone dots.

In every case, each minute area of the result is in a 'corrected' condition, proportional to the amount of light received from the original.

Retouching methods and masking systems will follow in an expanded form, while electronic scanning, because of its importance and ever-increasing use, will be the subject of the next chapter.

Retouching methods

It is very difficult to lay down any hard and fast rules for retouching procedures. This is a stage in the proceedings which relies heavily upon the experience, colour judgment and manual dexterity of the retoucher in question. The production of excellent reproductions revolves round this man's skill and always will, but this does not mean *carte blanche*. The retoucher must be prepared to work with colour keys and standard step-wedges, and not look upon a densitometer as some sort of mechanical monster. With the help of this instrument and systematic working, especially if the retoucher makes his own positives, retouching can become less complicated and more enjoyable. After taking a very general view of retouching on film, as this is the modern trend in every reproduction process, a four-step work-pattern can be seen.

1. Before starting any retouching work immediate checks must be carried out on the size, reversal and screen angle of the image. If these are satisfactory, assessment of the image's

Colour and tonal correction

tonal range against standard negative and positive stepwedges follows.

2. By referring to the original's colour key—a transparent overlay containing the negative and positive dot-sizes or density values predicted by the appropriate colour chart for the important subject areas of the original—the retoucher is given an immediate idea of how near the separation negative is to the required tonal range, what type and density of mask is necessary, and finally, the extent of manual colour correction required. The densities of the corrected negative can be recorded and used to give an accurate estimate of the exposure needed to give a positive image which will possess the densities or dot values predicted by the colour-key.

3. *Negative retouching.* If necessary, chemically reduce the tonal range and 'wanted' areas of the negative until they correspond to the standard negative range for the printing process in question. Increase the densities of the 'unwanted' colours, e.g. green filter negative—magenta printer. Carry on the work of the filter, using dye or opaque, increasing the density in green, cyan and yellow areas. Improve the drawing or detail of the picture. Dye-up and cut-in the register, centre and trim marks. Position the block-out mask or opaque round to ensure that the white areas do not receive a low-density or small dot.

Positive retouching. Once the overall appearance of the image is correct, increase the densities and importance of the 'wanted' colours. For example, in a magenta printer positive accentuate the red, magenta and violet areas by staging and dot-etching, applying dye or opaquing-up the solid areas. Use local retouching techniques to improve the definition of the picture.

4. Before the corrected positive is passed on to the platemaking department a final check must be made. Compare the dot sizes or density values of the important subject areas with the predicted figures on the colour-key. Assess the overall tonal range, especially the highlight, mid-tone and shadow areas, with the density capability of the printing process. Ensure that the image contains all the required information marks.

Although three colour reproductions are quite common, especially when the printing is entirely pictorial, i.e. devoid of any black typematter, it is usual practice to 'key' the picture with a black printing. The black printer performs a modifying role, producing deeper shadows, improving definition and retaining detail. Printing a black ink is also a comparatively

Colour reproduction

simple way of reproducing and maintaining neutral greys. There are numerous ways of exposing a black separation negative. The camera operator must select the method best suited to the original's content and character. These different methods may be tabulated.

(i) *K3 or gamma filters.* A coloured original photographed through either of these contrast-controlling filters is recorded at a brightness level comparable to the visual interpretation of a human eye.

(ii) *White light exposure.* The negative image is similar to one achieved by a heavy red-filter exposure.

(iii) *Split-filter exposure.* This method allows great flexibility. The exposures through each separation filter are adjusted in accordance with the predominant colours of the original.

(iv) *Infra-red emulsions.* Most pigments or dyes reflect or transmit infra-red light and record as white paper. Black and carbon pigments absorb infra-red in proportion to their visual density.

No matter what method was used at the camera stage the retoucher always has definite aims in view when he tackles the black printer. These may be listed as general directions.

(a) Colour reproduction needs to be darkened and the neutrality of grey areas increased.
(b) No black to print in clean colours.
(c) Outlines must be the correct strength.
(d) Contrast should be increased in the shadow areas.

The end result in every case must characterize the original picture. This means that possibly on one page containing numerous pictures of diverse appearance, the black printer may range from an image of high contrast with solid black outlines, reminiscent of a cartoon, to a subtle, ghost-like image producing muted greys and delicate shadows. The question of appropriate black printers is also associated with the method of printing. Printing each component ink layer on to a dry surface produces conditions vastly different from those encountered on a high-speed, wet-on-wet web-fed printing machine. This really revolves round the problem of tonal correction and is, therefore, discussed in the following section.

Mask strengths

Each separation negative records two facets of the original. The first is an image of its printing colour. The second is errors or imaginary contaminations of the other two printing

Colour and tonal correction

inks. Trichromatic yellow inks transmit red and green light so well that the red filter and green filter separation negatives need to be corrected for only one ink error. Cyan and magenta inks have a general inability to transmit blue light. This deficiency, apparent in both of these inks, must be corrected in the blue filter/yellow printer negative.

We require less yellow ink to print in magenta, red and violet areas of the original. To achieve this we must add density to these areas in the blue filter/yellow printer negative. This additional density needs to be in proportion to the amount of

Principle of colour correction by masking. An actual magenta printer negative contains the tones of an ideal magenta printer plus tones corresponding to an unwanted cyan printer negative. The mask compensates the values of the unwanted cyan, turning it into neutral density, so leaving the masked magenta printer negative only with the wanted magenta tone gradation.

Mask percentage. This is the ratio of the density range of the mask to the density range of the transparency, expressed in per cent

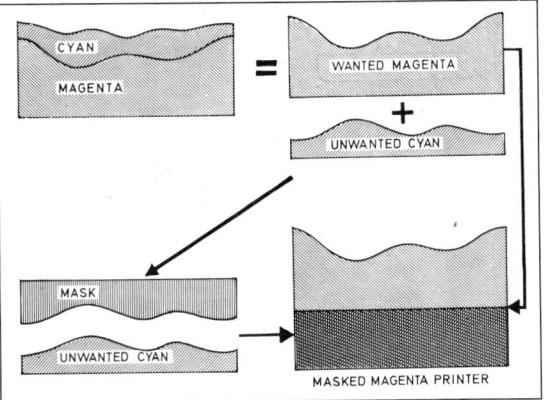

magenta ink to be printed. This colour-correcting density will be in the form of a positive image. The visual effect of adding this density to the magenta, red and violet areas of the original is to darken them.

This can be appreciated in a more accurate manner by measuring the density of these areas with a densitometer, firstly in their unmasked condition and then with the positive film containing the additional densities positioned in register on top of the blue filter/yellow printer negative. A considerable increase in density will be recorded. The final densities should be similar to the maximum highlight density of the negative. These higher densities in magenta, red and violet areas of the blue filter/yellow printer negative will produce a smaller dot or no dot at all on the final yellow printer positive.

After a careful study of the negative plus mask combination it will be seen that an improvement has been made in the colour separation. The improvement will be in proportion to the magenta printer positive. This relationship is required because we need less yellow ink to print in magenta areas of the original. This relationship is obtained by

Colour reproduction

making a positive from the green filter/magenta printer negative. This positive or 'mask' is a photographic density representing the imaginary amount of yellow ink contaminating the magenta ink. The mask will also contain areas of lower density which represent regions of the original that will be eventually printed in cyan ink. These areas of additional density are needed to compensate for the hypothetical amount of yellow ink contaminating the cyan ink.

A positive mask from the red filter/cyan printer negative will, after combination with the green filter/magenta printer negative, increase the cyan, green and to some extent the violet areas of the original. This masking will compensate for the analogous amount of magenta ink contaminating the cyan ink.

The all important strength or density range of a masking image is expressed as a percentage of the density range of the separation negative or coloured original with which it will be combined. The mask percentages are determined by measuring the density of the appropriate solid ink patch through primary filters and using masking equations such as those used in the Graphic Arts Technical Foundation system

Colour	Blue density	Green density	Red density
Yellow	1·00	0·06	0·01
Magenta	0·40	1·00	0·10
Cyan	0·20	0·40	1·30

Percentage mask strength to correct the blue filter/yellow printer negative:

$$\text{To correct solid magenta error} = \frac{M_b}{M_g} \times \frac{100}{1}$$

$$\textit{Example. To correct solid magenta error} = \frac{0 \cdot 40}{1 \cdot 00} \times \frac{100}{1}$$

$$= 40\%$$

Percentage mask strength to correct the green filter/magenta printer negative:

$$\text{To correct solid cyan error} = \frac{C_g}{C_r} \times \frac{100}{1}$$

$$\textit{Example. To correct solid cyan error} = \frac{0 \cdot 40}{1 \cdot 30} \times \frac{100}{1}$$

$$= 30\%$$

Colour and tonal correction

To ensure efficient masking the trichromatic inks should be 'balanced', that is the cyan and magenta inks are matched so that they have the same blue-green density ratio. This is important because when balanced inks are used the two imaginary deposits of yellow ink in the cyan and magenta inks are corrected with one mask.

A good working knowledge of colour correction can be obtained by learning the following masking systems, but a complete understanding is only acquired studying the methods of evaluating the efficiency of paper and printing inks. With this in mind reference should be made to the GATF publications and the excellent work of F. M. Preucil.

Colour masking systems

There are numerous colour masking systems, so in the interests of simplicity the explanations will be confined to the underlying principles of the more popular methods together with flow-charts of the four basic systems. Colour correcting masks may be positive images combined with separation negatives, or negative masks combined with the original image itself. Both systems can use a one-stage or two-stage masking approach. Positive masking systems are simple to apply, but the amount of correction tends to be limited by the filtration of the separation negatives. Negative masks, on the other hand, receive their correction properties from the original and if required these attributes can be altered considerably by filtration. When negative masks are being used with reflection originals precision equipment with pin-registration must be employed. Negative masking direct from the original is very popular at the present time, especially with multi-colour integral masking films.

One-stage positive overlay masking

Positive overlay masking is a simple system employed during the reproduction of reflection originals. The blue, green and red filter separation negatives are made on a suitable continuous-tone panchromatic emulsion. The black printer negative is produced by the 'split-filter' technique. The density range of each separation negative depends upon the percentage mask strengths required to correct the deficiencies of the magenta and cyan inks. An average value of 35% may be taken. The density range of the positive mask/separation negative combination must equal the halftone screen's reproducible density range if the required dot sizes are to be obtained without a flash exposure. The density ranges are calculated in this fashion.

Colour reproduction

$$\text{Unmasked negative range} = \frac{100 \times \text{Screen's reproducible range}}{100 - \text{Mask percentage}}$$

$$= \frac{100 \times 1 \cdot 3}{100 - 35} = \frac{130}{65} = 2 \cdot 0$$

Therefore the separation negative should be made to a density range of 2·0, e.g. shadow 0·3, highlight 2·3. This will allow for the flattening effect of the 35% mask.

$$\text{Mask density range} = \frac{\text{Unmasked negative range} \times \text{Mask percentage}}{100}$$

$$= \frac{2 \cdot 0 \times 35}{100} = \frac{70}{100} = 0 \cdot 7$$

The 35% colour correction positive mask will have a density range of 0·7, e.g. shadow 0·9, highlight 0·2. The combination of positive mask and separation negative will produce a density range of 1·3.

	Shadow	Range	Highlight
Separation negative =	0·3 +	2·0 −	2·3 +
Positive mask =	0·9	0·7	0·2
	1·2	1·3	2·5

An alignment chart for determining mask percentages is invaluable.

Two unsharp (see page 389) colour correcting masks are produced by contacting suitable positives from the red and green filter separation negatives. The positive masks are made on ordinary or orthochromatic film of low contrast. This emulsion must possess a characteristic curve which produces a 'toed' image after under-exposure and over-development and a 'shouldered' image resulting from over-exposure and under-development. A 'shouldered' positive mask increases the colour correction of light colours and two-colour overlaps, but also flattens the highlight gradation. A 'toed' positive mask increases the colour correction in the darker colours and does not flatten the highlight gradation so much, and because of this fact 'toed' masks are generally preferred. A contrast-reducing mask is sometimes made from the red filter negative with which it is combined at a later stage. This 'self' mask is strongly 'toed' because its function is to create

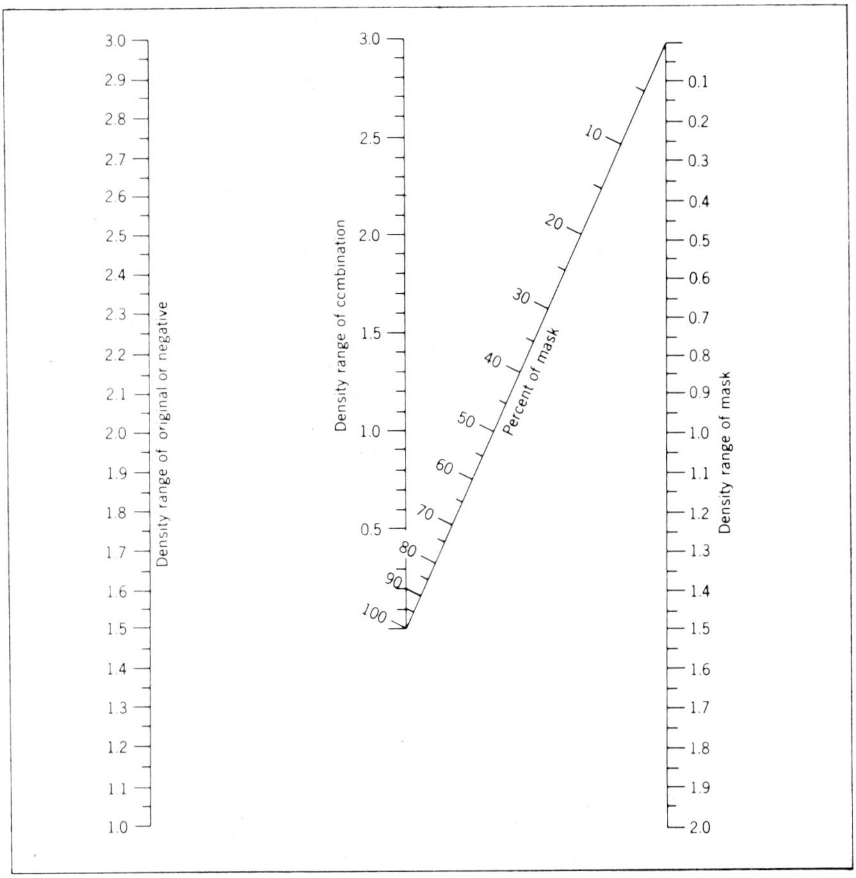

Colour and tonal correction

Alignment chart for determining mask percentages   (copyright Kodak)

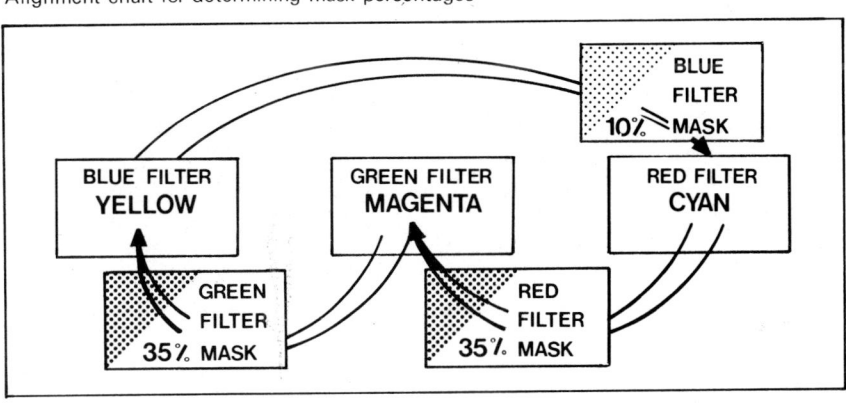

Basic Masking diagram

Colour reproduction

One stage positive overlay masking:

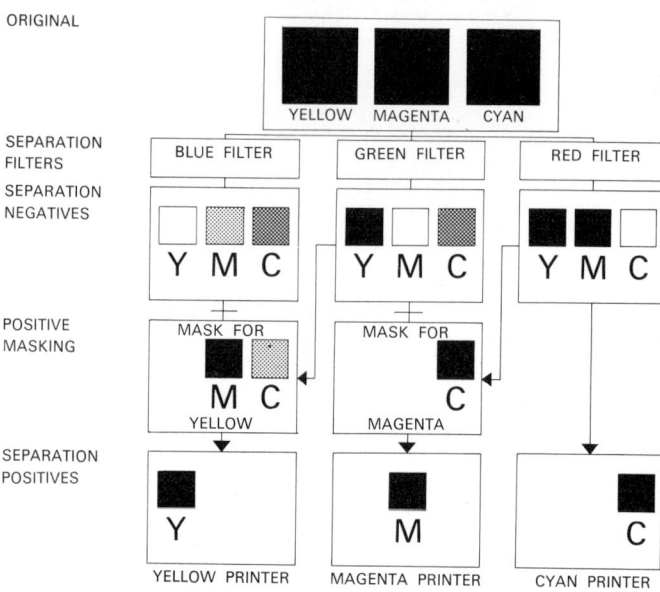

the adjustment or balance between the coloured inks which will produce a neutral grey scale.

The red filter positive mask is combined with the green filter/magenta printer negative. The green filter positive mask is combined with the blue filter/yellow printer negative. The degree of colour correction can be checked by taking densitometric readings of the 'unwanted' colours and comparing them with the density reading obtained from the white patch. If these densities are higher than the white reading the negative is being over-corrected and the positive mask should be remade using a more 'toed' approach. Densities of the 'wanted' colours are evaluated against the density of the 3-colour patch. If these densities appear higher than the 3-colour density, de-saturation of the printed colours will result. In this case remake the positive mask applying the 'shouldering' technique. The production of halftone screen positives has to be undertaken with care. The imbalance (predicted by the colour charts) between the positives necessary to print neutral greys must be obtained together with the required dot sizes in each important colour.

Two-stage positive overlay masking

This system works on the same correction principle as the previous one-stage overlay system, but isolates the grey scale, i.e. the tonal aspects of the original from the coloured

Colour and tonal correction

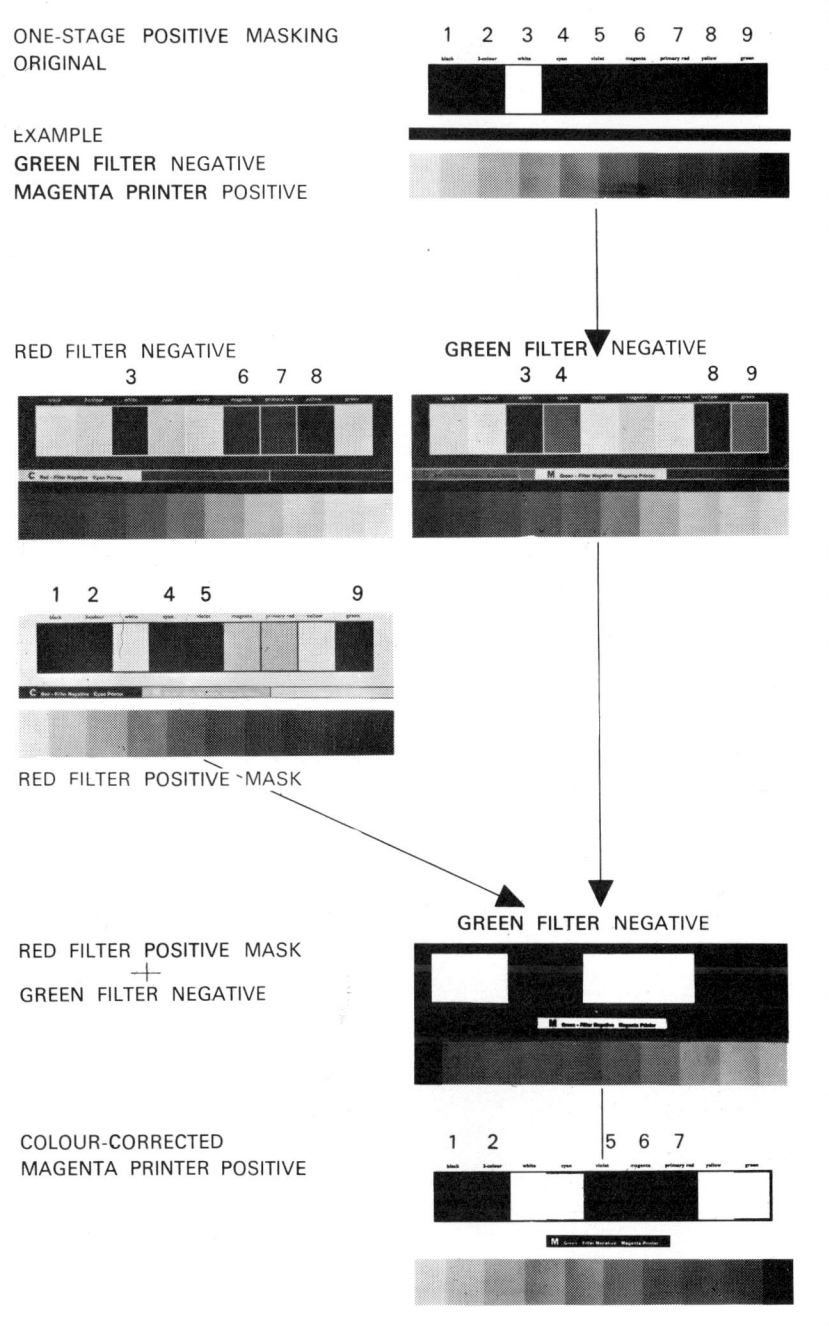

Colour reproduction

areas, allowing a greater degree of colour correction to take place. The effectiveness of the masks is reliant upon the filtration of the separation negatives. This masking system is an excellent method for complex reflection originals and can also be reversed into a negative system for correcting colour transparencies.

Blue, green and red filter continuous-tone separations are made. These negatives should have as near identical end densities and density ranges as is possible within the capability of the equipment. The black printer negative is exposed through red and green filters. Unsharp positives, which are termed premasks, are produced by contact from the separation negatives. The positive premask should have the same contrast and density range as the separation negative with which it will be combined, in order to make the final correction mask. This is easily checked when the premask is registered with the appropriate negative. The grey scale should be cancelled out, i.e. the shadow, mid-tone and highlight steps will be of equal density. This equalizing effect is of the utmost importance and is controlled by the gamma of the premask. If the separation negatives possess identical density ranges then the premasks are developed to a gamma of 1·0.

Unfortunately, more often than not there are slight discrepancies between the negatives. For instance, if the density range of the separation negative (1), from which the premask is being made, is correct, e.g., 1·30, but the density range of the separation negative (2) with which the premask is to be combined, in order to produce the final mask, is too high or too low, then the gamma to which the premask must be developed is found in the following manner.

$$\text{Premask gamma} = \frac{\text{Density range (2)}}{\text{Density range (1)}}$$

$$= \frac{1\cdot28}{1\cdot30} = 0\cdot98\,\gamma$$

$$\text{or} = \frac{1\cdot35}{1\cdot30} = 1\cdot03\,\gamma$$

When the density range of the separation negative (1) is too high, e.g. 1·35 instead of 1·30, its excess density of 0·05 is subtracted from the density range of the separation negative (2) (which may also have departed from the standard range). The figure obtained is divided by the range of the first separation negative.

Colour and tonal correction

$$\text{Premask gamma} = \frac{\text{Density range (2)} - \text{Excess density}}{\text{Density range (1)}}$$

$$= \frac{1 \cdot 28 - 0 \cdot 05}{1 \cdot 35} = 0 \cdot 92 \, \gamma$$

Conversely, in the case of the separation negative (1) having a low density range, e.g. 1·28 instead of 1·30, a deficiency of 0·02, the premask's gamma is found by the calculation:

$$\text{Premask gamma} = \frac{\text{Density range (2)} + \text{Deficient density}}{\text{Density range (1)}}$$

$$= \frac{1 \cdot 35 + 0 \cdot 02}{1 \cdot 28} = 1 \cdot 07 \, \gamma$$

The positive premasks are combined with the separation negatives in this manner, which is effectively the same as one-stage masking. The blue filter premask is registered with the green filter/magenta printer negative. The final mask from this combination will correct the blue filter/yellow printer negative. The green filter premask is combined with the red filter/cyan printer negative. The final mask will correct the green filter/magenta printer negative. The red filter premask is registered with the red-green filter/black printer negative. The final mask from this sandwich will correct the red filter/cyan printer negative. Finally, the blue filter premask is used again, this time in register with the red-green filter/black printer negative. The final mask will correct the red-green filter/black printer negative.

The final mask in each case is a colour-correction image and it should not reduce the contrast of the separation negative. Therefore it can be made to a higher density than one-stage masks, 50% or more not being uncommon. If the final mask puts a low density over the entire grey scale this can be compensated for when exposing the positive. The contrast of the final masks can be carefully controlled by exposing for the correction of 'wanted' colours and developing for the correction of 'unwanted' colours. Once again 'unwanted' colours are closely matched to the density of the white patch and the 'wanted' colours to the 3-colour density. If the percentage of the final correction mask is needed, this is found as follows:

$$\text{Final mask percentage} = \frac{100 \times \text{Final mask's gamma}}{1 + \text{Final mask's gamma}}$$

$$= \frac{100 \times 0 \cdot 95 \, \gamma}{1 + 0 \cdot 95 \, \gamma} = 48\%$$

Colour reproduction

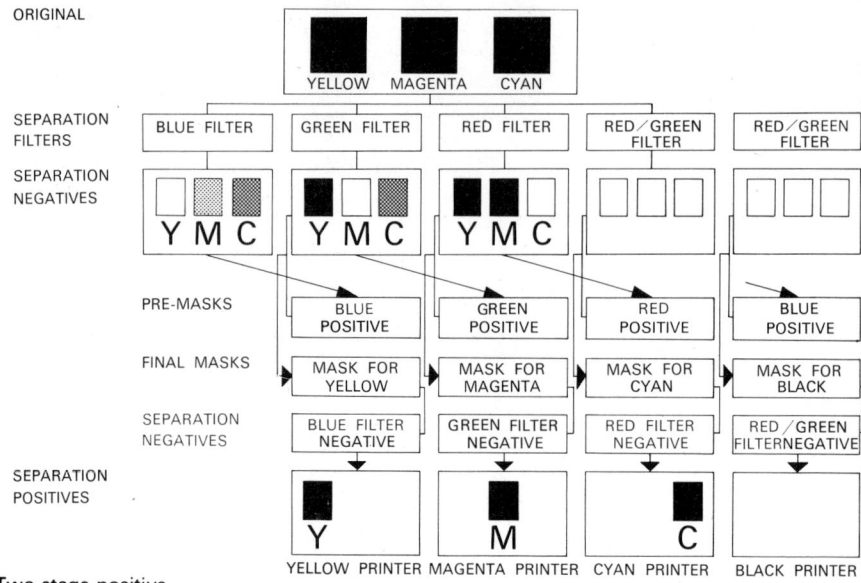

Two stage positive overlay masking:

The production of positives for each colour printer is fairly straightforward because the masked separation negatives remain at normal contrast levels. This also allows the imbalance between the positives for the printing of neutral greys to be obtained. Improved colour correction is achieved because of the non-linear nature of the final masks.

The registration of a negative mask on to a positive original achieves the same result as positioning a positive mask on a colour separation negative. The advantage of negative masks is that the original is their master image and they are the result of direct filtration which introduces greater flexibility in the degree of colour correction combined with improved definition. Negative masks can be employed to correct reflection originals using a camera-back masking system, as long as a precision camera with pin-registration is available but to begin with it is easier to understand negative masking on colour transparencies.

One-stage negative overlay masking

Three unsharp negative masks are made directly from the colour transparency by contact on a pan masking film, using a green filter, a magenta filter and an orange filter. The mask strength in each case is calculated from the printing inks and used as a percentage of the transparency's density range.

Colour and tonal correction

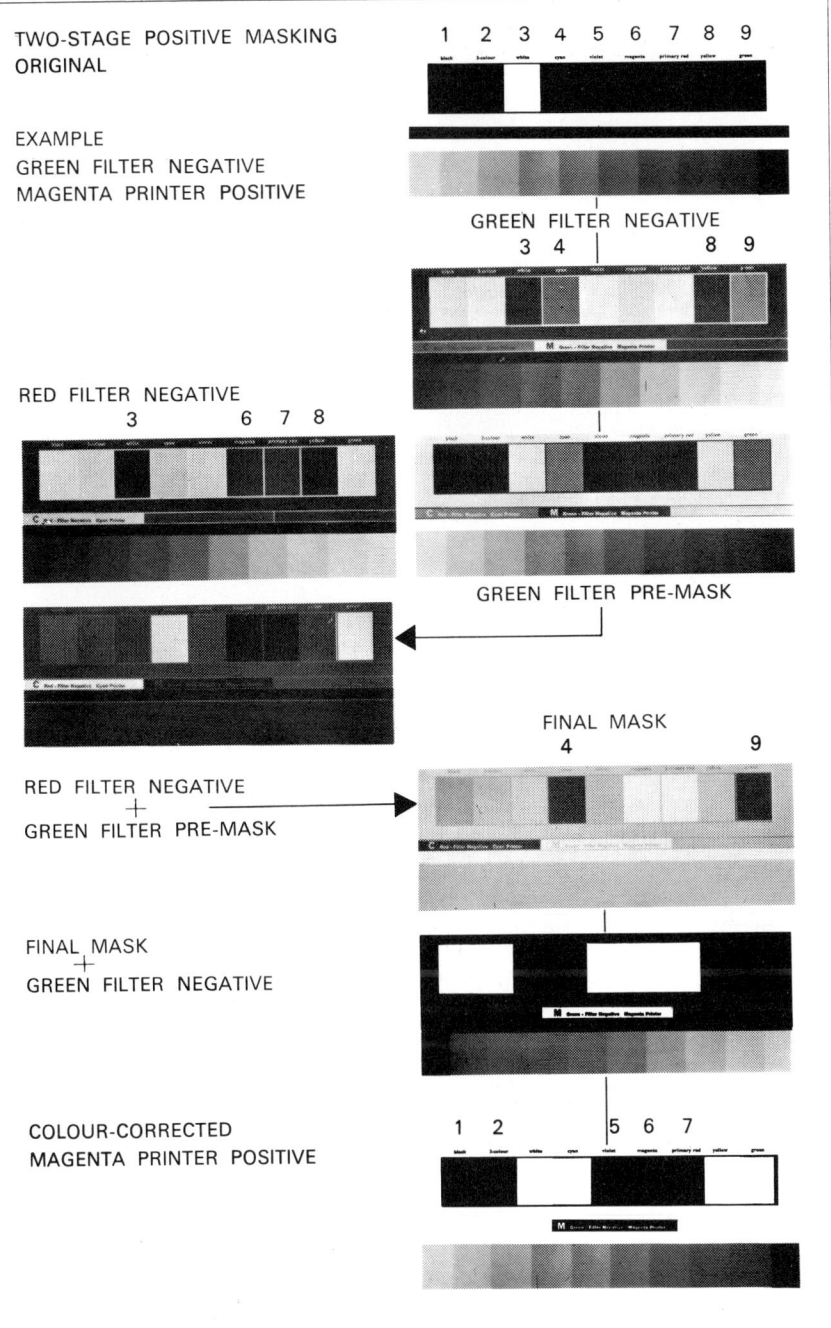

Colour reproduction

Masking a transparency. Taking the magenta image as an example, we have to remove from this a density equivalent to the unwanted cyan record of the green filter negative. This unwanted cyan is neutralized by a silver mask, bound up with the transparency while making the magenta printer negative through the green filter. This negative is then already corrected.

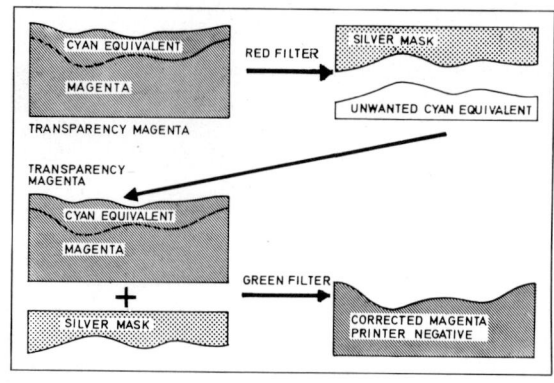

The density ranges of masks and combinations in question can be determined on the Kodak alignment chart.

Example

	Highlight	Range	Shadow
Density range of the transparency	= 0·4	2·0	2·4
Mask strength in each case	=	35%	
Density range of the masks	= $\dfrac{2 \cdot 0 \times 35}{100}$	= 0·7	

Combined density ranges	Highlight	Range	Shadow
Positive transparency	= 0·4	2·0	2·4
Negative mask	= 0·9⁺	0·7⁻	0·2⁺
	1·3	1·3	2·6

Range of final combination = 1·3

The colour-corrected blue filter/yellow printer negative is exposed from the transparency plus the green filter negative mask. This mask compensates for the blue absorption in the magenta and cyan inks. The green filter/magenta printer negative is made from the transparency re-registered with the magenta filter negative mask. A mask of this type corrects the green absorption of the cyan ink and for the slight green density apparent in the yellow ink. The red filter/cyan printer

Colour and tonal correction

One-stage negative overlay masking:

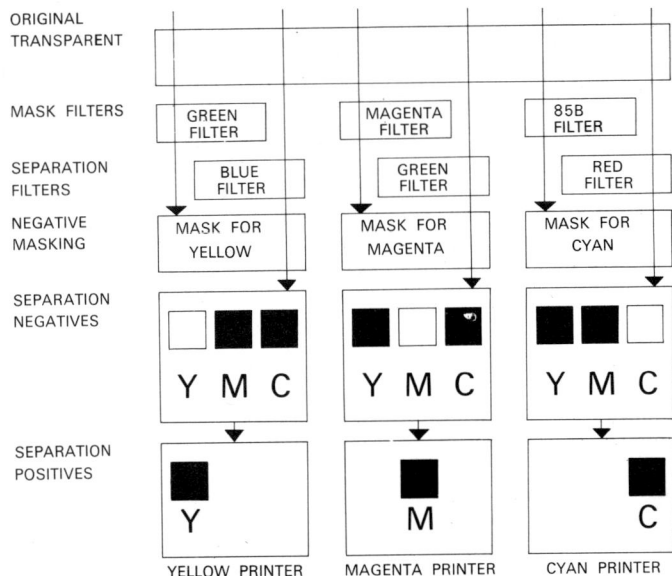

negative is produced from the transparency combined with the orange filter negative mask. This mask is similar to a red filter mask and therefore acts as a 'self' mask producing a reduction in contrast. Being orange the mask filter transmits green light to a certain extent which corrects for the small absorption of red light occurring in magenta inks. The black-printer negative may be produced by any of the normal procedures with the orange filter mask still in register, or a split mask and split filter technique can be used—for instance, a green filter exposure with the magenta filter mask in register followed by a split exposure through the red and blue filters with the green filter mask in combination with the transparency. Once again the extent and effectiveness of the colour correction is analysed on the colour patches. Positive making follows with all the usual references to grey balance.

Two-stage negative overlay masking

This system begins with the production of an orange filter contrast-reducing mask which after registering with the colour transparency will produce a more workable density range, e.g. 1·30. From this combination three uncorrected colour separation negatives are produced to a gamma of 1·0 plus a black-printer negative. The final masks are made by combining a separation negative with the transparency and exposing through the appropriate filter.

Colour reproduction

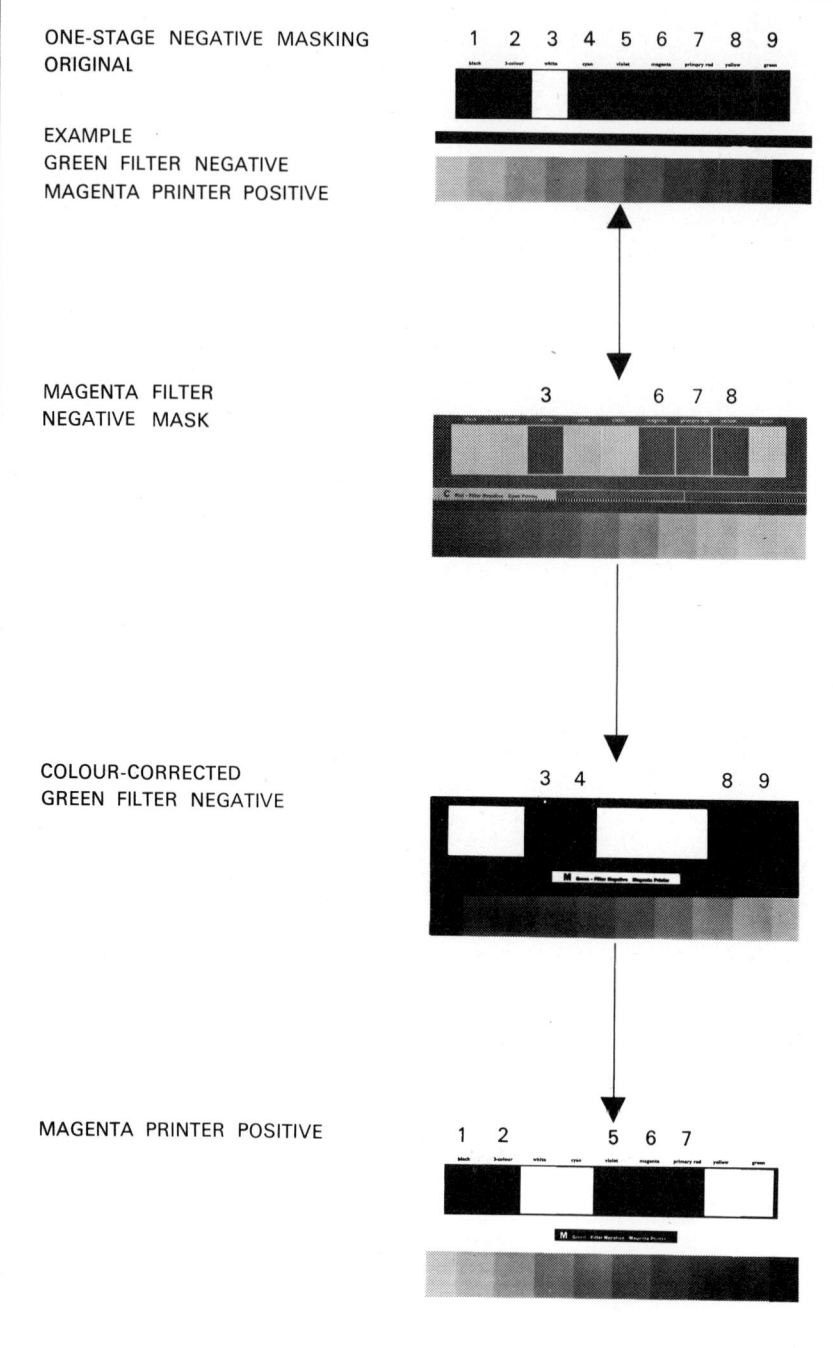

Colour and tonal correction

Blue filter	Green filter	Red filter
$\begin{pmatrix}\text{transparency}\\ \text{green filter}\\ \text{negative}\end{pmatrix}$	$\begin{pmatrix}\text{transparency}\\ \text{red filter}\\ \text{negative}\end{pmatrix}$	$\begin{pmatrix}\text{transparency}\\ \text{blue filter}\\ \text{negative}\end{pmatrix}$
↓	↓	↓
Final mask 40% for the blue filter/ yellow printer Negative	Final mask 40% for the green filter/ magenta printer Negative	Final mask 20% for the red filter/ cyan printer Negative

This system is also applicable to reflection originals. The uncorrected separation negatives developed to a gamma of 1·0 are replaced in register on the camera back and the final masks exposed in turn through each one using the appropriate filter. In both cases the final masks are registered with their corresponding separation negative to produce the colour-corrected positives.

Multi-colour negative masking

The fairly recent introduction of a single mask consisting of coloured images, by Gevaert (Multimask film) and Kodak (Tri-mask film) has revolutionized photographic masking. The mask is essentially a multi-layer colour film which produces negative dyed images after a white light exposure to the coloured original and processing in a colour developer and bleach/fix solution. It becomes unnecessary to remove and re-register numerous masks. The normal tricolour filters automatically select the appropriate coloured masking image, producing a colour-corrected negative.

The basic correction principle of dye-layer masking films may be appreciated in the following way. A green filter mask compensates for the blue absorption in the magenta and cyan inks. Both inks act as if they are contaminated with yellow ink, so if the image of the green filter mask was dyed yellow it would correct for these deficiencies. The yellow image would not affect the green filter/magenta printer negative or the red filter/cyan printer negative, but because it absorbs blue light it would colour-correct the blue filter/ yellow printer negative. A red filter mask compensates primarily for the green absorption of the cyan ink. Therefore, if the red filter mask was dyed blue, it would not affect the blue filter/yellow printer negative, but would colour-correct the green filter/magenta printer negative.

In Multimask film, layers 3 and 1 carry out colour correction similar to masks made through the green and magenta

Colour reproduction

filters, respectively; layer 2 performs a similar function to that of the orange filter mask. The gamma values of the yellow, magenta and cyan mask-images are 0.5, so that contrast reduction is balanced for each separation.

Masking procedure provided by Gevaert Multimask film.

Layer	Layer sensitive to	Mask colour	Separation affected	Mask gamma	Provides correction for	Function
1	Red and Blue light	Magenta	Green	0.5	Average Green absorption of Cyan and Yellow inks	Similar to Magenta Filter Mask
2	Red, Green and Blue light	Cyan	Red	0.5	Gamma of Cyan image, Red absorption of Magenta dye, and Red absorption of Yellow ink	Similar to an Orange Filter Mask
			Yellow filter layer			
3	Green light	Yellow	Blue	0.5	Blue absorptions of Magenta and Cyan inks	Similar to Green Filter Mask
			Film base			

In Tri-mask film the colour correction is more extensive. Layer 3 carries out the same function as a green-filter mask. Layers 1 and 5 perform the correction of a magenta-filter mask, but the two mask-images are made to different gamma values allowing the use of different mask strengths. Layers 2 and 4 simulate an orange-filter mask, but once again the two mask-images are adjusted independently to produce the correct gamma value. The Tri-mask film not only produces colour correction similar to green, magenta and orange filters, but provides photographic masking based on five coloured images instead of three images. The various mask-images add up to a total gamma of 0.5 in each case, so that contrast reduction is balanced for each separation.

Colour and tonal correction

Masking procedure provided by Kodak Tri-Mask film.

Layer	Layer sensitive to	Mask colour	Separation affected	Mask gamma	Provides correction for	Function
1	Blue light	Magenta	Green	0.1	Green absorption of Yellow ink	Similar to Magenta Filter Mask
		Yellow filter layer				
2	Green light	Cyan	Red	0.25	Red absorption of Magenta ink	Similar to an Orange Filter Mask
3	Green light	Yellow	Blue	0.5	Blue absorptions of Magenta and Cyan inks	Similar to Green Filter Mask
4	Red light	Cyan	Red	0.25	Gamma of Cyan image	Similar to an Orange Filter Mask
5	Red light	Magenta	Green	0.4	Green absorption of Cyan ink	Similar to Magenta Filter Mask
		Film Base				

The different manufacturers have their own colour-masking negative emulsions with suitable sensitizers and corresponding dye-couplers. A colour developer is a solution of chemicals capable of reducing exposed silver halide with simultaneous production of a coloured dye by the interaction of oxidation products of the developing agent with a colour coupler incorporated either in the emulsion or in the developing solution. A typical developing agent would be diethylparaphenylene diamine or its derivatives. A colour coupler is an

Colour reproduction

organic compound which reacts with the oxidation products of appropriate photographic developers to form an insoluble dye. For example, phenols give cyan dye, pyrazotones make magenta dye and ethylacetoacetate produces yellow dye. The unsharp colour mask may be used in contact with a transparency or used as a projected mask in a camera-back system. The dye-image, even in a dense area, will not scatter the light in the same way as a silver-grain mask and this improves the retention of detail. Generally speaking, the mask strength for a colour transparency using normal inks is usually in the order of 40%.

Example.

	Highlight	Range	Shadow
Density range of colour transparency	= 0·4	2·2	2·6
Mask strength	=	40%	
Density range of mask	= $\frac{2\cdot 2 \times 40}{100}$ = 0·88		

Combined density ranges	Highlight	Range	Shadow
Positive transparency	= 0·4	2·2	2·6
Negative mask	= 1·08 +	0·88 −	0·2 +
	1·48	1·32	2·8

Range of final combination = 1·32

For reflection originals the end densities of the mask are usually lower.

Example.

	Highlight	Range	Shadow
Density range of reflection original	= 0·05	1·5	1·55
Mask strength	=	40%	
Density range of mask	= $\frac{1\cdot 5 \times 40}{100}$ = 0·6		

Combined density ranges	Highlight	Range	Shadow
Positive original	= 0·05	1·5	1·55
Negative mask	= 0·80 +	0·6 −	0·20 +
	0·85	0·9	1·75

Range of final combination = 0·9

Colour and tonal correction

The final combination reduces the density range of the original so a high contrast emulsion is usually recommended to give the required negative density range, e.g. 1·30.

The use of multi-colour masking films has brought a new meaning to the term direct screening. In the past direct colour separations, especially from transparencies, were always associated with prolonged exposure times and poor halftone gradation in the separation negatives. This has now changed as a result of the work of F. R. Clapper, of Eastman Kodak, who introduced a direct screening system from colour transparencies. This system begins with an enlarger-type camera fitted with pulsed zenon illumination concentrated through optical condensers. The transparency is masked with a multi-colour negative masking film and exposed through grey negative-working contact screens on to panchromatic lith film. The three colour separated negatives are produced at standard halftone density ranges, the highlight areas being controlled by a 'no-screen' exposure and the shadows by a supplementary flash exposure. Originals with various density ranges are reproduced in a similar manner by making an appropriate black printer plate.

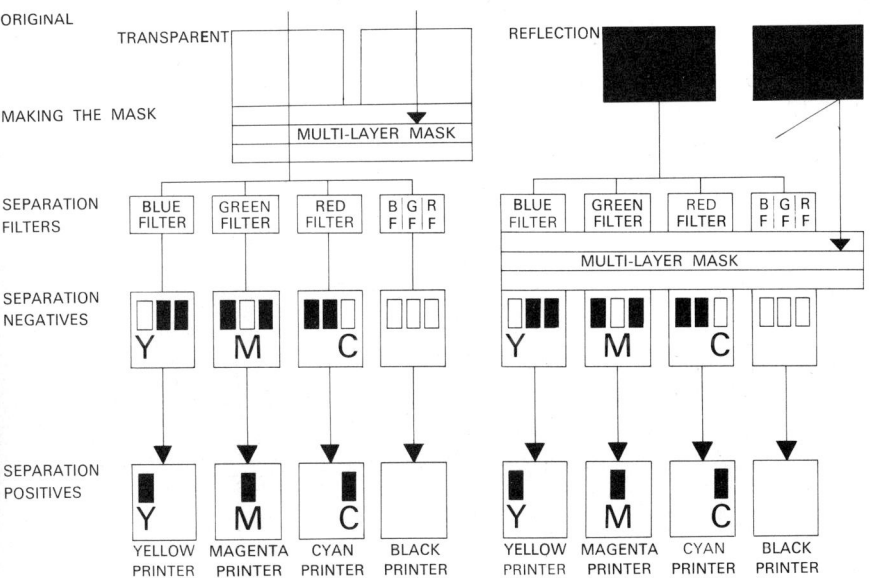

Multi-colour negative masking

Another way of employing a single negative mask throughout the colour separation procedure is to expose a sheet of low contrast pan masking film through an orange filter, or to use a similar emulsion which has this orange filter incor-

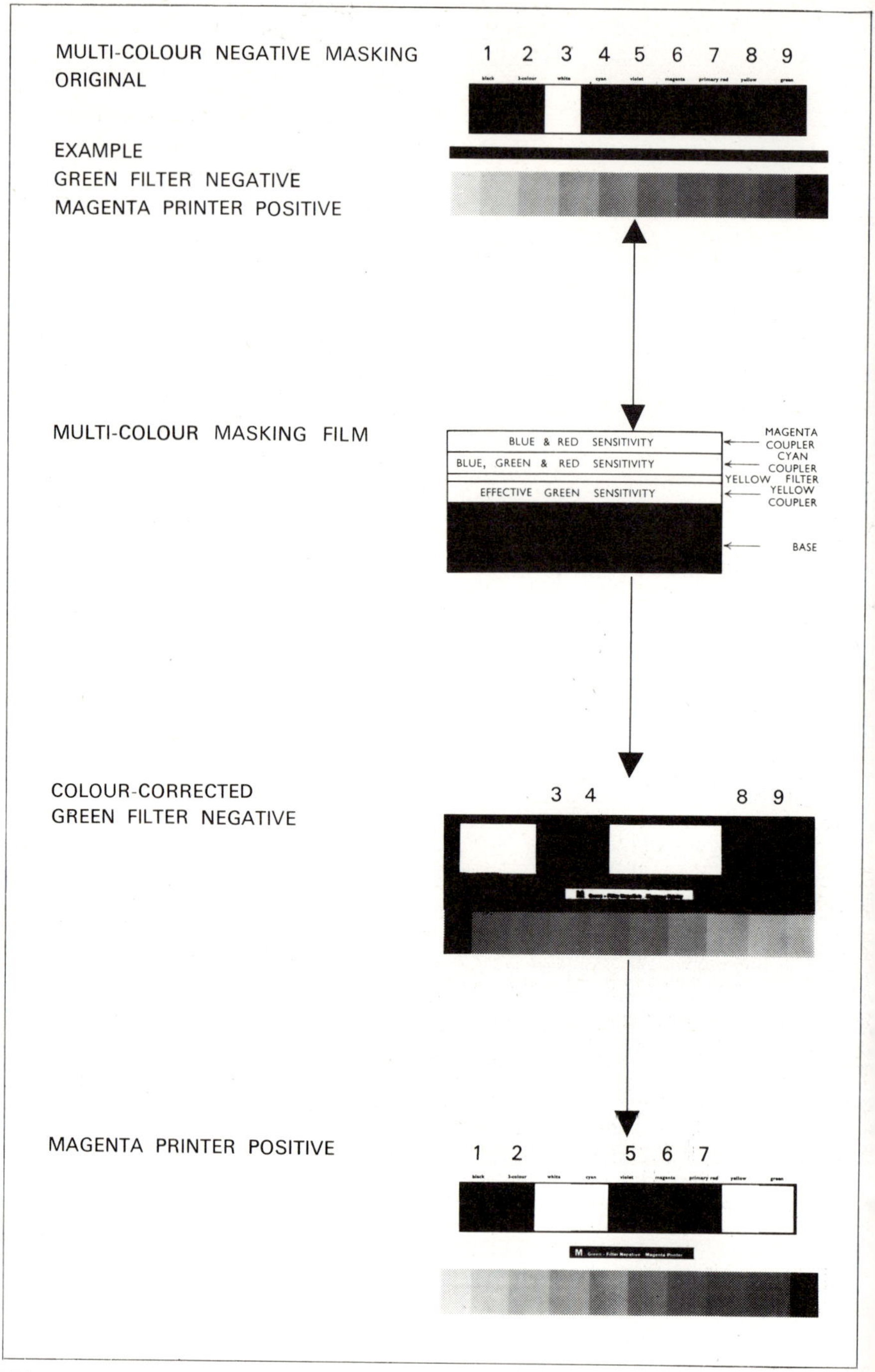

Colour reproduction

MULTI-COLOUR NEGATIVE MASKING
ORIGINAL

1 2 3 4 5 6 7 8 9
black koninur white cyan violet magenta primary red yellow green

EXAMPLE
GREEN FILTER NEGATIVE
MAGENTA PRINTER POSITIVE

MULTI-COLOUR MASKING FILM

BLUE & RED SENSITIVITY — MAGENTA COUPLER
BLUE, GREEN & RED SENSITIVITY — CYAN COUPLER
YELLOW FILTER
EFFECTIVE GREEN SENSITIVITY — YELLOW COUPLER
BASE

COLOUR-CORRECTED
GREEN FILTER NEGATIVE

3 4   8 9

MAGENTA PRINTER POSITIVE

1 2   5 6 7
black koninur white cyan violet magenta primary red yellow green

TABLE TOP: Uncorrected

Original

Uncorrected reproduction
conventional inks;

Uncorrected reproduction
Improved hue inks
movement towards the
original

TABLE TOP: Conventional build-up

3 colours plus
UCR black

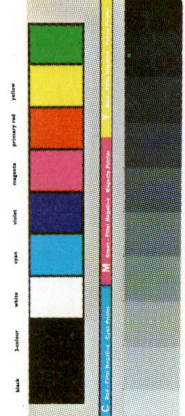 

3 colours with
under colour removal

UCR black

☐   *conventional black*
☐ ☐ *under colour
       removal black*

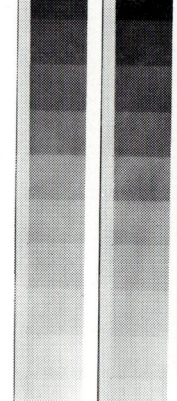 

Conventional black

3 colours plus
conventional black

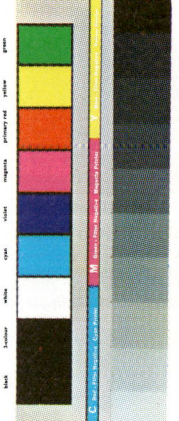 

3 colours: no under
colour removal on the
chair backs or seats

Colour and tonal correction

porated into its structure, e.g. Agfa Verimask film. A single silver mask of this type does not compensate for hue deficiencies—it corrects the relative brightness of the various colours and allows an increase in colour saturation to take place. This masking system lightens the blues, greens and cyans in the printed reproduction, all of which in an uncorrected result appear in a dark, degraded state. Although this simple system has the advantage of using normal emulsions and processing chemicals, to obtain the best printed results it is advisable to use printing inks of improved hue, such as those formulated for the Kodak short-run 'pleasing colour' process.

It has probably become quite clear by now that most masking systems may be employed by producing the colour correcting masks from colour transparencies or separation negatives using the contact method. Similarly the masking systems can be introduced on the vacuum camera back, or enlarger baseboard, as long as pin registration is available. These methods demand more care and precision while exposing the masks on the correct plane and re-registering them in turn over the next sheet of film. Colour masking is not just a matter of automation—its application requires intelligent thought.

Unsharp masks

Colour-correction and general contrast-reducing masks need to be rather unsharp in their appearance. The advantage of this is two-fold. Firstly registration is easier, and secondly the apparent sharpness of the reproduction is improved. The combination of sharp images of opposite signs reduces the contrast and the sharpness. An unsharp mask will decrease the contrast, but because of its unsharp nature it will mask critical detail without flattening it. An unsharp mask tends to accentuate the density gradient between two tones producing an emphasis in the form of a slight transitional hump of increased density. With the introduction of masking the contrast of the reproduction will be increased, especially in the highlight and lighter tones of the original. This emphasizes the fine detail and increases the apparent sharpness of each tonal movement. An unsharp mask can be made in three ways:

1. By exposing through the back of the film of transparency using a large, diffuse light source.
2. By introducing a 'spacer'—a matt surfaced diffusion sheet between the two surfaces, the matt side being positioned away from the emulsion on to which the mask image will be exposed.
3. Degrees of unsharpness can be controlled by the rotation method. A suitable contact frame is rotated, while a point

Colour reproduction

source of light, which is positioned away from the vertical centre, exposes the emulsion at an angle, thus spreading the mask image in an unsharp manner.

Masks:
1. Sharp mask
2. Reproductions (sharp mask)
3. Unsharp mask
4. Reproduction (unsharp mask)

Very unsharp masks should be avoided because they produce unpleasant outline effects, and unsharp masks are not required when two images of the same sign are being combined, e.g. a colour separated negative with a negative highlight mark.

Tonal correction

Original colours, irrespective of their hue and saturation, may be printed in a lightened or darkened condition in order to accentuate the highlight detail or improve the shadow definition appearing in the final reproduction. Tonal correction of this type is required because most graphic reproduction processes impose physical limitations on the printed density range and the degree of tonal movement in the highlight and shadow areas, together with definite tonal distortions which can be traced and recorded on 'Jones-type' quadrant curves. An attempt to overcome this distortion is to accentuate the highlight regions of the reproduction. This is more common than correction of shadow areas.

Tonal masks

A highlight mask is an underexposed, low density and short-range negative image of the highlight regions. This mask is usually made on panchromatic lith type film and developed in a dilute metol-hydroquinone developer until a density of 0·4 is reached in the lightest highlight. A highlight mask may be introduced during the exposure of negative overlay masks or combined with the separation negatives when the final positives are being exposed. The aim in negative masking is to flatten the mask in the highlight regions. This produces the

Colour and tonal correction

same result as increasing the highlight densities of the separation negatives.

A principal mask is usually a 35–40% general contrast-reducing mask made on a low-contrast, pan masking-type film. This mask is used to compress the density range of a colour transparency to a workable length. A principal mask should produce a flattening effect throughout the tonal range of the original; if a highlight mask is used the compression does not take place in the highlight regions.

A shadow mask is essentially a reversed highlight mask, both in appearance and function, but is rather more difficult to produce. A full strength no-filter negative is exposed on to a continuous-tone panchromatic emulsion of medium contrast. This is registered with the original or on the camera back. Its effect must be to block-out the highlight and middle-tone regions leaving the shadow areas in an apparently lightened condition. The panchromatic lith film is now employed to record a short-range negative image of the shadow areas. The full strength negative is removed and colour separation procedures performed in the normal way. The shadow mask is registered with the separation negatives when the final positives are being made.

Undercolour removal

There are two diametrically opposed methods of printing the black plate. The conventional method is associated with dry printing, when each ink layer is allowed to dry before another is printed. This approach facilitates the use of heavy colour printings and as a consequence the black printer is no more than a 'skeleton' monochrome image of the original producing contrast control in the shadow areas and maintaining the neutrality of greys.

Wet printing requires a different approach. The black printer needs to be made to a greater density, while each colour in the same area is printed at a correspondingly lower density. The heavy black printer plate produces the shadows and neutral grey areas, relieving the three colours of the task and therefore overcoming the 'piling-up' and drying problems. Generally speaking, this approach is the reverse of the conventional method.

When high-speed wet printing is employed it is not possible to apply the quantities of ink required to reproduce a black in three colours. Undercolour removal is a tonal correction which allows a dense black printing to be used by reducing the amount of the other three colours, yellow, magenta and cyan, in every areas where black is to be printed. Undercolour

Colour reproduction

removal prevents the accumulation of several ink layers, all trapping in an unsatisfactory manner. The balance of the three coloured component ink layers is less delicate, and relatively inexpensive black ink replaces the rather costly coloured inks.

It has been found in wet letterpress printing that a total ink density of 240% should not be exceeded in any area. This is achieved by adjusting the maximum dot size in each colour printer and adding their values, e.g. yellow 50%, magenta 55%, cyan 65% and black 70%. This is only an example and such adjustments can only be determined after exhaustive tests have been carried out on printing machines under prevailing conditions.

It is general practice to produce an undercolour removal (UCR) mask by contacting from the black printer negative. This positive mask is combined in turn with each separation negative while the final positives are being made. Densities in a UCR mask should only appear in the shadow areas; any highlight density will disturb the colour correction. This positive mask needs to cancel out the negative's lowest shadow tones into an even density over the last, say, four to five steps of the stepwedge. This will result in the saturated colours being retained, if not accentuated.

Colour transparencies may be approached in a different way. A black printer negative is made first and then becomes a UCR mask itself, being registered with the transparency for part of each separation exposure. The resultant separation negatives have the required amount of undercolour removal built in.

Assessment of coloured originals

A coloured original is defined as a picture presenting line and tone situations in colour, e.g. colour transparencies, prints, paintings and line drawings. Colour separation photography is only possible after a detailed study has been made of the original, resulting in an appreciation of its origination complemented by a complete knowledge of the colours, their make-up, saturation, harmonious or discordant nature, and the all important comparison with the appropriate colour chart printed on a stock similar to the paper selected for the printing. Awareness produces correct decisions and efficient printing.

A coloured image

Our eyes are stimulated by blue, green and red wavelengths of light energy. Cones, which are minute colour-sensitive

Assessment of coloured originals

light receptor cells positioned in the retinas of our eyes, act like blue, green and red filters separating and synthesizing pictures into nerve impulses which activate the brain, creating coloured sensations attributed to the source of the stimulus. The graphic reproduction photographer must duplicate this mechanism using a camera, blue, green and red filters, a panchromatic emulsion and finally positive images synthesized with blue, green and red-light absorbing inks on a sheet of white paper.

The photographer begins by finding the density range of the original. This is compared with the density capability of the printing process in question and evaluated with the tone standards previously mentioned. This indicates the amount of tonal compression or expansion required. With this in mind the highlight, middle-tone and shadow densities of the original are selected and transferred to a grey step-wedge on the control strip. These steps are referred to as the H, M and S reference points. These neutral areas are compared with the appropriate colour charts until a suitable neutral match is found for each one and the corresponding density values and dot sizes are recorded. Using the characteristic curve of the halftone screen to be employed a quadrant diagram, as previously illustrated, is drawn. From this the degree of imbalance between the three colour printers required to produce neutral greys becomes apparent.

Colour control strip

This should consist of the usual colour patches and a grey step-wedge masked-off and marked to indicate the H, M and S reference points. In addition, due notice must be taken of any limitations predicted by the tone standards. The next step is

Colour key:
Negative densities and positive dot sizes for each colour plotted

Colour reproduction

to produce a colour key on tracing paper of the most important colour areas of the original. By careful matching with the colour chart the required continuous-tone densities and halftone dot sizes relating to the yellow, magenta, cyan and black printing plates may be plotted in each area. These predicted values, which give a picture of the reproduction compared with the original, together with any non-reproducible or badly matched colours, are available to everyone at the very initiation of the reproduction.

Camera procedure

The camera procedures for colour reproduction vary slightly depending on which masking system is employed, but this can be helped enormously by the production of standard step-wedges.

To begin with these standard step-wedges must correlate closely to the continuous-tone densities and halftone dot sizes used to produce the colour charts. Once these have been established continuous-tone and halftone negative step-wedges and their corresponding positive step-wedges in halftone, continuous-tone, camera and contact screened form are made to a reasonable size, using the related piece of photographic equipment with standard exposure procedures and constant processing conditions.

Every one concerned with the reproduction process should have a set, so that a retoucher will know precisely what positive dot size he will receive from any given density. This is an immense help when masks are being registered with separation negatives. Specific dot sizes can be associated with the combined densities. This allows negatives and masks to be retouched locally in areas which present critical colours. The camera operator will be aware of his equipment's efficiency and the working characteristics of the halftone screens, emulsions, developers, etc. Furthermore, in every case he has a standard aim point—in short everyone knows what is going on and what to expect at each photographic stage—results become predictable.

Relevant information

Because of the fact that colour reproductions revolve around the necessity to expose through blue, green and red filters, the knowledge of correct filter ratios or factors related to the type of filter, colour temperature of the light source, camera, emulsion and processing conditions is essential. The ratio and factor numbers supplied by the emulsion manufacturer provide sensible starting points, but when

Camera procedure

balanced separation negatives are required containing similar end-densities, then appropriate tests must be carried out under the prevailing conditions.

These tests become rather lengthy if correct exposure increases with each filter are required for line, halftone and continuous-tone separation negatives, but in the end the results make it all worthwhile. As with most photographic tests average originals are simulated by a stepwedge. In this case the stepwedge is placed alongside the normal colour patches. To provide comprehensive tests these control strips need to be in reflection and transmission forms so that a complete simulation of both reflection originals and coloured transparencies is achieved.

We will now follow the test which will find the filter ratios for the medium band tricolour filters transmitting pulsed xenon light from reflection originals, using continuous-tone, medium contrast panchromatic film. The test for factor numbers is similar, commencing with a white light exposure as the base of 1·0. A series of exposures are given with each filter together with suitable blocking-out on the camera back. This produces one sheet of film containing, say, seven images of the step-wedge and colour patches. The exposures should start with the red filter as it is given the base of 1·0. Estimate an exposure, e.g. 16 light counts, make this the mean and give three exposures each side, increasing and decreasing by the square root of 2 (i.e. 1·4 approx.).

Example.

Red filter ratio 1·0 $\dfrac{5, 8, 11, 16, 22, 32, 45 \text{ light counts}}{\downarrow}$

Development γ = $\dfrac{\text{halftone screen's reproducible range}}{\text{average original density range}}$

Somewhere in this red filter separation negative one of the seven images will closely resemble the specifications. Occasionally it becomes necessary to repeat this with more closely associated exposure times. Looking at the series of test exposures above the precise exposure was 14 light counts related to a development gamma of 0·85. This negative image possessed a highlight density of 1·7, a shadow of 0·4 producing a range of 1·3.

The test is repeated with the green and blue filters. After inspecting the resultant negatives, two images are selected in each case, each one possessing density characteristics similar to those achieved in the red filter separation negative. The

Colour reproduction

development time may have to be altered for one of the negatives in order to maintain a constant highlight density. The filter ratio in each case is the proportional difference between the red filter 'base' and the green and blue filter exposures.

Example.

Red filter exposure	14 light counts	Filter ratio	1·0
Green filter exposure	18 light counts	Filter ratio	1·3
Blue filter exposure	10·5 light counts	Filter ratio	0·75

No matter what the actual red filter exposure becomes when originals of diverse nature are encountered, if the operating and processing conditions become standardized, then the filter ratios remain constant.

Colour separated negatives

The production of continuous-tone negatives is controlled by exposure and development times with reference to the densities of wanted and unwanted colours incorporated in a suitable density range. Separation negatives in line or halftone form on panchromatic lith film need to be controlled carefully at the processing stage. At times even automatic processing cannot bring about the delicate balance needed to retain an extremely fine line, while other colours are separated-out to record in an opaque manner. To overcome these problems with panchromatic lith film, Cosmocord have introduced the Acoscope viewer. This enables panchromatic film to be inspected during development using a special infra-red converter cell. When colour line separations are required the camera operator must think in terms of holding colours as black and losing others as white, assisting this with the correct choice of filter and sensitive emulsion. With this in mind a filter chart has been drawn up in Table 28, in the Appendix. For the more opulent exposure times, related to all types of negative making, can be predicted by electronic exposure meters such as the Densiprobe, Luxometer and Programma, to name just a few. These exposure computers usually employ three photo-cells, each one equipped with the appropriate colour filter. This enables accurate measurements of the light intensities within each spectral division, producing precise exposure times. All three photocells can be activated at once to assimilate the total light intensity for a white-light exposure. Exposure times integrated into light counts may be divided into three sections—flash, main and highlight units. Halftone images are controlled by dialling-in the screen's reproducible density range and relating it to the

Camera procedure

flash exposure unit; no-screen exposures are given through the highlight section, while colour compensating filters may be introduced during the main exposure. Continuous-tone emulsions can be given a short 'fogging' exposure with the flash exposure unit when tonal reproduction curves predict such measures. The dual-sensitivity of variable contrast films may be exercised to fine limits by using blue and yellow filters during the highlight and main exposure times.

Multi-light source:
1. Soft light and point source
2. Normal light
3. Hard light
4. Extra-hard and undercutting light

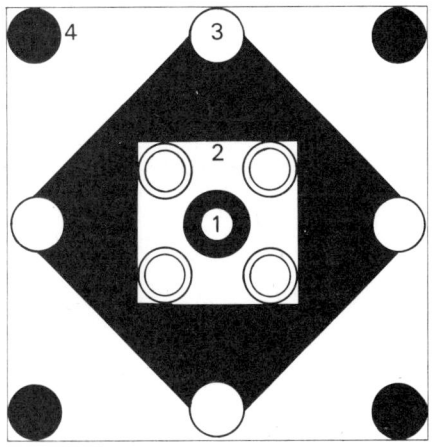

Screen angles

In four-colour work each colour is printed at a different screen angle. This is necessary to minimize moiré pattern. Moiré is the name given to the interference pattern which is set up when two or more images containing geometrical shapes are superimposed. With halftone dots moiré effects can be minimized by ensuring a 30° separation between the images. Therefore all halftone separation images, whether negative or positive, must be produced at a different screen angle or with due regard to the colour combination resulting from overprinting identical screen angles. The strongest colour is positioned at 45° because this is the angle least apparent to the eye. Yellow and pale tint colours may be placed at 15° from their adjacent colours, because of the fact that these colours are the least discernible to the eye and the resultant moiré pattern is not objectionable. Other colours can be set at the same angle as a darker colour as long as they do not print heavily in the same areas. When originals are extensively enlarged the retention of detail deteriorates and the original's surface or grain structure becomes noticeable. If the reproduction requires additional printings of pale tint colours, e.g. a flesh pink and a pale blue, then it is advantageous, when the

Colour reproduction

original contains portraits, to print the flesh colour at a finer screen ruling than the other colours. A reduced moiré pattern is produced on the important facial areas. The screen angles for most multi-colour printings can be tabulated:

Table 22. Screen angles for multi colour printings

2 Colours		3 Colours		4 Colours		6 Colours	
Strong	45°	Yellow	75°	Yellow	90°	Yellow	90°
Pale	75°	Magenta	15°	Magenta	15°	Dark Red,	
		Cyan	45°	Cyan	75°	Light Blue	15°
				Black	45°	Dark Blue,	
						Light Red	75°
						Black	45°

Elliptical-dot screens need to be angled in a slightly different way, because of the linking effect of the dots, which overcomes the 'halftone break' allowing dot sizes less than 50% to carry on the tonal gradation in a smooth and gradual manner. This continuous linking of dots at diagonal corners increases the probability of moiré pattern. Elliptical dot angles are arranged in the following order:

4 *Colours*

Yellow	90°
Magenta	15°
Cyan	165° (75 + 90)
Black	45°

⎡ The link between the diagonal corners of the cyan dot is turned to the opposite direction, minimizing the moiré pattern between the colour printers. ⎤

If a moiré pattern still persists it may be caused by any of the following faults:

1. An original containing uniform or geometrical patterns, or pictures of textiles, lace and embossed surfaces.
2. Two sets of rulings on a screen not crossing exactly at 90°.
3. An angular error in setting the screen.
4. A printing fault; dots distorting in wet printing, or a distinct lack of register.

When pre-printed coloured illustrations are presented as originals, precautions have to be taken during the reproduction photography to avoid a clash of dots resulting in moiré pattern. These precautions range from sophisticated optical equipment in the form of rotating glass discs in front of the lens and patterned filters to the crude approach of shaking the camera during the exposure. The aim in every case is to deflect the light and lose the shape of the printed dots making up the original. The most popular method seems to be the one in which the original is reduced to a very small size and a set of continuous-tone separation negatives produced. These negatives are placed in the camera or enlarger with a diffusion sheet positioned between their emulsion side and the glass holder. The required positive size is obtained by enlarging the image. The initial reduction plus the diffusion sheet usually obliterates the original dot pattern allowing re-screening to take place.

Colour separation for each reproductive process

The density range, end densities and overall appearance of the photographic image in a continuous-tone or halftone condition for each separation negative becomes apparent after a suitable colour key and a quadrant tonal curve have been produced. Specific aims can be stated for wanted and unwanted colours in each case.

For letterpress, lithography and screen process printing the end results must be in halftone. The specifications and general remarks which were made in the chapter on tone reproduction are just as applicable for colour reproduction once the colour and tonal predictions have been met.

Present day gravure colour printing is somewhat different. There has been a move away from conventional methods employing continuous-tone emulsions only. The 'invert halftone' system seems to be replacing the conventional methods. In essence this is a system of combining a halftone positive with a continuous-tone positive, gaining the advantages of both and at the same time producing ink cells which vary in width as well as depth.

The basic difference between conventional and invert halftone gravure is the way in which the tonal range of the image is constructed in the copper surface of the printing cylinder. The tonal range formed by the conventional method is composed of ink cells which are of the same width, but vary in depth.

Invert halftone systems produce ink cells which not only vary in depth, but also in width.

The advantages of invert halftone systems can be divided into two major groups—improved printing method and standardized cylinder-etching technique. In conventional gravure, the highlight regions of the tonal range are represented by very shallow ink cells, in the order of 0·5 µm in depth, separated by thin walls produced by the fish-net screen exposure. Ink retention in these shallow cavities is difficult and because of their thin dividing walls the friction and abrasion of the doctor blade and paper can cause rapid

Colour reproduction

Conventional gravure:
1. Stepwedge
2. Image construction
3. Etched ink cells

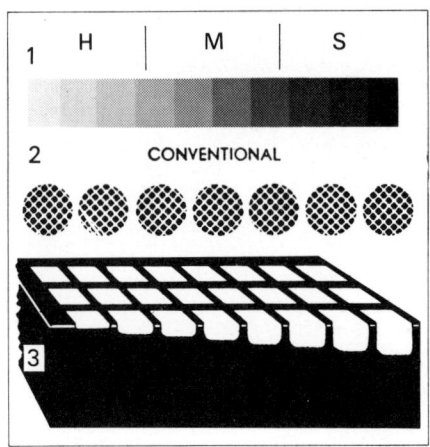

Invert halftone gravure
1. Stepwedge
2. Image construction
3. Etched ink cells

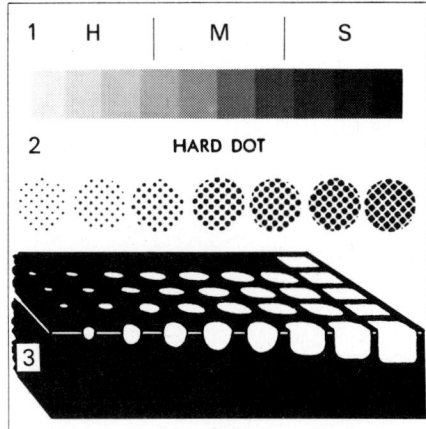

deterioration, resulting in these light tones disappearing from the reproduction. With an invert halftone construction these light tones are represented by comparatively deep ink cells divided by a considerable thickness of metal. This expanse of metal becomes a substantial support for the doctor blade and the deeper ink cells retain their ink when subjected to printing conditions. With regard to cylinder-etching the tonal range can now be expressed in dot sizes. This is an immense aid in standardization. In some companies the various etching baths of ferric chloride at different concentrations have been replace with one-bath techniques related to halftone and continuous-tone positives controlled to standard specifications by densitometric readings. Sensitometric working of the pigment paper completes the rationalization.

Halftone image:
A. Stepwedge
B. Etched ink cells
C. Required shadow cells

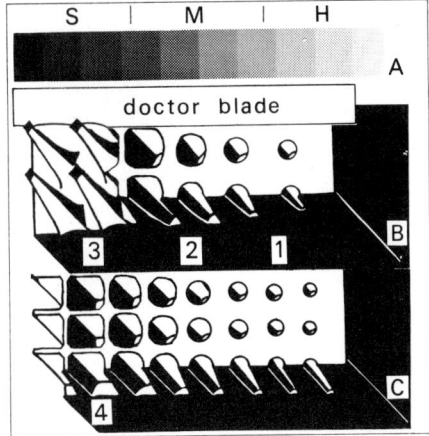

Colour separation for each reproductive process

halftone image. If a complete tonal range is translated into varying dot sizes and then etched into the metal surface of the printing cylinder, the result would appear as follows:

The etched tonal range may be divided into three sections, (1) highlights, (2) middle-tones and (3) shadows. The highlights and middle-tones are represented by correct-sized ink cells with considerable metal interspacing them. The shadow dots, however, are so large that the cell walls have become isolated metal 'stalagmites'. When printing commences the doctor blade runs smoothly over the highlight and middle-tone regions removing the surplus ink, but when confronted with the shadow areas, the lack of support and the 'stalagmite' structure will eventually result in the formation being bruised by the doctor blade producing an unsatisfactory ink transference.

To overcome this problem these isolated metal 'stalagmites' in shadow regions must be replaced by conventional walls separating the ink cells (section 4 in the diagram), while the middle-tone and highlight areas remain in a dot-structure. It is the different ways in which this combination of images can be achieved that has brought about the numerous invert halftone methods. The thin dividing walls in the shadow regions may be produced by special contact screens or by employing specially shaped diaphragms with the glass ruled screen.

The underlying principle in every halftone gravure process is in two parts.

1. The halftone positive is duplicated on to the pigment paper. This enables the subsequent ink cells to vary in *width*, with due regard to the construction of dividing walls in the

Colour reproduction

shadow regions. The halftone positive becomes the gravure screen, but the regular fishnet pattern is replaced with lines which vary in their separation depending upon the tonal value in question.

2. The continuous-tone positive is now copied on to the pigment paper, over the top of, and in register with the halftone image. The continuous-tone image, being essentially a wedge-shape, enables the subsequent ink cells to vary in *depth* in proportion to the particular tonal value.

For halftone gravure the graphic reproduction photographer is usually requested to produce the two positive images. A popular method is to make the screen positive from the continuous-tone positive. This ensures that all masked and retouched tone values are retained in the halftone image. The screen positive is produced by chemical reversal processing or on reversal film, through a gravure contact screen which provides a cell structure in the shadow regions and dot formation in the middle-tone and highlight areas. The usual specifications for each photographic result are as follows.

Table 23. Invert halftone specifications

	Shadow	Range	Highlight
Continuous-tone negative	0·4	1·1	1·5
Continuous-tone positive	1·6	1·3	0·3
Halftone positive	47% ± 4%	38%	9% ± 2%

These photographic images are part of an overall flow-chart.

With each reproduction system, most technicians and certainly the customers require a proof before the production

Halftone gravure:
1. Continuous-tone negative
2. Continuous-tone positive
3. Screen positive by reversal methods
4. Copying screen positive onto pigment paper
5. Copying continuous-tone positive onto pigment paper
6. Pigment paper to metal cylinder

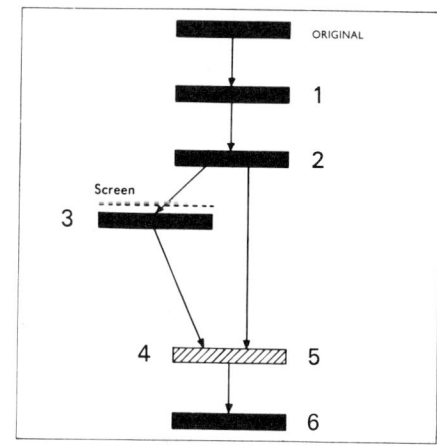

run commences. Usually this means that printing plates have to be made especially for proofing, involving a great amount of time and expense. For a long time now many companies have been experimenting with the available non-press colour-proofing systems, which produce a simulated printed reproduction from the final photographic images, whether negative or positive. Some of these methods employ dyed sheets of light-sensitive film in the complementary colours. These sheets are exposed under the appropriate separation negatives or positives. The resultant images in yellow, magenta and cyan dyes are superimposed in register on top of a sheet of paper taken from the stock selected for the printing.

A closer approach to actual printing conditions is achieved by electrostatic methods. The equipment consists of zinc-oxide coated paper, charging and exposing units, and toners which are matched to printing inks. The proofs are similar to colour photographic prints. These quick colour proofing systems are improving all the time and will eventually provide the much needed 'perfect' proof. In the meantime they are extensively used for checking the registration of 'white-out' or coloured lettering appearing within the pictorial boundaries of the reproduction.

Colour and line combination

Many coloured reproductions have type-matter appearing within their pictorial boundaries: for example, black type-matter printed across a light-blue sky and 'white-out' lettering against a dark shadow area. The incorporation of black lettering is essentially the same as the previously described procedure of line and tone combination. The inclusion of 'white-out' lettering is slightly different and is worth further investigation. Once again it is a wise move to simplify this work into four steps. To begin with the four colour-corrected separation negatives must contain pin-register holes. In this example a set of halftone positives with lettering appearing 'white-out' will be the end result.

1. *Making the line negative*

    Normal methods are used to produce a non-reversed line negative of the lettering to the required size indicated on the printing layout. This negative need only be on a minimum sized piece of film.

2. *Patching-up the master line negative*

    One of the separation negatives (the image in which most of the pictorial content is visible) is positioned on the registra-

Colour reproduction

Patching-up the master:

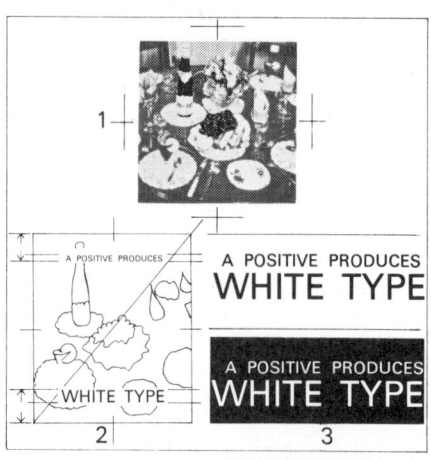

tion pins over a light-table. A sheet of clear film is punched with identical registration holes and laid over the negative using the pins. After consulting the layout, faint lines are drawn on to this clear film to indicate the precise position of the lettering. Using great care the line negative is taped into position and the rest of the clear film covering the separation negative blocked-out with black paper. In doing this a master line negative is produced.

3. *Producing the positive*

The next step is contacting. A sheet of lith film is punched so that it contains the same register holes as the master line negative and the separation negatives. This film is laid over the master line negative in register and a positive image of

Colour and line combination

The type positive:

the lettering produced. When the lettering on this positive is checked for position over the separation negative, the rest of the separation negative must be covered with clear film.

4. *Exposing the final screen positives*

Whether the final positives are to be made in the camera, enlarger or contact frame the procedure is the same. Using the registration pins the positive containing the lettering in its exact position is placed on the back of each separation negative in turn. When each screen positive is made the positive image of the lettering will block the light, therefore rendering the lettering 'white' amidst the dot values of the final positives.

Sometimes the 'white-out' effect is produced throughout the separation negatives so that the printed result contains a section of 'white' lettering which can be overprinted with a

Final positive:
1. Negative plus type-positive
2. Screen positive "white-out" type

Colour reproduction

special colour. This would mean an extra printing plate containing a positive image of the particular lettering, which on printing would fall exactly over the 'white' type with sufficient overlap all round to ensure perfect register. To achieve this another positive is made from the master line negative, but this time a 0·202 mm spacer of clear film is inserted between the two films to 'spread' the positive image so that it possesses the required overlap.

12. ELECTRONIC REPRODUCTION

Introduction

In recent times technological advances have been prolific and the rate of these new discoveries is rapid. The field of communication is no exception. Photography is primarily a means of expression and communication, so it was not unnatural that pictures of current events were incorporated into a radio transmitting system. News pictures from all over the world and now, of course, from the moon, can be received and built up line by line, using a scanning system. Television is a rapid scanning technique employing photo-electrical and radio processes which enable an actual or recorded scene to be reproduced at a distance, point by point on a screen.

These photo-electrical scanning systems have been modified for graphic reproduction processes. Their function is to separate the original into input signals representing the amounts of blue, green and red light being reflected or transmitted from any minute area of the original. These colour-separated signals are fed into an analog computer which calculates the amount of complementary printing ink required in each case. The output for these colour-corrected signals can be either a vibrating stylus which will engrave the appropriate dot pattern, or a variable intensity glow lamp exposing continuous-tone or halftone images on film.

The calculations carried out by the computer are based on the deficiencies of the printing inks. The appropriate modifications resulting in colour correction were approached, to begin with, through two different avenues—firstly by employing colour-correction (Neugebauer) equations as the foundation for the computer, and secondly by using the speed of electronics to apply modifications based on photographic masking systems, the mask strengths, degree of undercolour removal and the characteristics of the black printer being calculated by the analog computer. The former approach has not been successful, whereas the latter method is the basis of the electronic scanners available at the present time.

Scanning methods

Electronic scanners have been built round three different

Electronic reproduction

Automatic engraving.
Here a half-tone block is engraved by scanning the original with a scanning head. This contains photo-cells which measure the intensity of a light beam reflected from the original photograph. The photo-cell output controls the depth to which the engraving head digs into the metal of the printing plate.

movements, each one resulting in a picture being scanned line by line. These movements may be listed.

1. Reciprocal motion

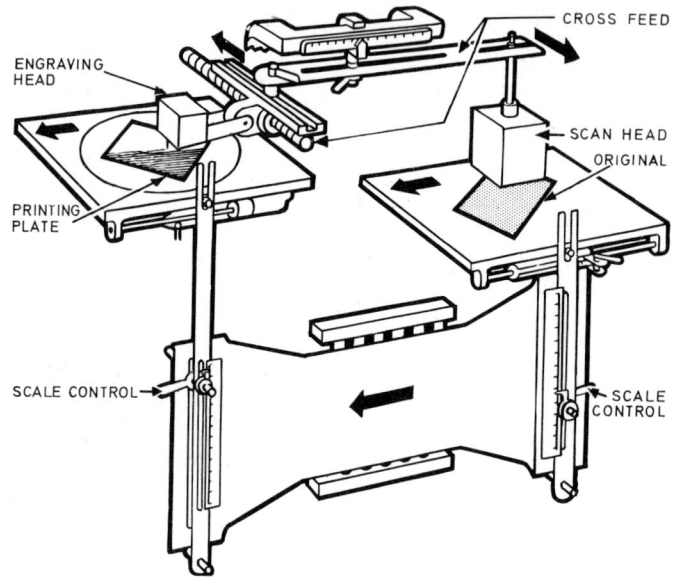

2. Rotating cylinder

Diagram of the Time-Life Scanners. These produce fully masked continuous-tone separation positive or negatives from a colour transparency. The latter is mounted on a glass cylinder, and light impulses from the lamp system reflected through the image to three filtered photo-cells. The cell outputs through a computer system which controls printer beams to expose the separation images on films wrapped around the same cylinder. The transparency and the separations are scanned in unison.

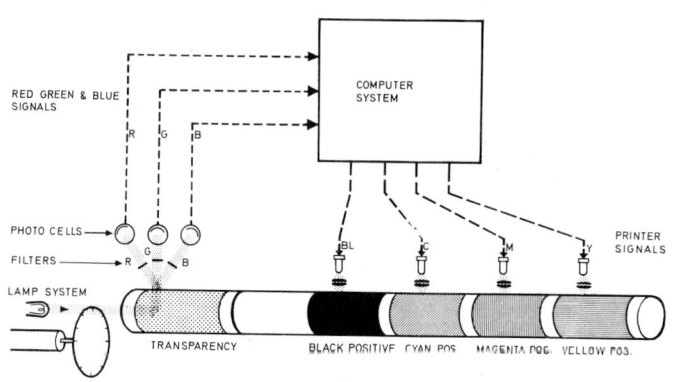

Scanning methods

3. Electronic flying-spot

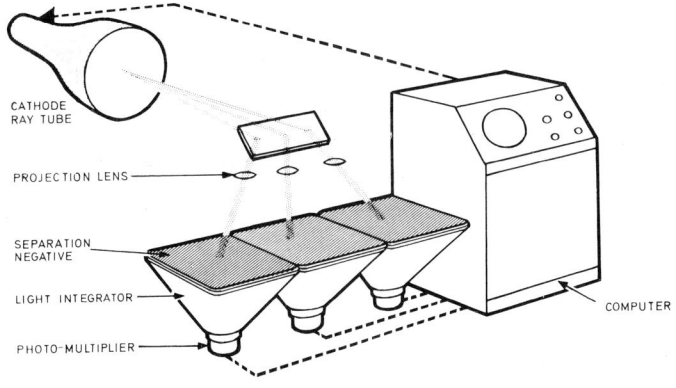

Diagram of the Scanatron. A flying spot projected from a cathode ray tube via a mirror simultaneously scans all three separation negatives over light integrators. At the same time one negative is contact printed to produce a separation positive. The photo-multiplier output from all three light integrators is fed into the computer system. This controls the brightness of the flying spot from the cathode ray tube to produce a fully corrected continuous-tone colour separation positive. The three positives are printed in succession in this way.

Drum scanners

As in television, the details of a picture may be converted, point by point into electrical signals by scanning the picture in successive lines. It is necessary to scan at a speed which produces a high degree of definition. During the electrical stage the signals can be amplified, attenuated and manipulated; so that they correspond to algebraic equations equivalent to photographic colour and tonal correction. After correction has taken place the picture is reconstituted into either continuous-tone or halftone separations.

The drum or rotating cylinder scanning system is emerging as the most popular method of producing electronic separations. Large drum scanners typified by the Crosfield Magnascan 550 and the Hell Chromagraph DC300 exposing four colour separations simultaneously, enlarged or reduced and screened in a single step from transmission and flexible reflection originals are forming the basis of add-on packages which are being continuously developed, such as page composition systems, electronic retouching, colour television monitors for display proofs and radio-transmitting methods.

Electronic reproduction

The main components of the Chromagraph DC 300:
1. Lamp compartment
2. Halogen lamp housing
3. Feed motors
4. Transparency arm
5. Scanning drum (interchangeable)
6. Scanning head
7. Mask scanning head
8. Mask drum
9. Recording space
10. Daylight cassette
11. Colour computer with control unit and extended selective correction
12. Base frame

By courtesy of Dr-ing. Rudolf Hell

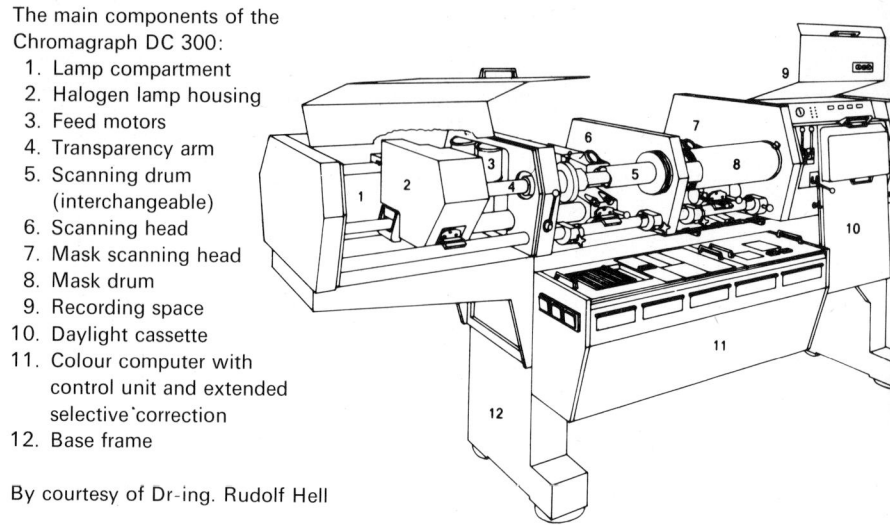

Scanning magnification

A large diameter input drum 61×51 cm (24×20 ins) provides easier mounting of flexible reflection originals, while a smaller 19×13 cm (7.5×5 ins) input drum allows big enlargements from small originals. The output drum 61×51 cm (24×20 ins) permits four separations up to 30×23 cm (11.8×9 ins) to be produced in one scanning pass. Separations up to 50×30 cm (19.6×11.8 ins) are produced two at a time and separations up to 61×50 cm (24×19.6 ins) are produced singly including register marks. A complete A4 four colour set of screened separations can be scanned in less than six minutes. One disadvantage of drum scanners is the need for flexible originals, but most rigid originals can be copied on to colour transparency film and this copying stage provides an opportunity to correct colour casts, alter the sizes of originals and mount individual pictures together or assemble them in their page positions before scanning commences.

Magnification

An important development has been the facility to obtain a wide range of different magnifications and reductions between the original picture and the separations. A range of magnifications from 0.3 to 20 times is quite common. Original colour transparencies are usually smaller than the separations required, so by making the output drum twice the diameter of the input drum and the pitch of the larger output drum leadscrew twice that of the smaller input drum, the final separations

Scanning magnification

would be twice the size of the original picture. Degrees of enlargement greater or less than 2 in the axial ←→ direction may be obtained by driving the two lead-screws at different speeds, but the same degree of enlargement must be obtained in the transverse ↕ direction. This is done by taking a complete line of picture signals from the input drum and placing them in a digital store. The position of these signals along the picture line is also stored. They are then accurately read out onto the output drum at such speed as will enlarge or reduce the separation. The more slowly it is played back the greater the degree of enlargement. Conversely, a fast play-back speed will produce a reduction.

Digital scale alteration in the circumferential direction

The adjacent diagram shows digital scale alterations in the circumferential direction to 50% and 200%, which are effected by extending or shortening the time (t) of the picture signal.

The signal (a) from the colour computer is digitalised in the scanning cycle (b). For this purpose the entire density range (D) is subdivided into a large number of density stages (only 25 of these stages are shown in the diagram) and stored in a core memory as picture lines (c). The upper row of figures denotes the memory address, the lower one the density stage. The recording cycle (d) altered by the scale alteration factor in relation to the scanning cycle (b) reads out the stored density values more slowly (enlargement) or faster (reduction) from the memory (e). After conversion the analogue recording signal (f) is obtained. The scale is altered in the axial direction by appropriate control of the scanning head feed rate.

By courtesy of Dr-ing. Rudolf Hell

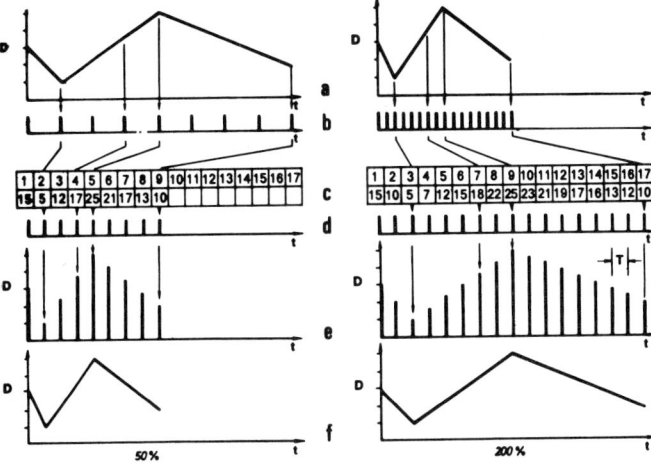

Electronic reproduction

Computer functions

Electronic scanners use both analogue and digital computers. An analogue recording is the simulation of physical variables by another set of equivalent variables. For example, light intensities converted into electrical signals or the luminance pattern of an original picture recorded as different photographic density values. A digital recording is the simulation of physical variables in numerical values (digits) which represent the variations. A binary record being made up from high and low photographic densities in an array of image cells.

The objective of computerisation is to automate the many repetitive calibration and image balancing operations. The introduction of a digital computer is a large step in this direction. Calibration of a scanner begins with the machine automatically exposing an electronically generated grey scale. After processing, these steps are measured and the actual density readings obtained fed back into the computer and calibration is complete—the scanner and film processor are correctly matched to provide accurately predicted densities on the separations. Balancing the input circuits to zero is carried out automatically in ten seconds by the scanning head measuring two density patches fixed permanently to the input drum. The movement of the scanning head and the adjustments are controlled by the computer. Standard 'facsimile' programmes for transmission and reflection copy, continuous-tone or screening operations, positive or negative separations and a range of colour and tonal reproduction requirements for different press and paper conditions may be stored in the built-in magnetic tape unit and can be called into use when required. The original's characteristics are measured by the machine. The computer calculates the correct exposure, tone and colour values for optimum reproduction. The computer reduces the chances of error, but for originals requiring special treatment, there are ample 'overriding' facilities for editorial changes.

Although the computer is built to function on basic principles it needs an operator to feed in specific instructions with regard to the original in question. A proficient approach is essential. A colour key of the original with important coloured areas analyzed and translated into the scanner's language, e.g. dot sizes as meter readings, digital density readings or curve shapes, is a necessity, together with all the previous references to printing standards and methods of prediction. Instructions relating to scanning pitch (lines per cm), amount of undercolour removal, degree of unsharp masking, enlargement, reduction and special treatments must be known before scanning commences.

The electronic computer is extremely effective in the following sections of colour separation.

1. Optimum tonal control with the inclusion of unsharp masking effects.
2. Highlight accentuation.
3. Generation of a corrected black printer image together with associated undercolour removal.
4. Duplication of colour-corrected separations with the utmost accuracy after long periods of time.

Colour correction is usually a masking system simulated with separation signals and correction signals. A masking method is basically the addition of two photographic densities. The result is that the combined optical densities of the colour separation image and the colour-correcting image affect the intensity of the transmitted light logarithmically. The addition of density in unwanted areas of a negative separation results in a reduction of the transmitted light and its product on a positive image. For example, an unwanted density of 1·00 transmits 10% of the incident light; an additional density of 0·30 would reduce the transmitted light to 5%. The wanted areas, on the other hand, will appear in an apparently less dense condition, thus allowing the transmitted light to increase the density of the positive image in these areas.

The electrical signal from a photo-cell is proportional to the light intensity received from the original. Electrical signals representing colour separations—the amount of blue, green and red light present in a particular area of the original—are amplified or attenuated (reduced) to give the same effect as photographic masking. The number of signals used to modify or colour-correct the separation signal determine the complexity and sophistication of the scanner. One-stage masking is simulated by the Hell Vario-Klischograph electronic scanner. A green filter photo-cell generates the separation signal for the magenta printer positive. The red filter photo-cell produces the colour-correcting signal.

Numerical set-up

A numerical read out meter indicates original input density, or output separation density, or dot percentage. Having mounted the coloured original onto the input drum and

Electronic reproduction

selected the required programme the original's highlight and shadow areas are positioned under the optical viewer. The required densities or dot percentages are dialled directly into the scanner by the digital switches. These are typically 0.3 and 1.6, or 5% and 95% respectively and are not normally changed from one original to another. The computer now takes over the calculation of tonal compression, cast removal, correct middle tone, end densities, grey balance and colour values, whilst the unexposed film is being loaded.

Colour correction set-up

The electrical input to the computer consists of three signals proportional to the blue, green and red transmissions or reflections from the original. The signals are generally converted from linear to a logarithmic form so they become proportional to the densities of the original. Then the signal range is usually compressed to suit the limited luminance range of the printed result. Colour correction is related to three simultaneous equations, which add and subtract different proportions of the logarithmic signals; similar to photographic masking when densities are added to unwanted colours at the negative stage or large halftone dots are subtracted by etching at the positive stage. Colour controls on a scanner relate to wanted and unwanted colours. A standard masking profile may be programmed, then its masking strength increased or decreased to affect the saturation of wanted colours and the reduction of unwanted colours. Each separation may be adjusted by six independent colour controls. In a magenta separation for instance, three controls for the wanted colours—magenta, primary-red and violet blue; three controls for unwanted colours—green, yellow and cyan. Each colour control has a centre-zero position enabling a standard correction to be set for a given ink and paper combination.

Tonal correction set-up

On average, original pictures have longer density ranges than the printing process chosen to reproduce them and tonal compression must be carried out. This loss of tonal gradation is also aggravated by the fact that at each stage of the reproduction process, some change in tonal gradation takes place, the worst effect being the loss of highlight detail. Ink gain has a similar effect on the shadow detail, but in most pictures the shadow detail is less important since it has less visual impact on the eye than the highlight detail. The separated signals may be passed through curve-shaping circuits to produce high contrast in highlight and shadow regions, but low con-

Contact screening and electronic laser screening

trast and tonal compression in the middle tones. This high-low-high gamma characteristic is extremely difficult to achieve on conventional photographic emulsions. There are four individual tonal controls on a scanner providing full adjustment of the tonal curve. Three controls operate in the highlight, middle tone and shadow areas. The fourth control operates in the catch-light region. In order to simplify the operation of the scanner, when all the tone controls are set to a centre-zero position a standard tonal curve is produced for a given printing process and an average ink and paper combination.

Contact screening and electronic laser screening

Tonal gradations in most printing processes can be represented only by halftone dots. The output of a drum scanner can be a xenon light source exposing the separation film through an electro-optic crystal modulator and fibre-optics, creating an intensity high enough to penetrate a grey contact screen and produce elliptical or square halftone dots. The size of a dot depends on the quantity of light that reaches the

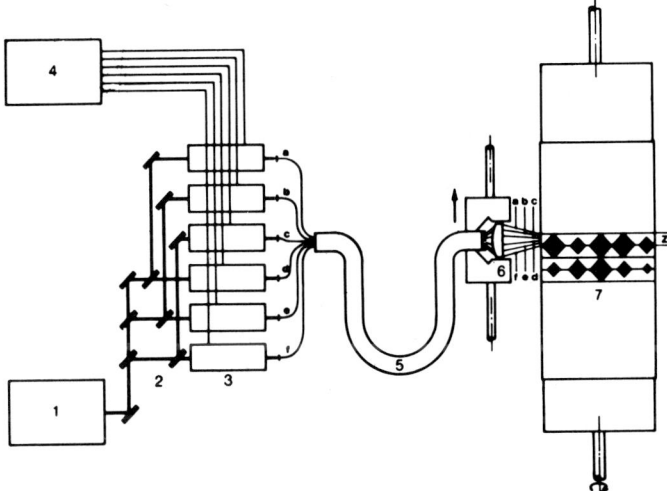

Recording principle of the electronic screen:
The light beam from the laser (1) is broken down into six separate beams with the same intensity by a mirror system (2) consisting of ten mirrors. These separate beams pass to the modulation unit (3) consisting of six modulators, where they are switched on or off individually by a digital control signal from the screen computer (4). The modulated separate beams (a–f) are fed to the recording head (6) via a light cable (5) and are projected by a zoom lens on to the recording drum (7) and expose the film mounted on the drum on one recording line (Z).
Electronic screening is effected line by line. Two drum revolutions are required for recording a screen dot. A 60 line screen, for example, is recorded with 120 lines/cm.

By courtesy of Dr-ing. Rudolf Hell

Electronic reproduction

film. As the contact screen absorbs a certain quantity of light, a higher number of recording lines is necessary. Alternatively, halftone dots may be recorded on film by laser beams. The final picture signal is not fed directly to the output head, but to a screen computer containing programmes for different dot densities and screen angles. A digital signal modulates a laser beam divided into six separate beams. Six dotlets are recorded onto the output scanning line and form half a screen dot. The other half is exposed during the next drum revolution. The digital construction of the half-tone dots improves dot definition and the dot profile is less sensitive to varying emulsion characteristics and unintentional deviations during photographic processing. The formation of laser dots is two to four times faster than present day contact screen method. All commonly used screen rulings from 34 dots per cm (85 d.p.inch) to 80 dots per cm (200 d.p.inch) can be produced electronically.

Electronic circuits use small and simple calculations, but by employing rapid interplay and combination are capable of solving complicated equations. An electronic scanner based on two-stage masking must be capable of isolating the colour-correcting signal from the grey tonal aspects of the original. The electronic circuits will then produce the required tonal correction with due regard to highlight accentuation and undercolour removal. An output signal representing complete correction emerges after the colour-correction signal is re-combined with the corrected tonal separation signal.

With each new scanner the degree of modification increases. To indicate the flexibility of these corrections a brief list is in order.

1. Inversion—the immediate conversion from negative to positive images. Tonal compression or expansion—scanned images fit the required reproduction density range.
2. Highlight 'drop-out' and 'peaking' which emphasize tonal movements. Shadow detail boost—an asset when reproducing 35 mm colour transparencies.
3. Limitation—density of highlight areas controlled to a predetermined level—a necessity for photo-engraving reproductions. Shadow areas produced at a standard density—a requirement desirable for gravure reproductions.
4. Selective properties—when the black printer is being calculated the highest filter signal in each case is taken as the separation signal.
5. Under colour removal—By using an electronic computer a black printer of any required type can be made with great

Contact screening and electronic laser screening

accuracy. Difficulties experienced in printing on high speed presses may be reduced if only two of the three coloured inks are printed heavily in the same area of a reproduction, any darkening of a colour being achieved by overprinting black ink, rather than the third colour, a method termed under-colour removal. This approach lowers ink costs, as cheap black ink replaces expensive colour ink. Registration of fine detail is less critical. Neutral greys are maintained and the heavier black printer improves the appearance of the reproduction because impressions of sharpness and resolution are controlled by differences in luminance are more apparent in a single black image than three coloured images superimposed. The under-colour removal controls are independent of all other controls and can be adjusted to give a removal percentage anywhere between 0 and 100%. The point on the tonal curve where under-colour removal begins may also be selected. The compensatory amount of black ink is automatically computed for the heavier black printer separation.

6. Unsharp masking and peaking—Apparent sharpness of a picture may be increased, if a fringe is produced around the edges of contrasting tones. The fringe is produced by a special photomultiplier scanning through a larger aperture than the other three photomultipliers. Catchlights for instance can be emphasised by unsharp masking and enlarged photographic grain structures may be minimised. Image definition is also visually improved by electronic peaking which accentuates sharpness without introducing fringes.

7. Register system—A pin register system is provided on the output drum and register marks for subsequent planning are automatically exposed on each separation. Tabulators may be provided to isolate an area within a picture and surround this area with a tint border.

8. Picture distortion—An original picture may be slightly distorted to fit a particular format which is not in proportion, or alternatively grossly distorted to produce artistic effects. Distortion is achieved by using differing degrees of enlargement in the horizontal and vertical axes.

9. Monochrome scanning and customer programmes—Black and white bromide prints may be wrapped round a large input drum and the colour scanner programmed for monochrome or sepia-toned reproductions, thus utilising the scanner to its full capacity. Special operating programmes relating to tonal curves and colour correction requirements for individual customers or different printing processes can be stored on magnetic tapes and called into use by simply pressing a recall button.

Electronic reproduction

10. Television image display—In colour reproduction, the opportunity of examining a coloured proof or a predicted sample, before the sequence of operations takes place, is highly advantageous to both the customer and printer. This predicted coloured proof may be shown on a television monitor using the original picture as the input. By including electronic circuits which duplicate the stages in the printing process, the coloured image on the television screen can be adjusted to obtain the desired result. A zoom lens permits close examination of selected image areas. After viewing the original picture and setting-up the appropriate corrections, the original is mounted onto the drum scanner and separated. The scanning programme being in accordance with the corrected television image.

Construction of the Chromaskop
1. Original plane
2. Reflection lighting left and right (pushed in)
3. Transmission lighting
4. Mirror
5. Zoom lens
6. Camera head
7. Camera amplifier
8. Unit of the colour camera (5-6-7)
9. Colour computer
10. Colour convertor
11. Colour monitor
12. Measuring rod

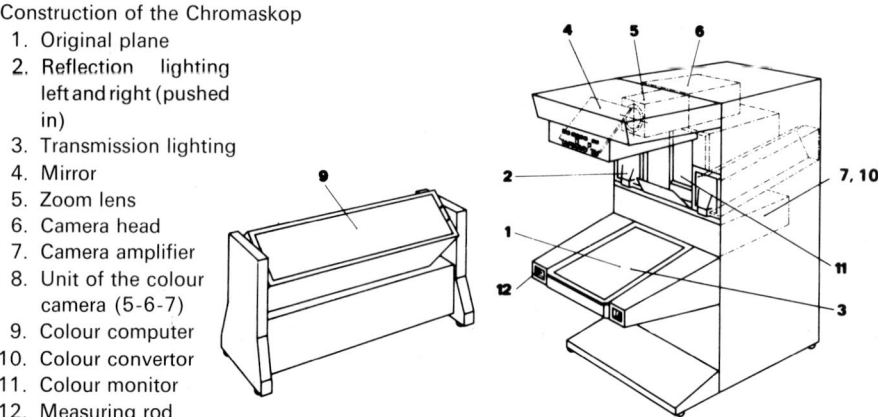

By courtesy of Dr-ing. Rudolf Hell

Page assembly system and electronic retouching

An electronic page assembly system works in conjunction with a drum scanner producing a complete set of separations with all the illustrations and text components planned into their correct printing positions. At the same time all geometric shapes, such as picture borders, rounded corners, coloured backgrounds and pictures inserted into one another are also produced electronically.

Accuracy and registration surpass conventional methods which use hand-cut or painted masks visually sellotaped into register. Originals are mounted on to the input drum and scanned to their final sizes. This picture information is stored on a magnetic disc. Disc storage allows individual pictures to be displayed on a colour monitor, so they may be arranged

Page assembly system and electronic retouching

into their page position. Furthermore, pictures can be retouched electronically either entirely or selectively. The lay-out is drawn with an electronic pen on a digitising table with x/y co-ordinates. Using the electronic pen, the planner moves the pictures into their correct printing position.

Once the whole page, including all the pictures and text matter has been assembled on the colour monitor, the digital information is transferred to another disc, which is used to control the output of the drum scanner so that the full page is exposed onto a large film in a simple operation, employing continuous tone or electronic screening.

In the near future complete copy preparation, sophisticated page make-up, spatially localised tone and colour retouching will be incorporated into a direct plate-making system with the final images being exposed onto flat or curved printing surfaces using the electronic "flying-spot" principle.

Everyone involved in graphic reproduction photography should try to obtain operating experience on an electronic scanner. Although these machines cannot always cope with the more complicated reproduction the mere sharpness of the result because of the absence of lens flare is at times enough to swing the balance in favour of purchasing such expensive pieces of equipment. Most people with an experienced eye for analyzing colour combined with a sensible attitude towards priorities find the transition easy.

13. REPRODUCTION PROCESSES

Introduction

Printing is a marvellous vocation. It is full of creative thought and stimulation. The graphic reproduction photographer is in a unique position, being at the very initiation of the printing processes, and it is because of the importance of his early contribution—the accurate photography of all manner of designs, pictures and illustrations—that it is imperative that he fully understands and appreciates all the subsequent stages of the four major printing processes.

Letterpress printing

Letterpress relief printing is the earliest known printing process. Non-image areas are recessed while the image areas are raised. It is interesting to note that literacy was far more prevalent in China during the fifth century than it was in Europe nine centuries later. So it is not surprising that the invention of paper is usually attributed to the Chinese, in fact to Ts'ai Lun, a member of the court of Emperor Ho Ti, in the year A.D. 105. The art of printing progressed and it was not long before the Chinese and Koreans were using stone seals and even moveable type faces cast in copper as their printing formes.

During the seventh and eighth centuries letterpress techniques were being used in Europe to print designs in gold on leather. The early monks and scribes cut recurring words into wood and 'pressed' the inked 'letters' to save time. This work led to whole-page woodcuts of type and the use of pictorial woodcuts. The printed image formed an outline for delicate hand-painting techniques.

The work of cutting or etching pictorial images in relief goes back to early times. Its use in printing terms began in earnest after Johann Gutenberg (1395–1468) of Mainz, Germany, introduced into Europe the art of printing from moveable type faces employing the letterpress method in the year 1450. Gutenberg was in some ways more of a printing technologist than a creative typographer. Consequently he was very interested in the mechanics of printing. His method of cutting lettering punches and striking them into soft metal

Letterpress printing

Johann Gutenberg (1395–1468)

surfaces to form dies for moulding was probably an extension of coin manufacture, an art well known to Gutenberg's uncle who was a master of the Mainz mint. While writing to Johannes Fust, his financial partner, Gutenberg spoke of his 'type tools and printing equipment for the work of the books'. The Gutenberg 42-line Bible was a triumph and illustrated vividly the potential of letterpress printing.

Photo-engraving

Photo-engraving is a process for producing a relief image on a sensitized metal plate by placing a photographic negative of the original picture on the coated plate in front of a light source. The non-image areas of the coating which are not rendered insoluble by light are washed off and etched with acid until the correct depth is obtained. Let us follow the reproduction process step by step.

1. *Preparation and coating*

A sheet of metal is scrubbed clean with the aid of a felt pad and pumice powder; zinc, copper and magnesium plates are the block-makers' main choice of metals in that order. Zinc is used for linework and coarse-screen halftones, and copper, because of its fine crystalline structure, is reserved for halftones containing fine screen rulings. Magnesium emerged as a suitable metal with the introduction of powderless etching.

Once the sheet of metal is clean it is placed in a whirler. This piece of equipment is essentially a turntable rotated by a variable-speed motor with a heating element and fan housed in a hinged lid. When the motor is switched on the metal

Reproduction processes

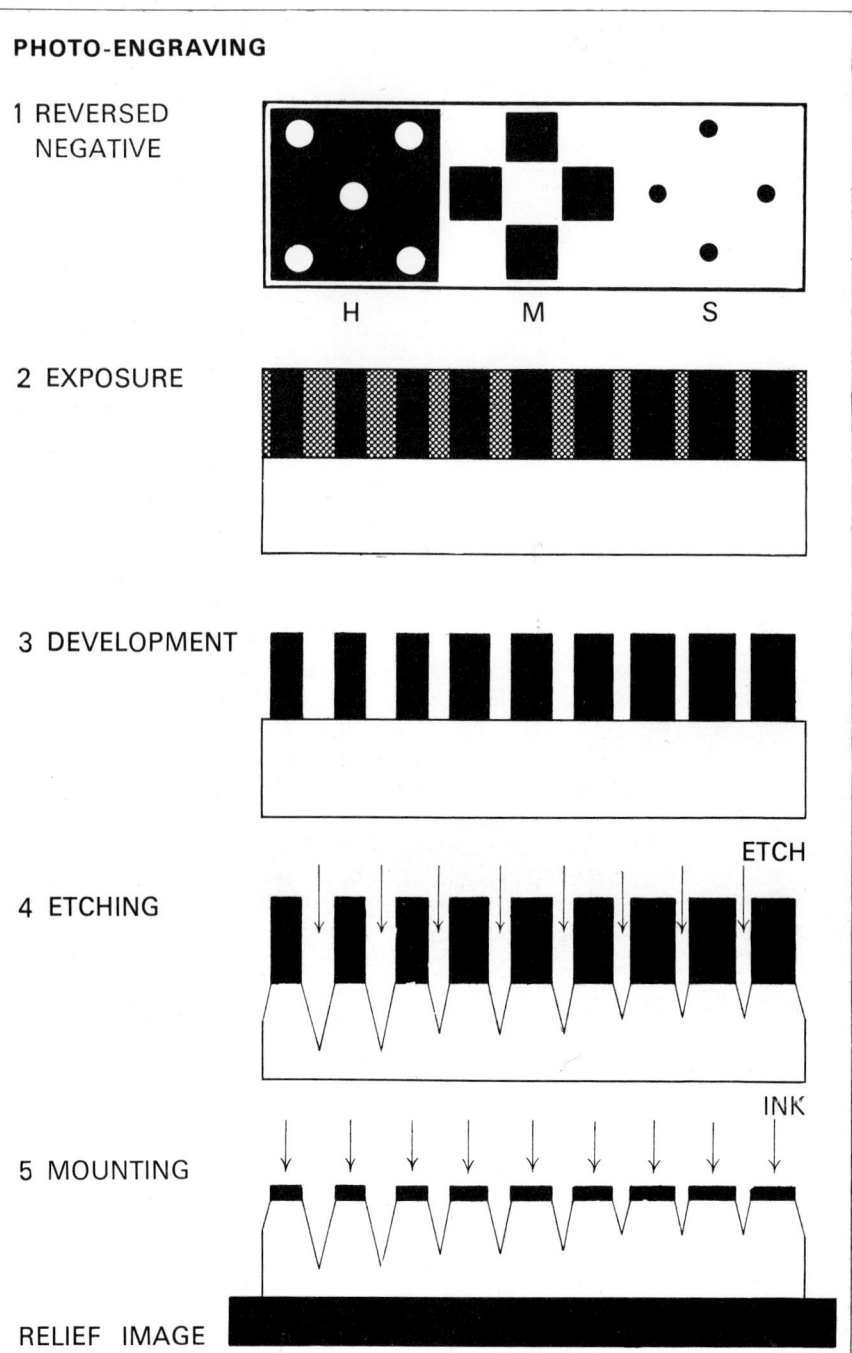

dichromated colloid is poured on to the centre of the plate, where the centrifugal force causes the solution to spread uniformly over the surface of the metal. The lid is lowered and the heat switched on in order to dry the light-sensitive solution into a thin coating.

2. Exposure and development

The dichromated colloid has two roles to perform. Firstly, it provides a light-sensitive coating so that the photographic image may be transferred to the metal surface; and secondly, it becomes a strong acid resist covering the image areas. These dichromated colloidal solutions consist of two main parts: (a) light-sensitive ammonium or potassium dichromate and (b) a colloid or organic substance, such as albumen, fish glue or more recently polyvinyl alcohol (PVA).

Printing down the photographic image to metal is carried out in a vacuum frame. The reversed negative is placed on top of the coated metal plate. The transparent areas of the negative will ultimately become the relief image areas, while the opaque portions represent the non-image areas. Dichromated colloids are less sensitive to light than photographic silver halides and an arc lamp is used to pass intense, actinic light rays through the transparent areas of the negative and harden the dichromated coating underneath. This reaction is the first stage of forming an acid resist.

The unexposed non-image areas remain in a soluble state. Development takes place removing these areas of soluble albumen or fish glue with a copious supply of running water. With PVA coatings a suitable developer is used to remove non-image areas. This is followed by washing and immersion in a chemical hardening bath.

3. Forming an acid resist

To ensure that the hardened coating covering the image areas does not break down during the etching process heat is applied to the coating. This is called 'burning-in'. PVA coatings need to remain in the burning-in oven until a temperature of 230°C is reached: this treatment forms an extremely strong acid-resisting enamel, the image being sufficiently protected, while the non-image areas remain bare and vulnerable. To aid a uniform etch the surface of the prepared plate is descrummed and the back of the plate coated with an acid-resisting paint.

4. Etching process

The plate is clamped on a turntable housed in the lid of a

Reproduction Processes

powderless etching machine. This slowly turns the face-down plate over a tank containing a set of paddles or spray nozzles which throw the etching solution up on to the plate's surface. The acid contained in the solution will attack the bare, non-image areas, gradually producing a raised, relief image. The acid certainly etches in a vertical direction, but left to its own course it will etch laterally as well, gradually 'undercutting' the raised image areas.

With conventional etching 'undercutting' was prevented by etching in stages or 'bites'; as each new depth was gained the walls of the image areas had to be banked and fused with a red resinous powder known as 'dragon's blood'. This lengthy operation of etching and powdering prevented 'undercutting', but produced unwanted 'shoulders' on the image walls. The removal of these bumps was again a time-consuming business.

Powderless etching machine (spray type)

It was certainly an advancement when the Dow Chemical Company of America introduced 'powderless' etching machines. The basis of their success is the oil-like additives which are mixed with the acid and water to form the etching solution. When the etching solution is thrown up on to the surface of the metal plate, the acid attacks the bare metal leaving the additive on the surface as an oily film. When the next splash or spray hits the surface its vertical thrust divides the oily film and pushes it aside. As the acid begins to etch vertically, so the additives build an oily film bank on the image walls. Thus, the oily film takes the place of the conventional dragon's blood and prevents 'undercutting' while producing smooth image walls in strong pyramidal shapes.

5. Finishing

After etching, a number of finishing operations are necessary to produce a relief plate ready for printing. These begin with 'routing' when a machine similar to a drill is used to cut away superfluous metal and deepen any large non-image areas. If the image is a rectangular halftone then the edges may be 'bevelled' to produce a flange for drilling holes into, so that pins can be used to secure the metal plate to a suitable base or mount. This mount should bring the surface of the plate to exactly type high 23·32 mm (0·918 in.) Metal bases, with plates held with double-sided adhesives, are also a popular method of mounting. The final stage is to take a proof from the relief image. The printed image will indicate the quality of the reproduction and the necessity for any hand corrections. The proof also provides a guide for subsequent letterpress printing.

6. Printing and identification

The letterpress printing method is a precise, rather mechanical way of transferring ink to paper. A thin layer of ink is deposited on to the relief image areas. This ink layer, now a representation of the image, is transferred directly on to the paper under pressure on a platen or cylinder type of printing machine.

The image bearing surface may be autographically produced linocuts and woodcuts, photo-engraved plates combined with moveable metal types, duplicate plates made by the stereotyping and electrotyping processes, or wrap-around photopolymer plates containing illustrations and typematter all on the same surface for direct or offset rotary printing.

Letterpress printing:
A. Platen type
  1. Image
  2. Ink
  3. Paper
  4. Impression
B. Cylinder type
  1. Image
  2. Ink
  3. Paper
  4. Impression

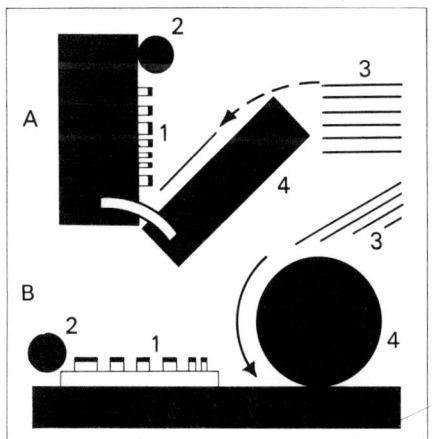

Reproduction Processes

Identification: Letterpress
1. Ink squash
2. Impression mark
3. Increased contrast
4. No dot drop-out

It is the precise, mechanical nature of relief printing that produces the instantly recognizable characteristics of inked images printed by the letterpress method. With the help of a magnifying glass these characteristics may be identified in four ways.

1. *Ink squash* is seen in printed typematter and halftone dots. Heavy ridges of ink appear around the outside edges of the images.
2. An *impression mark* may be seen on the reverse side of the paper.
3. Halftone images are often interlayed or bumped to increase the pressure on the darker tones. This produces *increased contrast* between shadow and highlight areas.
4. The absence of vignettes merging imperceptibility into the white paper background. *No dot drop-out* in extreme highlight areas.

Letterpress printing has always been associated with work that requires constant alteration, the facility of moving the type faces and illustration blocks within the forme on the printing machine being the advantage. This is indeed a key reason for many newspapers, books and much business literature being printed by the letterpress method. Overriding this fact is the sharpness, precision and quality of the printed result. Although modern economy demands rotary printing and the use of wrap-around plastic and nylon plates complemented by photo-typesetting, resulting perhaps in the widespread use of 'offset letterpress', the letterpress printing method will always earn a place in the production of good quality printing.

Gravure printing

The image areas in gravure intaglio printing are recessed. Letters or designs are engraved or etched into the metal surface. The metal surface is inked and then wiped, so that ink remains only in the recessed image areas from which printing takes place. This method of printing is directly linked with ancient crafts. Early weapons and suits of armour were engraved with intricate designs and embellishments. Armourers occasionally took patterns from popular designs by rubbing the 'intaglio' images with soft blackened wax and then pressing and rubbing thin pieces of leather over them in order to release the image.

These early printings led to the Mezzotint and Aquatint methods of producing an intaglio printing image in a metal surface. With the discovery of photography, Joseph Nicéphore Niépce presented the 'Heliogravure' printing plates and Fox Talbot the 'Photoglyphy' process, with its use of various concentrations of ferric chloride, for the production of gravure plates. Gravure printing received its greatest impetus when Karl Klič (1841–1926), a son of a Czechoslovakian chemist, became interested in its development. Klič was a talented artist with a comprehensive knowledge of photography. He wanted to invent a printing process capable of recreating the tonal appearance of paintings, sketches and photographs. Karl Klič expressed this desire while writing an explanation of his first attempts. 'The stroke of an artist's brush is a beautiful motion, the result of this act should not be lost when printing takes place. The combination of a carbon image and grain etching provides a manipulation which avoids this loss.' This meant the printing of different ink

Karl Klic (1841–1926)

thicknesses and therefore the printing plate needed to hold and impart these various amount of ink. Klič achieved this and rotary printing in 1890 when he replaced the aquatint grain with a cross-line screen to form the walls of minute ink cells which varied in depth and made up the image.

Photogravure

Photogravure is a method of etching an intaglio image into a metal cylinder by positioning a photographic positive of the original picture on light-sensitive pigment paper in front of a light source. The non-image areas become hardened and after being transferred to the surface of the metal cylinder act as partial acid resists, controlling the depth of the etched ink cells. The steps in reproduction are as follows.

1. *Preparation and sensitizing*

The copper surfaced cylinder is cleaned, buffed and polished, while the pigment paper is being sensitized. Pigment paper is essentially pigmented gelatine coated on to a paper base. The brown pigment gives visual indication of the effect of exposure and development. The pigment paper is rendered sensitive to light by soaking it in a solution of potassium bichromate. The bichromate penetrates the colloidal coating and forms a light-sensitive dichromated colloid. The paper is now in a sensitized condition and can be dried.

2. *Exposure*

The non-reversed halftone positive in the case of invert halftone, or the gravure 'fish-net' screen for conventional etching, is placed on the pigment paper in a vacuum frame and exposed to an arc lamp. The gelatine layer is hardened throughout its depth in the exposed areas. After this exposure the non-reversed continuous-tone positive is positioned on the pigment paper and a second exposure is given. Underneath the highlight regions of the continuous-tone positive the gelatine layer is hardened to a considerable depth, while the shadow areas will hold back the light resulting in only a thin layer of gelatine being hardened. The pigment paper now contains two images—the screen combined with the illustration.

3. *Mounting and development*

The exposed pigment paper is now transferred to the copper surface of the printing cylinder. This is carried out on a type of rolling machine, the gelatine surface of the pigment paper being wetted and rolled under pressure on to the metal

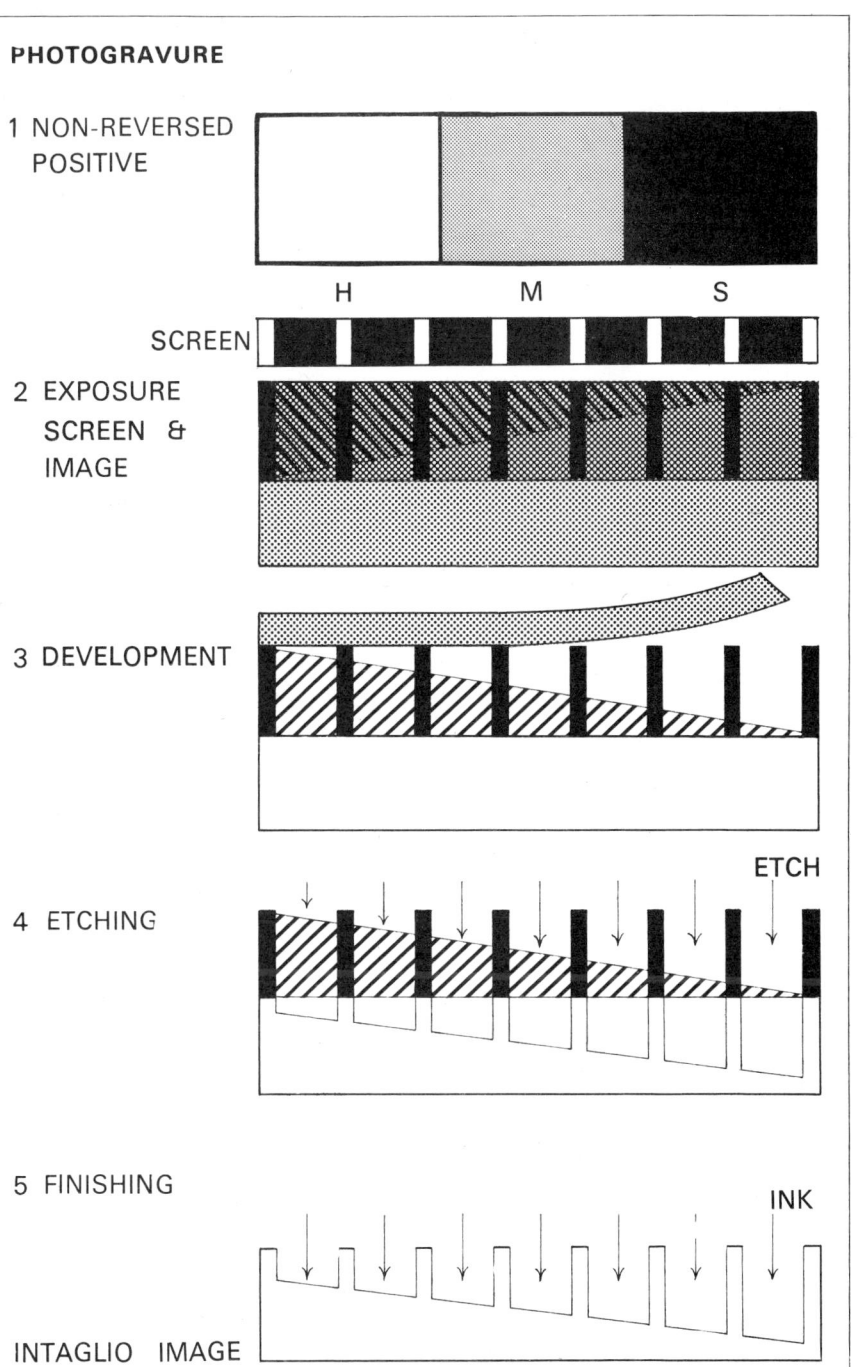

surface. The gelatine adheres to the copper and development can now take place. The paper base is washed with alcohol to increase its porosity and then the whole cylinder is rotated in a bath of warm water. The paper base will detach itself and the unhardened areas of gelatine will dissolve away. The copper surface has received a negative image of the screen and original in hardened gelatine which will serve as a partially acid-resisting mould. The surface of the cylinder is dried and any unwanted areas painted-out with a varnish impervious to the etching solutions.

4. *Etching process*

The image is etched into the copper surface with aqueous solutions of ferric chloride. During the etching the ferric chloride penetrates the gelatine resist more rapidly where it is thinnest (shadow areas), while the thicker regions (highlights) are penetrated later. The concentration of etching solutions

Etching procedure:

INTAGLIO IMAGE

may be varied to control the penetration and eventual ink cell depth. A high concentration, e.g. 41° Baumé, will penetrate the gelatine at a much slower rate than a solution of 35° Baumé. When each tonal region of the original picture is represented by ink cells of the correct depth the cylinder is washed, removing the resist.

5. *Finishing*

A proof is pulled from the intaglio image to indicate the need for hand corrections. Once the cylinder is correct its copper surface is chromium-plated to prolong its printing life by withstanding the wiping action of the doctor blade which is peculiar to gravure printing.

Gravure printing:
1. Image
2. Ink
3. Paper
4. Impression
D. Doctor blade

Gravure printing

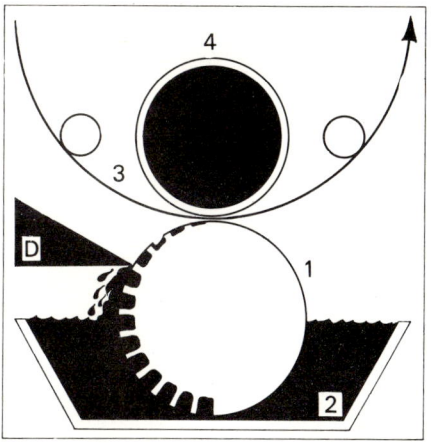

6. *Printing and identification*

Gravure printing is an almost continuous-tone method of transferring ink. An intaglio image is used and in this respect is the opposite of relief printing. The recessed image areas in the form of minute ink cells are flooded with ink. After this the surface of the image cylinder is wiped clean by the doctor blade of surplus ink before it comes into contact with the paper, so that ink transference may take place under pressure.

The image bearing ink cells may be the result of conventional etching, invert halftone processes or an electronic engraving system, making high-speed, good quality rotary printing possible.

It is the transference of different weights of ink that reproduces the continuous-tone effect that gives gravure printing its individuality. This individuality may be recognized by looking with a magnifying glass for the following identification factors.

1. The printed image has a *continuous-tone appearance* even on the cheaper machine finished papers. Typematter contains a *serrated edge* caused by the introduction of the 'fish-net' screen.
2. Conventional gravure breaks the picture up into *small squares of equal size*, but of differing density according to the tones of the image.
3. Invert halftone images contain small highlight dots at the lighter end of the tonal scale and *heavy ink density* in the shadows.
4. Very occasionally a *wipe mark* is seen. A minute piece of

427

Reproduction Processes

Identification: gravure
1. Continuous-tone appearance
2. Serrated edge
3. Small squares
4. Heavy ink density

grit becomes trapped underneath the doctor blade and leaves either score marks or ink streaks.

Gravure printing has become synonymous with vivid, pictorial printing, presenting this attribute in magazines, periodicals and packaging—in fact, most printing which requires high speeds and long runs with the accent on coloured illustrations. In the past the production of gravure cylinders has been associated with high cost, but with the introduction of electronic automatic-etching machines and invert halftone methods speeding up the cylinder's progress to the reel-fed rotary printing machines, gravure printing is expanding, especially in the packaging field.

Lithographic printing

Lithography is a planographic process—the image and non-image areas lie on the same plane. The printing action relies on the fact that grease repels water. The image areas are made grease receptive-water repellent, while the non-image areas become water receptive-grease repellent.

Alois Senefelder (1771–1834), the German inventor of lithography, was born in Prague. He was the son of Peter Senefelder, an actor. Because of his father's occupation Alois's early life was spent travelling with theatrical groups and it was love of the theatre that inspired him to become a playwright. Alois spent most of his working life in Munich and Offenbach on the Main. He felt the need to print his own plays and experimented with both letterpress and gravure techniques, but found them difficult and expensive. He tried to replace the copper plate with relatively inexpensive blocks of limestone, using the principle of copper engraving to

Lithographic printing

Alois Senefelder (1771–1834)

produce an intaglio printing image. What happened next is best related in Senefelder's own words.

'I had just cleanly polished a stone plate in order to then cover it with etching base and continue my exercises in mirror-writing, when my mother asked me to write her a laundry bill. The washerwoman was waiting for the laundry, but there was no piece of paper at hand and my own stock had just been exhausted in the experiments; there was also no ordinary ink left, as this had dried up, and since there was nobody to send out for new writing materials I did not hesitate, but simply wrote the laundry list on the polished stone plate using my stone drawing ink made of wax, soap and charcoal, so that I could copy it when some fresh paper had arrived. When I afterwards wanted to wipe this writing off the stone it suddenly occurred to me what would happen to this writing with wax ink on to the stone, if I etched the plate with aqua fortis, and whether it might not be possible to ink the stone and print from it in the manner of type characters or woodcuts.'

Using a slightly relief image in stone led Senefelder to improve the printing image. By striving for perfection his thoughts began to crystallize and he related, 'but should it not be possible to prepare the stone plate itself, in a manner so that it would take ink only in the parts where it had been marked with the greasy writing ink, and repel it in the damp parts? That was the obvious and simple idea which had to occur to me'. Thus, in the year 1798 Alois Senefelder had invented lithography.

Photolithography

Photolithography is a means of preparing a planographic printing image on a sensitized metal plate by laying a photographic positive (in the case of deep-etch) of the original

Reproduction Processes

PHOTOLITHOGRAPHY

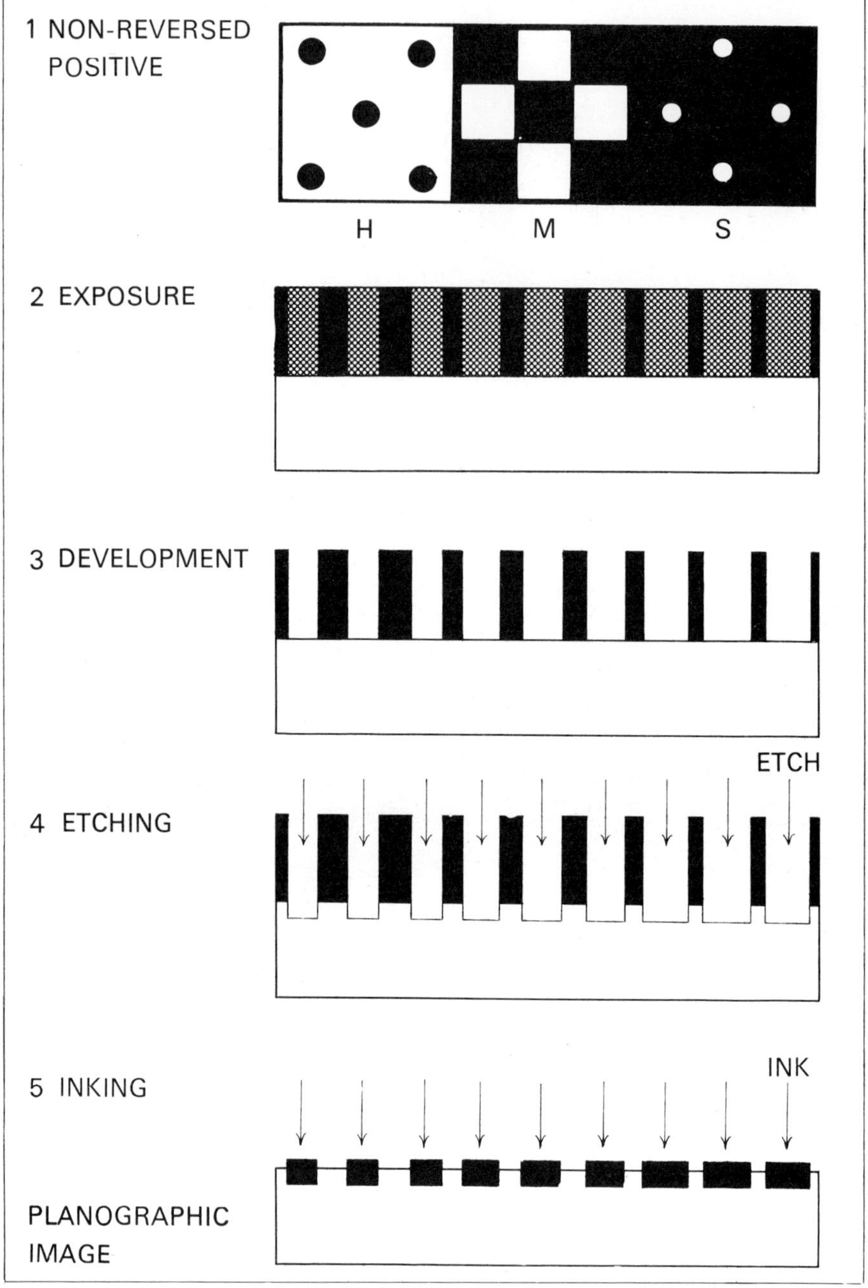

1 NON-REVERSED POSITIVE

H   M   S

2 EXPOSURE

3 DEVELOPMENT

4 ETCHING — ETCH

5 INKING — INK

PLANOGRAPHIC IMAGE

Lithographic printing

picture on the coated plate in front of a light source. The non-image areas become hardened allowing the image areas to be developed, etched and inked, ready for accepting the greasy ink during printing operations. We will now take a close look at the reproduction procedure.

1. *Preparation*

The metal plates for lithographic printing are usually sheets of grained zinc or anodized aluminium. The anodic layer on the grained aluminium is responsible for three desirable lithographic printing properties—porosity, non-corrosiveness and hardness. The first operation is to chemically clean the metal plates with counter-etch, e.g. a weak solution of acetic acid. With surface plates, such as albumen, pre-sensitized and wipe-on, the plate-making procedure is very similar to exposing photographic bromide prints. A non-reversed negative is used and the resulting positive image is produced slightly in relief on the metal surface.

2. *Coating*

When the deep-etch process is required, the metal plate is fastened on the turntable of the whirler. A solution of dichromated gum arabic is poured on to the centre of the spinning metal plate, so that a thin, uniform light-sensitive coating is formed.

3. *Exposure*

The dried metal plate with its light-sensitive coating is laid in the vacuum frame. A non-reversed photographic positive is placed on top of the coating in its printing position and held there under vacuum pressure. The combination of plate and positive is exposed to the rays of an arc lamp. The action of the light hardens the coating in the non-image areas of the plate producing a negative image in hardened gum arabic.

4. *Development*

The unexposed gum still present in the image areas is now developed-out with a solution of calcium chloride and lactic acid. Development continues until all the gum has been removed from the image areas, leaving the image areas as bare metal surrounded by a hardened gum coating.

The metal in the image areas is now vulnerable and may be attacked by a suitable etchant, e.g. a mixture of calcium chloride, iron perchloride and hydrochloric acid. With anodized aluminium plates the etchant is allowed to penetrate halfway

Reproduction Processes

Deep-etch plate:

through the anodic layer and this may be as little as 0·0178 mm (0·0001 in.), illustrating how misleading the title of deep-etch can be when first encountered.

Nevertheless it is the slightly recessed image produced by the etching that reduces the amount of image wear caused by friction on the printing machine. After etching, the plate's surface is thoroughly cleaned with anhydrous spirit. A vinyl resin in the form of a lacquer base is now applied to the image areas. When dry this lacquer base becomes the foundation of the image areas, its ink-receptive properties providing half of the lithographic principle. The image areas are finally covered with a thin film of black ink. The hardened gum stencil has served its purpose and can now be washed away in a stream of tepid water.

5. *Desensitizing*

The plate is dried and its surface desensitized by applying a thin layer of gum arabic. The image and non-image areas are protected in this way until they are required for lithographic printing.

6. *Printing and identification*

The lithographic printing method uses a substantially flat or planographic surface to hold and transfer ink. This gives the printed result its appearance of uniformity and smoothness. The division of image and non-image areas, resulting in the selective deposition of ink, relies on the natural antipathy of grease and water. With every printing revolution the non-image areas receive a thin layer of moisture. The presence of moisture in the grained metal surface prevents these areas

Lithographic printing

Lithographic printing:
1. Image
2. Ink
3. Paper
4. Impression
5. Blanket
6. Water

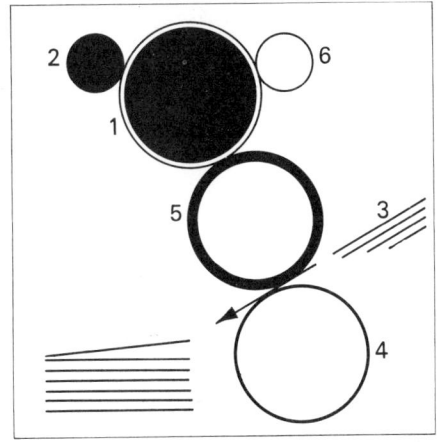

from accepting ink when the inking rollers pass over the plate surface, charging the image areas with ink. It is the balance between ink and water that determines the quality of the final reproduction.

The image bearing plate may be made from paper, zinc, aluminium or of a bi- or tri-metal structure. Printing can take place using a direct or offsetting principle on rotary machines. It is the planographic printing surface in conjunction with the action of the offsetting blanket that creates the distinguishing features of lithographic printing. These features, listed below, can be recognized with the aid of a magnifying glass.

1. A *uniform* and *smooth image*, no trace of ink squash. The ink coverage over typematter and halftone dots is perfectly even.

Identification: Lithography
1. Uniform, smooth image
2. Fine-screens—coarse surfaced papers
3. Dot drop-out
4. Identical ink density

Reproduction Processes

2. *Fine-screen rulings* on *coarse surfaced papers* and boards.
3. Vignettes and highlight regions merging into the white paper background, extreme highlights containing *no dots at all*.
4. All ink images transferred to the paper, whether lines, highlight dots or shadow dots, are of *identical ink density*.

Lithographic printing has been related to the reproduction of pictures since its invention, the ease of preparing large printing plates complemented by the use of the rubber 'offsetting' blanket being the main reasons. By interposing a rubber blanket between the printing plate and paper, it becomes possible to transfer a smooth, uniform layer of ink on to coarse surfaced papers and hard materials such as tin-plate. With the expansion of reel-fed rotary machines, lithography aided by photo-typesetting, electronic scanning machines and automatic plate-processing is moving into the newspaper, book and magazine printing fields. Small offset printing has become a well established part of most office and in-plant printing departments. The use of offset rotary machines is becoming widespread and is responsible for the ever-present coloured picture even on the most meagre of printed matter.

Screen process printing

The screen process printing method is based on the simple principle of using a stencil. If coloured ink or any other medium is rubbed or drawn across a stencil, the ink will pass through the unprotected image areas on to the paper underneath. The non-image areas are represented by the solid or masked portions of the stencil which stop the transfer of ink in these areas.

Silk screen printing and now the printing method of screen process are perfections of the early stencil printings used since ancient times. It has been proposed that primitive man gained the art of stencilling by dabbing coloured dye round his hands and through his fingers on to a flat surface underneath. Another theory is that information was learnt from watching insects eating leaves, finishing only when the fine vein 'stencil-like' structure was left. Early studies of native life on the Fiji Islands have shown that the Islanders cut stencils from banana leaves and then applied vegetable dye through the image areas on to cloth.

Oriental artists employed stencilling techniques to print beautiful designs on to textiles and fibrous paper. The Japanese became extremely adept at cutting stencils. The centres of

Screen process printing

circular letters and other loose design shapes were held in position by ties of fine human hair. The product of the silkworm was used by the craftsmen to weave a fine mesh of silk, which formed a support for these delicate stencils.

It is difficult to isolate one man's work and equate it with the monumental achievements of Gutenberg, Klič or Senefelder. Nevertheless, the use of a silk screen stretched across a frame to form a support for a 'tieless' stencil in Europe, is generally attributed to Samuel Simon of Manchester, England, who introduced this method in 1907. In his application for a patent Simon wrote:

'This is an advancement on stencil-writing, the result of applying, in an up and down movement, a stiff brush charged with ink on top of a cut-out stencil laid over the surface to be marked. The taut silk-mesh becomes a ground for the stencil, which is placed on and held without the use of ties or bridges. The ink or paint can be brushed, rolled or wiped through the stencil and silk-mesh on to the material underneath.'

It was the use of this silk mesh, now replaced by nylon, wire or man-made fibres, that has led to the development of silk-screen printing. Under the new title of screen process this method of transferring ink to paper is now regarded as the fourth major printing process.

Photo-stencil

Photo-stencil is a procedure for making stencils for screen process printing by placing a photographic positive of the original picture on the light-sensitive stencil film in front of a light source. The non-image areas become hardened and after being transferred to the underside of the mesh become the means of blocking the ink, so that it only penetrates through the image areas which remain open. The reproduction sequence is carried out stage by stage in this manner.

1. Stencil film

The mesh is stretched across the frame and thoroughly cleaned with powdered soap and water. There are many proprietary brands of light-sensitive stencil films, which are essentially the same, consisting of a film base coated with an emulsion whose sensitivity and reaction is similar to dichromated colloids.

2. Exposure

A sheet of light-sensitive stencil film is placed with its base-side in contact with the emulsion of a reversed photographic positive. This combination is positioned in a vacuum frame and exposed to an arc lamp. The light rays pass through the transparent non-image areas of the positive and harden the stencil film forming a negative image.

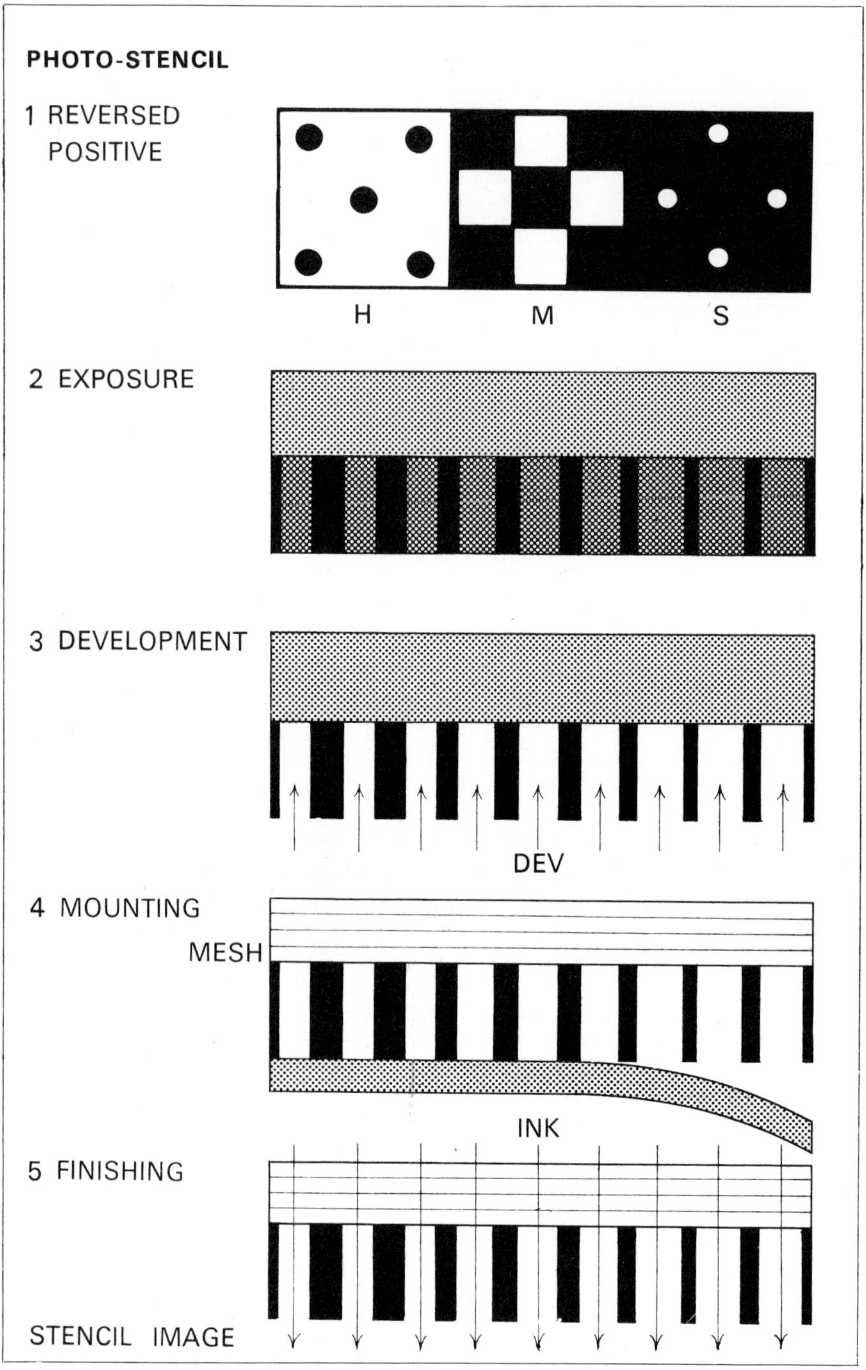

Reproduction Processes

PHOTO-STENCIL

1 REVERSED POSITIVE

H   M   S

2 EXPOSURE

3 DEVELOPMENT

DEV

4 MOUNTING

MESH

INK

5 FINISHING

STENCIL IMAGE

Screen process printing

Mounting the stencil

3. *Hardening*

Although the light-sensitive stencil film has become light-hardened it must go through a process of chemical hardening. This is carried out in a bath of hydrogen peroxide and ensures that the non-image areas will stand up to the developing process.

4. *Development*

The soluble image areas are now ready for development. Warm water is sprayed on the emulsion side of the stencil film. The image areas gradually dissolve, leaving the hardened non-image areas on the film base.

5. *Mounting*

While still in a moistened state the emulsion side of the stencil is placed firmly on the underside of the mesh. By applying gentle pressure and warm air the stencil becomes firmly adhered to the mesh. Once this has been established the stencil's base film is carefully peeled off, leaving the stencil—that is the non-image areas—blocking out the mesh. Finally, the surrounding areas are masked out with tape, so that the only openings through the mesh are in the image areas. The frame containing the mesh and stencil can now be positioned on the printing press and screen process printing may commence.

6. *Printing and identification*

Screen process printing relies on a direct, stencil-like method of transferring ink on to paper. Ink is held in the top of a frame which has the fine, nylon mesh stretched tautly

Reproduction Processes

Screen process printing:
1. Image
2. Ink
3. Paper
4. Impression
M. Mesh
S. Stencil

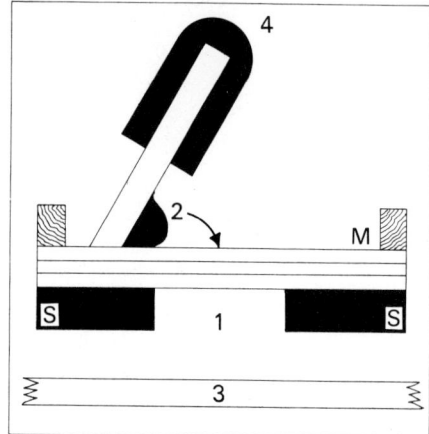

across its underside and secured around the outside edges with clamps or staples. A sheet of paper is positioned at the side lays, and the frame, mesh and stencil lowered on to its surface. With one wipe of a squeegee blade the ink is forced through the mesh and stencil on to the paper underneath.

The image bearing stencil may be produced by painting-out the unwanted areas of the mesh, hand-cut on special laminated foil or exposed on to photographic stencil film. Printing is usually based on the simple squeegee-wiping method, while the frame can be made to fit most shapes, flat or curved.

It is the printing method that allows such a thick layer of ink to be deposited and this causes a bold edge to build up around the printed areas which is responsible for the 'well-defined' look of screen process printing. This type of printing may be identified by looking for the following characteristics with a magnifying glass.

1. The *thickness of the printed ink image* detected by the touch of the fingers. Fairly coarse halftone screens.
2. Occasionally the *mesh pattern* seen in the printed image.
3. The overprinting of *white ink* on top of other colours.
4. The printing of bold, brilliant colours on *all manner of surfaces*.

Screen process printing could be renamed 'versatile' printing because it is capable of printing so many different types of work on all manner of flat and curved surfaces. The direct application of ink through the stencil and fine mesh on to the surface of the material is the key factor. The work can range from large posters containing huge letters printed

The future scene

Identification: screen process
1. Thickness of printed image
2. Mesh pattern.
3. White ink.
4. All manner of printed surfaces.

in fluorescent ink from hand-cut stencils to the finest printed circuit for some complicated electrical device. Screen process with its recent influx of finer meshes, photographic stencils and automatic printing machines positioned in production lines, is developing at a rapid pace. It takes in its stride the printing of such things as glass dials, bottles, containers, textiles, and laminated plastics.

The future scene

It is extremely difficult to forecast the future of graphic reproduction and printing. Perhaps the best approach is to say 'possible trends' because with man's ingenuity everything is possible. It does not take a crystal ball to see that rotary printing must develop together with the use of nylon and plastic printing plates. Phototypesetting and electronic colour scanning will expand and become less expensive. Zerography and electrostatic printing on to synthetic paper is in its infancy, but will grow into fruition. Unsewn book-binding combined with casing-in machines is already with us.

The major printing processes will gradually merge, with the advantages of each being highlighted in one efficient process. Computers will definitely be employed to aid calculations and control the production of printing. The future with its murmurings of automated graphic reproduction may be rather disquieting to some. This need not be, because there will always be a place for intelligent analysis and a cool voice when things start to go wrong. Future printing will demand more than ever before the application of creative thought and control. The unknown is always a stimulating challenge.

APPENDIX

*Table 24. Lens flare factor 1%**

D.O	D.C	D.O	D.C	D.O	D.C
3·00	1·96	2·00	1·70	1·00	0·96
2·95	1·95	1·95	1·67	0·95	0·91
2·90	1·95	1·90	1·65	0·90	0·87
2·85	1·94	1·85	1·62	0·85	0·82
2·80	1·94	1·80	1·59	0·80	0·77
2·75	1·93	1·75	1·56	0·75	0·73
2·70	1·92	1·70	1·52	0·70	0·68
2·65	1·91	1·65	1·49	0·65	0·63
2·60	1·90	1·60	1·45	0·60	0·58
2·55	1·89	1·55	1·42	0·55	0·53
2·50	1·88	1·50	1·38	0·50	0·49
2·45	1·87	1·45	1·34	0·45	0·44
2·40	1·85	1·40	1·30	0·40	0·39
2·35	1·84	1·35	1·26	0·35	0·34
2·30	1·82	1·30	1·22	0·30	0·29
2·25	1·81	1·25	1·18	0·25	0·24
2·20	1·79	1·20	1·14	0·20	0·19
2·15	1·77	1·15	1·09	0·15	0·14
2·10	1·75	1·10	1·05	0·10	0·09
2·05	1·72	1·05	1·00	0·05	0·05

*Original density compared to camera density, e.g. original = highlight 0·1; range 1·8; shadow 1·9; camera = highlight 0·09; range 1·56; shadow 1·65.

Table 25. Opacity ⟷ density

Opacity	Density	Opacity	Density	Opacity	Density
1·00	0·00				
1·12	0·05	11·2	1·05	112	2·05
1·26	0·10	12·6	1·10	126	2·10
1·41	0·15	14·1	1·15	141	2·15
1·58	0·20	15·8	1·20	158	2·20
1·78	0·25	17·8	1·25	178	2·25
2·00	0·30	20·0	1·30	200	2·30
2·24	0·35	22·4	1·35	224	2·35
2·51	0·40	25·1	1·40	251	2·40
2·82	0·45	28·2	1·45	282	2·45
3·16	0·50	31·6	1·50	316	2·50
3·45	0·55	35·4	1·55	354	2·55
3·98	0·60	39·8	1·60	398	2·60
4·47	0·65	44·7	1·65	447	2·65
5·01	0·70	50·1	1·70	501	2·70
5·62	0·75	56·2	1·75	562	2·75
6·31	0·80	63·1	1·80	631	2·80
7·08	0·85	70·8	1·85	708	2·85
7·94	0·90	79·4	1·90	794	2·90
8·91	0·95	89·1	1·95	891	2·95
10·00	1·00	100·0	2·00	1000	3·00

Table 26. Density ↔ transmission

D	T%	D	T%	D	T%
0·00	100				
0·05	89·2	1·05	8·92	2·05	0·89
0·10	79·4	1·10	7·94	2·10	0·79
0·15	70·0	1·15	7·09	2·15	0·71
0·20	63·6	1·20	6·33	2·20	0·63
0·25	56·2	1·25	5·26	2·25	0·56
0·30	50·0	1·30	5·00	2·30	0·50
0·35	44·6	1·35	4·46	2·35	0·45
0·40	39·8	1·40	3·98	2·40	0·40
0·45	35·5	1·45	3·55	2·45	0·36
0·50	31·6	1·50	3·16	2·50	0·32
0·55	28·2	1·55	2·82	2·55	0·28
0·60	25·1	1·60	2·51	2·60	0·25
0·65	22·4	1·65	2·24	2·65	0·22
0·70	20·0	1·70	2·00	2·70	0·20
0·75	17·8	1·75	1·78	2·75	0·18
0·80	15·9	1·80	1·59	2·80	0·16
0·85	14·1	1·85	1·41	2·85	0·14
0·90	12·6	1·90	1·26	2·90	0·13
0·95	11·2	1·95	1·12	2·95	0·11
1·00	10·0	2·00	1·00	3·00	0·10

Table 27. Density ↔ dot percentage

Integrated halftone density	Percent dot area	Integrated halftone density	Percent dot area
0·00	0	0·36	56
0·01	2	0·38	58
0·02	5	0·40	60
0·03	7	0·42	62
0·04	9	0·44	64
0·05	11	0·46	65
0·06	13	0·48	67
0·07	15	0·50	68
0·08	17	0·54	71
0·09	19	0·58	74
0·10	21	0·62	76
0·11	22	0·66	78
0·12	24	0·70	80
0·13	26	0·74	82
0·14	28	0·78	83
0·15	29	0·82	85
0·16	31	0·86	86
0·17	32	0·90	87
0·18	34	0·95	89
0·19	35	1·00	90
0·20	37	1·10	92
0·22	40	1·20	94
0·24	42	1·30	95
0·26	45	1·40	96
0·28	48	1·50	97
0·30	50	1·70	98
0·32	52	2·00	99
0·34	54		

Appendix

Table 28. *Filter chart*

To print as black	Filter	Emulsion
Red	Green Green None	Pan Ortho Blue
Orange	Green Green None	Pan Ortho Blue
Yellow	Blue Blue None	Pan Ortho Blue
Yellow-green	Blue Blue None	Pan Ortho Blue
Green	Red	Pan
Blue-green	Red	Pan
Blue	Red Yellow	Pan Ortho
Violet	Green Deep yellow	Pan Ortho

To appear as white	Filter	Emulsion
Red	Red	Pan
Orange	Red	Pan
Yellow	Red Yellow	Pan Ortho
Yellow-green	Green Green or yellow	Pan Ortho
Green	Green Green or yellow	Pan Ortho
Blue-green	Blue or green Blue or green	Pan Ortho
Blue	Blue Blue None	Pan Ortho Blue
Violet	Blue Blue None	Pan Ortho Blue

INDEX

A
Aberrations in lenses 108, 114
Absorption of light 68, 70
Accelerator, developer 170, 167
Acid fixing bath 183
Actinometer 188
Additive colour principle 40, 346
  mixture 347
Additivity failure 354, 355
Absorption 139, 149
After-ripening 146
Agitation of developer 177, 179
Airbrush 59
Ammonium thiosulphate 183
Angle screen 393
Antifoggant 171
Antihalo backing 151
Aperture, actual 116
  effective 117
  for halftone 319
  lens 312
  shape 305
Aquatint method 29
Archer, F. Scott 137
Arc lamp 81
Argentosulphates 183
Artwork 44
A.S.A. speed rating 152
Assessing exposure 229
Atoms 138
  structure 155
Atomic weight 139
Autochrome process 346

B
Bar Gee curve 335, 336
Baryta coating 151
Berwick, Thomas 27
Bi-metal plates 428
Black Printer 363, 386
Blood, dragon's 415
Box camera 15, 94
Brightness, luminance 191
  range 189
British Standard 92, 356
  colour filters 354
  colour inks 355
Bromide, acceptor 143
  paper 149
  prints 60
  reproduction of 295
Bunsen-Roscoe law 141

C
Cadmuim sulphide cell 221
Callier coefficient 220

Calotype process 136
Camera 94
  back projection 325
  dark room 123, 124
  extension 229
  gallery 122
  operating 226
  reversible 119
  roll film 125
  scales 103
  screen gear 297
  vertical 123
Camera obscura 13
Candela 72
Carbon process 294
Catalysis 165
Cathode ray oscilloscope 406
Chalk drawings 57
Characteristic curve 201
  exposure on 206
  interpretation of, 210
Characteristic screening curve 327
Chemical change 138
  compounds 139
  development 166
  equations 139
  intensification 185
  introduction 140, 163
  reduction 184
  reversal 286
  transfer 184
C.I.E. system 351
Clapper, F. R. 381
Coating 141
  colloid, dichromated 412
  emulsion 149
  lens 114
  P.V.A. 414
  spectral sensitivity of 153
Collodion process 136
Collotype process 291
Colour, additive 40, 346
  appraisal 387
  bars 353
  chart 258
  compensating filters 333
  correction 360, 415
  couplers 348
  density 349
  electronic correction 403
  filters 246, 354
  key 388
  line reproduction 54
Colour masking 364, 367
  choice of method 366
  for black printer 374

443

for under colour 387
negative masking 374, 377
one-stage masking 367, 370
positive masking 367, 370
two-stage masking 370, 377
with coloured mask 379
Colour matching lamp 93
measurement 349
original 61, 387
pattern 393, 394
Colour photographs 63, 64
photography 344, 345
printing, wet 386
proofing 399
proportionality 355
psychological effect 345
sensitivity 153
separation 353
subtractive colour principle 40, 347
retouching 362
Colour temperature 78
theory 40, 346
Coloured image 387
control strip 388
transparencies 61, 62
Combination work 399, 200
Complementary colours 39
Computer functions 407
Computer programs 329
Concentration of fixer 183
Conjugate foci 105
Contact screen 298, 299, 411
Continuous-tone photography 334
Contrast 211, 335
Correct exposure 208, 209
Crystal lattice 139, 155

D

Darkroom, lay-out 128, 132
Daguerre, L. J. M. 135
Daguerreotype 136
Daylight 79
Deep-etch 426, 427
Definition and resolution 36
Densitometer 218
photoelectric 224
visual 223
Densitometry 217
Density 194
diffuse 220, 221
measurement 194, 223
Density range 204
bromide print 60
continuous-tone 280, 281
colour separation 395
halftone 330
Density, reflection 193
specular 220, 221
Developer(s) 163, 166
accelerator 170
action of 173
agitation 177

formulae 172
lith 275, 278
M.Q. 168
preservative 169
replenishment 178
restrainer 171
solvent 171
Developing agents 167
Development 163, 166
automatic 181
centres 173
chemical 166
effective 166
lith 278
mechanism 167
physical 163
temperature 176
time 175
Deville, E. 307
Diffraction of light 308
Diffraction halftone theory 307
Diffuse, density 221
light 219
Diffusion of light 383
Digestion, emulsion 147
Dispersion of light 67
Distortion of picture 413
DIN speed 152
Direct halftone method 325
screen separations 381
Doctor blade 427
Doctoring, emulsions 148
Dot area 292
Dot-etching 362, 363
Dot-formulation 296, 301
Dragon's blood 420
Drop-out highlight 337, 385
Drum scanners 405
du Hauron, Louis Ducos 347
Dufay colour process 347

E

Eastman, George 22
Electronic colour correction 362
Electronic flash 87
Electronic laser screening 411
Electronic scanning methods 403
Elements, chemical 138
Emulsion 142
after-ripening 146, 147
colour sensitivity 149, 153
data 152
doctoring 148
double-layer 335
manufacture 142
ordinary 140, 153
orthochromatic 140, 153
panchromatic 140, 153
pre-screening 302
ripening 146
super-coating 150
speed 152

variable-contrast 336
Enlargers 124
Exhaustion, developers 178
  fixing baths 183
Exposure 154
  control 229
  correct 208
  flash 319
  latitude 211
  meter 391

F
Farmer's reducer 184
Ferric chloride 421
Ferrous sulphate 162
Film base 149
  grain size 144
  Multimask 379
  Tri-pack 380
Filter 246
  factor 247
  neutral density 228, 320
  ratio 247
Fixation 182
  acid 183
  exhaustion 183
  hardening 183
  mechanism 183, 184
  rapid 183
  time 182
Flare, lens 89
  calculation 91
  factor 92
Flash exposure 319, 331
Fluorescent lamps 85
F/numbers 117
Focal length 100
Focusing 103
Fog, development 171
Fox Talbot, William Henry 136

G
Gallery camera 122
Gamma 215
Gamma/time curve 216
G.A.T.F. 367
Gelatine 143
Gillot, Charles 272
Gillot, M. F. 272
Gillotage 272
Gutenberg, Johann 416
Graphic reproduction 23
  photography 26, 186
  processes 416
Gravure printing 423
Grains, silver halide 144
  size distribution 146
Gurney, R. W. and Mott, N. F. 154

H
H and D curves 188
  speed 189

Halation 151
Halftone 291, 295
  direct 325
  for different papers 256
  indirect 325
Halftone screen 295
  screen compensation 341
  screen distance 312, 313
  screen theories 304
Halides, silver 142
Herschel effect 159
Herschel, Sir John 162
Highlight, density 204
  exposure 319
  mask 385
Hue 43
Hurter & Driffield 188
Huygen's wave Theory 65
Hydroquinone 168
Hypo—see sodium thiosulphate 182

I
Identification of printing processes:
  gravure 418
  letterpress 416
  lithography 428
  screen process 434
Illuminant 77
  carbon arc 81
  colour matching 92
  electronic flash 87
  fluorescent 85
  mercury vapour 84
  photoflood 83
  requirements 77
  tungsten 82
  xenon 87
Illumination, control 73
  intensity 228
  terminology 72
Image, formation 103
  magnification 105
  sharpness 116
Ink, gravure 423
  letterpress 416
  lithographic 428
  screen process 434
Ions 138
Incident light 69
Indirect halftone 325
Instructions, for reproductions 50
Intensification 185
Intensity scale sensitometer 204
Intermittency effect 158
Inverse exposure system 233
Irradiation 151
Ives, F. E. 292

J
Joly screen 346

K
Kelvin temperature scale 78
Klič, Karl 418
Kodak 22

L
Lamp (see Illuminant) 77
Latensification 157
Latent Image 154
Latitude, exposure 276
Layout, printing factory 129, 130
   reproduction department 132
le Bron, J. C. 345
Lens 97
   aberations 108
   aperture 116
   aperture tables 231, 232
   coating 115
   compound 113
   convergent 99
   divergent 99
   evolution 95
   flare 89
   flare factor 92
   focal length 100
   focus 103
   image formation 104
   manufacure 98
Letterpress printing 416
Levy, M. 293
Light 65
   absorption 69
   action of 154
   diffraction 308
   diffuse 220
   dispersion 67
   interference 393
   inverse square law 74
   luminous intensity 66
   rectilinear propagation 68
   reflection 68
   refraction 70
   transmission 192
   waves 66
   white 67
Light, general 86
   for viewing and printing 92
Light meter, integrating 226
Lightness 43
Line reproduction 270
   image 272
   results 273
Line colour combination 399
Line and tint combination 228
Line and tone combination 340
Lith emulsion 275
   development 278
Lithography 423
Log of exposure 200
Logarithms 195
Lumiére brothers 346
Luminous flux 73
Luminosity 43
Lummer-Brodhum cube 218

M
Maddox, Dr R. L. 140
Magnification factor 103
Mask, contrast reducing 386
   highlight 384
   multilayer 379
   one-stage negative 374
   one-stage positive 367
   for reflection originals 368, 374
   for transparencies 376, 379
   strengths 364
   two-stage negative 377
   two-stage positive 370
   unsharp 385, 413
Masking, camera back 383
   double-overlay 372, 373
   single-overlay 367, 368
Max Levy 293
Maxwell, James Clerk 345
Measurement, density 194
Meisenbach, G. 292
Metol 168
Mezzotint process 28
Molecules 138
Moiré pattern 393
Multimask 379
Mungo Ponton 271
Munsell system 349

N
Negative, by electronic flash 87
   colour separation 391
   continuous-tone 334
   halftone, direct 325
      indirect 325
   line 275
Negative-working screen 299, 328
Neutral density filter 228
Newton, Isaac 65
Newton rings 302
Nièpce, Nicéphore 17, 135

O
Offset printing 428
Opacity, definition of 193
Ordinary emulsion 140, 153
Original, characteristics 44
   coloured 54, 61
   density range 60
   instructions 50
   line, 51
   overlay, 54
   preparation 44, 58
   scaling 46, 47
   tone 56
   transparencies 63
Orthochromatic emulsion 140, 153
Ostwald ripening 146
Oxidation 165

P

Panchromatic emulsion, 140, 153
Paper, bromide 149
    coated 256
    proofing 399
    uncoated 256
Pencil sketches 59
Penumbral halftone Theory 305
Perfect negative 209
    reproduction 258, 259
Petit, C. G. 292
Petzval, Joseph 95
pH, definition 164
Photo-cell, efficiency 223
    spectral sensitivity 222
Photo-chemical requirements 141
Photo-electric densitometer 224
Photo-engraving out-line 418
Photographic, emulsion 142
    equipment 122
    filters 246
    originals 60, 63
    procedure 274, 324
    processing 161
    processes 134
    reduction 184
    speed 152
    studio 132
Photography, basic factors 186
    history 11
Photolithography out-line 430
Photometry 73
Photopolymer-plates 439
Phototypesetting 439
Physical development 163
Pinhole camera 96
Pinhole halftone Theory 305
Plant lay-out 128
Polyvinyl alcohol 414
Pointevin, A. 272
Ponton, Mungo. 271
Positive, continuous-tone 286
    duplicate 287
    halftone 330
    in line 284
    reversal 286
Positive-working screen 299, 328
Potassium, dichromate 271, 419
    ferricyanide 184
    metabisulphite 170, 183
Precipitation 144
Preservative, developer 169
Preucil, Frank 367
Primary colours 39
Printing, gravure 423
    letterpress 416
    lithography 428
    screen process 424
Printing standards 252
    colour 258
    line 252
    tone 256

Processing, photographic 161
    automatic 181
Proofing colour 399
Proportionality failure 355

Q

Quantitative evaluation:
    of photographic results 273, 321
    of printed results 256
Quality control 266
    strip 252

R

Rapid access papers 184
Rapid fixation 183
Reciprocity failure 157
Reducer, photographic 173
Reducing agent 167
Reduction 184
Reflection 68
    critical angle 71
    densitometer 223
    laws of 69
    total internal 71
Refraction 70
    index 70
    laws 71
Register, marks 49
    pins and punch 289
    system 413
Relief image 23
Replenisher 179
Restrainer, developer 171
Retouching 361
    by dot-etching 362
    electronic 414
    original 58
    photographic 363
Reversal, chemical 286
    film 287
    optical 118
    straight-line systems 120

S

Sala, Angelo 134
Saturation 43
Scale of reduction 46
Scanning methods 403
    drum scanners 405
    electronic fly-spot 406
    monochrome 413
    reciprocal motion 404
    rotating cylinder 405
Scott-Archer, F. 137
Scheele, C. W. 134
Schulze, H. J. 16
Scraper board 52
Screen angle 394
Screen compensation 341
Screen, contact 298
    chain-dot 301
    distance 312, 313

grey 301
highlights exposure 319
magenta 298
negative-working 299, 328
positive-working 299, 328
Screen cross-line 291
Screen negatives 328
Screen positives 330
Screen process printing 432
Screen rulings 324
Selenium cell 221
Senefelder, Alois 429
Sensitivity, of emulsions 144
  specks 148
Sensitometer 203
  intensity scale 204
  time scale 204
Sensitometric terms 191
Sensitometry 189
Sharpness 36
Silver, atom 138
  bath 137
  bromide 138, 142
  chloride 142
  halides 142, 184
  iodide 142
  ions 138
  nitrate 142
Spectrophotometry 350
Spectrum, visible 66
Specular density 219
  reflection 68
Speed of emulsions 152
Stabilization system 184
Standardization 225, 249
Step-and-repeat 127
Studio, lay-out requirements 132
Subtractive colour processes 40, 347
Super-proportional reducer 185
Swan, J. W. 76

T
Taylor, H. D. 96
Temperature, colour 78
  control 175
  developer 177
Three-colour reproduction 363
Time, development 175
Tint, laying 288

Tonal correction 360, 410
Tonal control strip 256
Tonal key 323
Tonal masks 384
Tonal results 321
Tone, originals 56
  reproduction 291
Transparencies, colour 63
  masking 374, 379
  preparation 63
  selection 64
  viewing 93
Trichromatic reproduction 363
Tritton, F. J. 314
Type, repro proofs 53

U
Under-colour removal 386
Unsharp masks 385

V
Vacuum frame 126
Valency 139
Viewing conditions 93
Visible spectrum 66
Vision 33
Visuals and roughs 31
Vogel, H. 140
V/ratio exposure system 229

W
Wash drawings 59
Washing 183
Waterhouse stop 116
Wedge-spectrogram 78
Wedgewood, Thos. 17
Wet collodion process 137
Wet-on-dry printing 386
Wollaston menicus lens 95

X
Xenon lamp 88
Xerography 439

Y
Young, Thomas 345
Yule, J. A. C. 351

Z
Zinc plates 426